Composites in Energy Harvesting, Conversion and Storage

Also of interest

Advanced Energy Materials.
Natural Polymers, Renewable Energy, Sustainable Development
Verma, Sonika, Sharma, Rajput (Eds.), 2025
ISBN 978-3-11-128890-1, e-ISBN 978-3-11-128907-6

Hybrid Aerogels.
Energy and Environmental Applications
Gangawane, Panda (Eds.), 2024
ISBN 978-3-11-125135-6, e-ISBN 978-3-11-125200-1

Surface-Functionalized Nanomaterials.
Environmental, Energy Storage, Energy Conversion Applications
Rasheed, Verma (Eds.), 2025
ISBN 978-3-11-138164-0, e-ISBN 978-3-11-138180-0

Smart Alloys.
In Energy, Biomedical, Engineering, Aerospace and Automotive Industries
Lakshmi, Vallimanalan (Eds.), 2025
ISBN 978-3-11-163005-2, e-ISBN 978-3-11-163020-5

Composites in Energy Harvesting, Conversion and Storage

Biocomposites, Nanocomposites and Bionanocomposites

Edited by
S. M. Sapuan, Mohd Roshdi Hassan,
Eris Elianddy Supeni and Azizan As'arry

DE GRUYTER

Editors
Salit Mohd Sapuan
Department of Mechanical and Manufacturing Engineering Faculty of Engineering
Universiti Putra Malaysia
43400 UPM Serdang Selangor
Malaysia

Mohd Roshdi Hassan
Department of Mechanical and Manufacturing Engineering
Faculty of Engineering
Universiti Putra Malaysia
43400 Selangor
Malaysia

Eris Elianddy Supeni
Department of Mechanical and Manufacturing Engineering
Faculty of Engineering
Universiti Putra Malaysia
43400 Selangor
Malaysia

Azizan As'arry
Department of Mechanical and Manufacturing Engineering
Faculty of Engineering
Universiti Putra Malaysia
43400 Selangor
Malaysia

ISBN 978-3-11-123448-9
e-ISBN (PDF) 978-3-11-123476-2
e-ISBN (EPUB) 978-3-11-123499-1

Library of Congress Control Number: 2025947824

Bibliographic information published by the Deutsche Nationalbibliothek
The Deutsche Nationalbibliothek lists this publication in the Deutsche Nationalbibliografie; detailed bibliographic data are available on the Internet at http://dnb.dnb.de.

Cover image: Just_Super/iStock/Getty Images Plus
Typesetting: TNQ Tech Private Limited, Chennai, India
Printing and binding: CPI books GmbH, Leck

www.degruyter.com
Questions about General Product Safety Regulation:
productsafety@degruyterbrill.com

Contents

List of contributing authors

Alhadi A. Abosbaia
Mechanical Engineering
Faculty of Engineering
University of Zawia
Zawia
Libya

Mohd Ridhwan Adam
School of Chemical Sciences
Universiti Sains Malaysia
Minden, Penang 11800
Malaysia

Mohammed Hassouna Alaaeddin
Advanced Engineering Materials and Composites Research Centre (AEMC)
Department of Mechanical and Manufacturing Engineering
Universiti Putra Malaysia
43400 UPM Serdang, Selangor
Malaysia
and
Laboratory of Biocomposite Technology
Institute of Tropical Forestry and Forest Products (INTROP)
Universiti Putra Malaysia
43400 UPM Serdang, Selangor
Malaysia

Faris M. AL-Oqla
Department of Mechanical Engineering
The Hashemite University
13133 Zarqa
Jordan

Belal Alemour
Department of Mechanical Engineering
Bahrain Polytechnic University
Isa Town
Bahrain
E-mail: belalemour@gmail.com
https://orcid.org/0000-0002-7346-934X

Munir Faraj Alkbir
Advanced Facilities Engineering Technology Research Cluster (AFET)
Plant Engineering Technology (PETech) Section
Universiti Kuala Lumpur Malaysian Institute of Industrial Technology
81750 Masai, Johor
Malaysia
E-mail: munir@unikl.edu.my

Azizan As'arry
Department of Mechanical and Manufacturing
Faculty of Engineering
Universiti Putra Malaysia
Serdang, Selangor 43400
Malaysia
E-mail: zizan@upm.edu.my

Nazrin Asmawi
Institute of Power Engineering
Department of Electrical and Electronics Engineering
College of Engineering
Universiti Tenaga Nasional
Jalan IKRAM-UNITEN
Kajang 4300, Selangor
Malaysia
and
Fakulti Kemanusiaan
Pengurusan Dan Sains
Universiti Putra Malaysia Kampus Bintulu Sarawak
Jalan Nyabau
97008 Bintulu, Sarawak
Malaysia

Nuha Awang
Advanced Facilities Engineering Technology Research Cluster (AFET)
Plant Engineering Technology (PETech) Section
Universiti Kuala Lumpur Malaysian Institute of Industrial Technology
81750 Masai, Johor
Malaysia

https://doi.org/10.1515/9783111234762-201

Faieza Abdul Aziz
Department of Mechanical and Manufacturing Engineering
Universiti Putra Malaysia
43400 UPM Serdang, Selangor
Malaysia

Omar Badran
Department of Mechanical Engineering
Al-Balqa Applied University
FET
Amman
Jordan
E-mail: omar_badran@bau.edu.jo

Adnan Bakri
Advanced Facilities Engineering Technology Research Cluster (AFET)
Plant Engineering Technology (PETech) Section
Universiti Kuala Lumpur Malaysian Institute of Industrial Technology
81750 Masai, Johor
Malaysia

Joseph Selvi Binoj
Institute of Mechanical Engineering
Saveetha School of Engineering
Saveetha Institute of Medical and Technical Sciences (SIMATS)
Chennai, 602105
Tamilnadu
India

Haider Jaafar Chilabi
Department of Mechanical and Manufacturing
Faculty of Engineering
Universiti Putra Malaysia
Serdang, Selangor 43400
Malaysia
and
Midland Refineries Company (MRC)
Ministry of Oil
Baghdad 10022
Republic of Iraq
E-mail: haider.chilabi@gmail.com

Anil Kumar Dussa
Department of Electronics and Communication Engineering
BMS Institute of Technology & Management
Yelahanka, 560064
Bengaluru
India

Abdul Habib
Faculty of Engineering
Advanced Engineering Materials and Composites Research Centre (AEMC)
Universiti Putra Malaysia
Serdang 43400, Selangor
Malaysia
https://orcid.org/0009-0008-0648-548X

Muhammad Firdaus Abdul Halim
Faculty of Engineering
Advanced Engineering Materials and Composites Research Centre (AEMC)
Universiti Putra Malaysia
Serdang 43400. Selangor
Malaysia

Nor Salwa Hamdan
Department of Skills Development
Ministry of Human Resources Malaysia
Level 7-8, Block D4, Complex D
62530 Putrajaya, Malaysia
E-mail: salwa.hamdan@mohr.gov.my

Moklis Muhammad Harussani
Department of Transdisciplinary Science and Engineering
School of Environment and Society
Tokyo Institute of Technology
Meguro
Tokyo 152-8552
Japan
and
Advanced Engineering Materials and Composites Research Centre (AEMC)
Department of Mechanical and Manufacturing Engineering
Universiti Putra Malaysia
43400 UPM Serdang, Selangor
Malaysia
E-mail: harussani.m.ab@m.titech.ac.jp
https://orcid.org/0000-0003-4475-8764

Wan Zuha Wan Hasan
Department of Electrical Engineering
Universiti Putra Malaysia
43400 UPM
Serdang, Selangor
Malaysia
https://orcid.org/0000-0003-4691-066X

Mohammad Zaid Hasan
Department of Mechanical and Manufacturing Engineering
Advanced Engineering Materials and Composites (AEMC) Research Centre
Universiti Putra Malaysia
43400 UPM Serdang, Selangor
Malaysia
E-mail: ham123zaid@gmail.com

Shukur Abu Hassan
Centre for Advanced Composite Materials
Office of Deputy Vice-Chancellor (Research & Innovation)
Universiti Teknologi Malaysia
Johor Bahru
81310, Johor
Malaysia

Mohd Roshdi Hassan
Department of Mechanical and Manufacturing Engineering
Universiti Putra Malaysia
43400 Serdang, Selangor
Malaysia
E-mail: roshdi_hassan@upm.edu.my

Rushdan Ahmad Ilyas
School of Chemical and Energy Engineering
Faculty of Engineering
Universiti Teknologi Malaysia
81310 UTM Johor Bahru
Johor
Malaysia
and
Centre for Advanced Composite Materials (CACM)
Universiti Teknologi Malaysia
81310 UTM Johor Bahru
Johor
Malaysia

Zafar Iqbal
Applied Sciences and Humanities
Jamia Millia Islamia
New Delhi
India

Yusuf Jameel
Faculty of Engineering
Advanced Engineering Materials and Composites Research Centre (AEMC)
Universiti Putra Malaysia
Serdang 43400. Selangor
Malaysia

Fatihhi Januddi
Advanced Facilities Engineering Technology Research Cluster (AFET)
Plant Engineering Technology (PETech) Section
Universiti Kuala Lumpur Malaysian Institute of Industrial Technology
81750 Masai, Johor
Malaysia

Mohamad Haafiz Mohamad Kassim
Bioresource Technology Division
School of Industrial Technology
Universiti Sains Malaysia
Minden, Penang 11800
Malaysia
and
Green Biopolymer, Coatings & Packaging Cluster
School of Industrial Technology
Universiti Sains Malaysia
Minden, Penang 11800
Malaysia

Lim Wei Keong
Department of Mechanical and Manufacturing Engineering
Universiti Putra Malaysia
43400 UPM
Serdang, Selangor
Malaysia

Abir Khan
Advanced Engineering Materials and Composites Research Centre (AEMC)
Department of Mechanical and Manufacturing Engineering
Universiti Putra Malaysia
43400 UPM Serdang, Selangor
Malaysia
and
National Institute of Textile Engineering and Research (NITER)
Nayarhat, Savar
Dhaka, 1350
Bangladesh
E-mail: akhan@niter.edu.bd

Mmhamad Shahrul Effendy Kosnan
Advanced Facilities Engineering Technology Research Cluster (AFET)
Plant Engineering Technology (PETech) Section
Universiti Kuala Lumpur Malaysian Institute of Industrial Technology
81750 Masai, Johor
Malaysia

Hong Ngee Lim
Department of Chemistry
Universiti Putra Malaysia
43400 Serdang, Selangor
Malaysia
E-mail: ganet@upm.edu.my

Mageswari Manimaran
Bioresource Technology Division
School of Industrial Technology
Universiti Sains Malaysia
Minden, Penang 11800
Malaysia
and
Green Biopolymer, Coatings & Packaging Cluster
School of Industrial Technology
Universiti Sains Malaysia
Minden, Penang 11800
Malaysia

Zulhaimi Mohammad
Advanced Facilities Engineering Technology Research Cluster (AFET)
Plant Engineering Technology (PETech) Section
Universiti Kuala Lumpur Malaysian Institute of Industrial Technology
81750 Masai, Johor
Malaysia

Murshiduzzaman
Department of Mechanical and Manufacturing Engineering
Universiti Putra Malaysia
43400 UPM
Serdang, Selangor
Malaysia
E-mail: gs55223@student.upm.edu.my
https://orcid.org/0009-0002-5389-6552

Mohd Nurazzi Norizan
Bioresource Technology Division
School of Industrial Technology
Universiti Sains Malaysia
Minden, Penang 11800
Malaysia
and
Green Biopolymer, Coatings & Packaging Cluster
School of Industrial Technology
Universiti Sains Malaysia
Minden, Penang 11800
Malaysia
E-mail: mohd.nurazzi@usm.my

Mohd Nor Faiz Norrahim
Research Centre for Chemical Defence
Universiti Pertahanan Nasional Malaysia
Kem Perdana Sungai Besi
Kuala Lumpur 57000
Malaysia

Agung Nugroho
Research Center for Advanced Materials
National Research and Innovation Agency (BRIN)
South Tangerang
Banten, 15314
Indonesia

Sathishkumar Palanisamy
Department of Electronics and Communication Engineering
BMS Institute of Technology & Management
Yelahanka, 560064
Bengaluru
India

Lakshminarasimhan Rajeshkumar
AU-Sophisticated Testing and Instrumentation Centre
CoE-Advanced Materials Synthesis
and
Department of Mechanical Engineering
Alliance School of Applied Engineering
Alliance University
Bangalore 562106
Karnataka
India

Md Rubel
Department of Transdisciplinary Science and Engineering
School of Environment and Society
Tokyo Institute of Technology
Meguro
Tokyo 152-8552
Japan
E-mail: rubel.m.aa@m.titech.ac.jp

Sapuan Mohd Salit
Faculty of Engineering
Advanced Engineering Materials and Composites Research Centre (AEMC)
Universiti Putra Malaysia
Serdang 43400, Selangor
Malaysia
E-mail: sapuan@upm.edu.my
https://orcid.org/0000-0002-4598-6358

Suhad D. Salman
Mechanical Engineering Department
College of Engineering
Mustansiriyah University
Baghdad
Iraq

Ganesan Shyni
Department of Electrical and Electronics Engineering
University College of Engineering
Konam, Nagercoil
629004, Tamil Nadu
India

Vasi Uddin Siddiqui
Faculty of Engineering
Advanced Engineering Materials and Composites Research Centre (AEMC)
Universiti Putra Malaysia
Serdang 43400. Selangor
Malaysia
https://orcid.org/0000-0003-3427-8943

Eris Elianddy Supeni
Department of Mechanical and Manufacturing Engineering
Universiti Putra Malaysia
43400 UPM
Serdang, Selangor
Malaysia

Anitha R. Vaddinuri
Department of Electronics and Communication Engineering
BMS Institute of Technology & Management
Yelahanka, 560064
Bengaluru
India
E-mail: anithavr@bmsit.in

Mohd Zuhri Mohamed Yusoff
Advanced Engineering Materials and Composites Research Centre (AEMC)
Department of Mechanical and Manufacturing Engineering
Universiti Putra Malaysia
43400 UPM Serdang, Selangor
Malaysia
and
Laboratory of Biocomposite Technology
Institute of Tropical Forestry and Forest Products (INTROP)
Universiti Putra Malaysia
43400 UPM Serdang, Selangor
Malaysia
https://orcid.org/0000-0002-1069-7345

Hanafiah Yussof
School of Mechanical Engineering
College of Engineering
Universiti Teknologi MARA
Shah Alam
Malaysia

Edi Syams Zainudin
Advanced Engineering Materials and Composites Research Centre (AEMC)
Department of Mechanical and Manufacturing Engineering
Universiti Putra Malaysia
43400 UPM Serdang, Selangor
Malaysia
and
Laboratory of Biocomposite Technology
Institute of Tropical Forestry and Forest Products (INTROP)
Universiti Putra Malaysia
43400 UPM Serdang, Selangor
Malaysia

Vasi Uddin Siddiqui, Yusuf Jameel, Muhammad Firdaus Abdul Halim, Mohd Sapuan Salit*, Nazrin Asmawi and Zafar Iqbal

1 Composites in battery casing and energy storage

Abstract: Battery casing is a vital part of the electrical vehicles. The material used in it plays an important role in deciding the safety of the vehicle. This chapter refers to the polymer composites, which gives a good replacement for the currently used materials. The chapter explains the concept of battery pack along with its safety issues. It also explores various features and characteristics behind the vitality of polymer composites. The usage of polymer composites as an energy storage gives this material an upper hand.

Keywords: composites; battery casing; energy storage; EV's

1.1 Introduction

The ever-expanding global society necessitates a vast number of new energy devices to keep up with the rapid advancement of technology [1]. Polymer composites offer intriguing advantages in the design and construction of electric vehicle battery casing. Their popularity escalated along with the rise of electric vehicle (EV) in recent decades. The ubiquity of polymer composites is found in multidisciplinary sectors, which cover every part of human daily life from household appliances to industrial equipment. Automotive and aerospace industries had been implementing robust polymer composites materials in their products for better durability, efficiency, and productivity. The substitution of lightweight polymer composites for metals significantly reduces energy consumption, thus increase the distance travelled by the EV. Exposure to corrosive substances pose little to no threat owing to their high chemical stability. Furthermore, they can be easily moulded into complex shape and design allowing wide range of flexibility to be integrated into the structure of vehicle. The selection of polymer composites for battery casing also based on their capability to act as an energy storage

***Corresponding author: Mohd Sapuan Salit**, Faculty of Engineering, Advanced Engineering Materials and Composites Research Centre (AEMC), Universiti Putra Malaysia, Serdang 43400, Selangor, Malaysia, E-mail: sapuan@upm.edu.my. https://orcid.org/0000-0002-4598-6358
Vasi Uddin Siddiqui, Yusuf Jameel and Muhammad Firdaus Abdul Halim, Faculty of Engineering, Advanced Engineering Materials and Composites Research Centre (AEMC), Universiti Putra Malaysia, Serdang 43400, Selangor, Malaysia. https://orcid.org/0000-0003-3427-8943 (V.U. Siddiqui)
Nazrin Asmawi, Institute of Power Engineering, Department of Electrical and Electronics Engineering, College of Engineering, Universiti Tenaga Nasional, Jalan IKRAM-UNITEN, Kajang 4300, Selangor, Malaysia
Zafar Iqbal, Applied Sciences and Humanities, Jamia Millia Islamia, New Delhi, India

As per De Gruyter's policy this article has previously been published in the journal Physical Sciences Reviews. Please cite as: V. U. Siddiqui, Y. Jameel, M. F. Abdul Halim, M. S. Salit, N. Asmawi and Z. Iqbal "Composites in battery casing and energy storage" *Physical Sciences Reviews* [Online] 2024. DOI: 10.1515/psr-2024-0019 | https://doi.org/10.1515/9783111234762-001

application particularly in storing and releasing heat energy. To dissipate heat generated by battery, good heat conductivity is crucial as a way to prolong the longevity and performance of battery pack. Typically, polymer itself has low heat conductivity but through the addition of good conductive filler, highly conductive pathways can be established. A variety of fillers can be incorporated into polymer matrix to create conductive polymer composites such as carbon-based nanoparticles (graphene, carbon nanotubes) [2], metal oxides (aluminium oxide, zinc oxide) and silicate (boron nitride) [3]. Manipulating the filler loading will modify the characteristics of the developed material for better or worse. Table 1.1 indicates the thermal conductivity of various kind of fillers.

Table 1.1: The thermal conductivity of various fillers [4].

Type of filler	Thermal conductivity (W/mK)	Density (g/cm^3)
Aluminium	234	2.7
Aluminium nitride	100–319	3.235
Boron nitride	185–400	2.27
Carbon fibre	10–1,000	1.5–2.0
Copper	386–400	8.9
Carbon nanotubes	2,000–6,000	2.1
Graphene	4,000–7,000	1.06
Graphite	100–500	2.25
Nickel	88.5	8.9

1.2 Battery casing

Polymer composites are the prospective alternative in manufacturing EV battery casing due to their distinct properties of high strength, lightweight, corrosive-resistant and ease processability [4, 5]. The combination of these traits is critical to accomplished safe and high performance of EV, which will propel the vehicle [6]. Thermal runaway is one of the vital issues in developing battery casing. It is a major safety concern for energy storage devices, which can lead to flames and even an explosion. Besides adaption of solid-state batteries, excellent thermal management through battery casing design and technology also contribute to the prevention of hazard to occur. Phase change materials (PCMs) are vastly used in electronic appliances due to its ability to store thermal energy relative to their small volume. In a study by Huang et al. [7], flexible composite phase change material (CPCM) embedded at lower part of battery pack was able to effectively drop the battery temperature by 18 °C at 10 C discharge rate. The CPCM is made up of olefin block copolymer (OBC) doped with graphite as thermal conductivity enhancement additive. Flexible CPCM recorded lower thermal contact resistance of 461 mm^2 K/W compared to conventional CPCM of 1818 mm^2 K/W indicating better thermal conductivity to dissipate

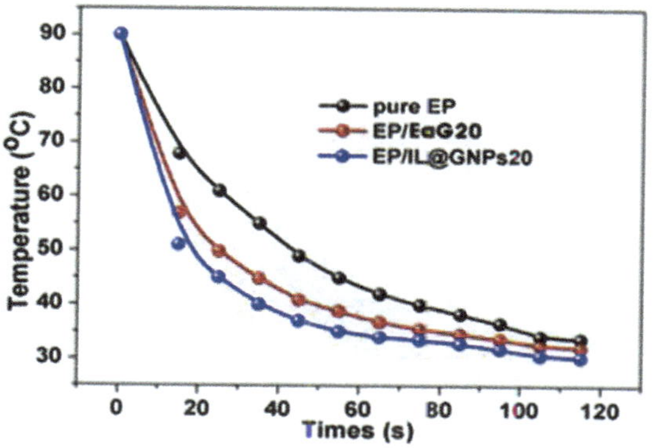

Figure 1.1: The cooling trends from 90 °C of pure epoxy, epoxy/graphene and epoxy/graphene-imidazolium ionic liquid [3].

heat energy. The flexibility feature eases the installation, prevents brittle failure and improves poor surface contact that caused high thermal contact resistance. Meanwhile, Lee et al. [8] investigated the air cooling system on different aspect ratios of PCM battery casing. The aspect ratio of PCM and air velocity was manipulated to evaluate the effectiveness of the cooling system. The lowest average temperature of PCM chamber and minimum volume fraction of molten PCM was recorded at an aspect ratio of 0.2 ,whereas the highest average temperature of PCM chamber and maximum volume fraction of molten PCM was recorded at an aspect ratio of 0.35. The increment in air velocity hastens the solidification of PCM as greater heat transfer occurred between the air and the PCM area. Zhang and Park [3] prepared graphene-imidazolium ionic liquid (G-IL)/epoxy nanocomposites for electroactive devices. Functionalisation using IL contributes to a well-exfoliated conductive network of graphene and optimised thermal interface between matrix and filler to develop high thermal performance material. The heat dissipation performance demonstrated quicker cooling time when compared with nanocomposites without IL incorporation as indicated in Figure 1.1.

1.3 Energy storage

Polymer composites as an energy storage material had been widely used in place of liquid electrolytes as they offer several advantages including improved safety, enhanced stability and reduced cost. These composites constitute of polymer matrix and one or more energy storage fillers acting as an electroactive particles or redox-active species. Lithium-ion or li-ion battery is the most commonly found application of polymer electrolyte through the mechanism of conducting ions. Unlike battery casing mechanism, energy storage materials do not require critical high thermal conductivity. It prioritised on the energy or power density to store more energy for longer usage or greater performance. Though, thermal energy storage demands good thermal conductivity for efficient heat transfer and thermal management. Nowadays, multifunctional approaches for energy

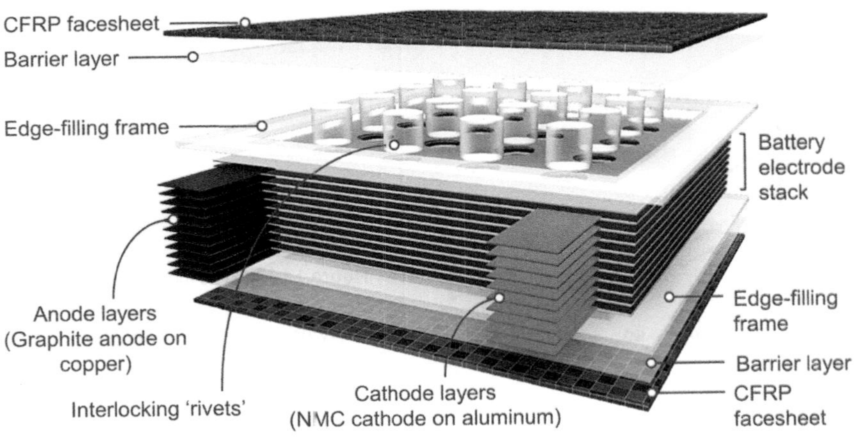

Figure 1.2: The 3D illustration diagram of multifunctional energy storage composite [9].

storage devices are sought to maintain energy density and battery safety. Besides having high ionic conductivity, polymer electrolytes can be fabricated into thin and flexible batteries of high energy density. However, it is difficult to balance between high ionic conductivity and mechanical stability of the polymer electrolytes. If surplus loading of lithium salts added, the polymer will end up too brittle thus leading to crack and lowered performance. Hence, development of multifunctional structures can be integrated to offset or compensate the flaws of the individual battery structure. In a study by [9], an innovative multifunctional energy storage composite (MESC) structure comprising stack of li-ion battery electrodes presented an improvement of mechanical stiffness and strength over conventional pouch cells. The encapsulated electrodes were protected by a high-strength carbon fibre composites and stabilised by interlocking polymer rivets. The structural design enables efficient load transfer between battery layers contributing to the load-bearing performance while storing electrical energy without fluctuating battery constituent. Figure 1.2 elucidates the multifunctional energy storage composite structure. Also, the implementation of thin solid electrolyte is likely to raise risk of short-circuit in battery. In that manner, [10] adapted ultrathin nanoporous polyimide (PI) film filled with polyethylene oxide/lithium bis (trifluoromethanesulfonyl) imide (PEO/LiTFSI) for a safe solid polymer electrolyte. The study indicated that the nonflammable and mechanically robust of PI film helps to prevent short-circuit occurrence of the PI/PEO/LiTFSI. All-solid-state li-ion batteries designed using PI/PEO/LiTFSI solid electrolyte demonstrated good cycling performance of 200 cycles at C/2 rate at 60 °C and endure harsh tests of bending, cutting and nail penetration. Meanwhile, [11] incorporated lithium aluminium titanium phosphate (LATP) nanoparticles into PEO/lithium trifluoromethanesulfonate (LiOTf) to enhance the performance of li-ion battery. The presence of LATP ceramic resolves the lack in thermal stability and quickly evaporated solvents in conventional li-ion batteries and liquid electrolyte when operating at elevated temperature of 60 °C for a long duration. Additionally, higher temperature induced the ionic conductivity of PEO/LiOTf/LATP solid electrolyte.

Yusuf et al. [12] used graphene nanoplatelets (GNPs) as nano-fillers; it was found that mechanical and thermal properties of green polymer matrices were improved. The low percentage of loading of GNPs demonstrated an enhanced thermal stability of the polymer matrix with 0.25 wt% loading, which caused the thermal degradation temperature (Td) of GNPs curve with 0.25 wt% loading to move toward a higher temperature (from 336 °C to 342 °C). Overall, the findings demonstrated notable enhancements in the performance of green nanocomposite materials with low weight percentages of GNPs, which can be used as a material for battery casing.

1.4 Challenges and future prospective

Both difficulties and strong future potential are presented using composites in battery casings and energy storage devices. Assuring the mechanical strength and longevity of composite materials used in battery casings is one of the key problems. The operating conditions that battery casings must be able to tolerate include temperature changes, mechanical strains and vibrations. Thus, it is essential to create composites with improved mechanical qualities, such as high strength and stiffness, while preserving lightweight attributes. Most battery pack enclosure designs took mechanical tests like vibration loading and deformation into account. However, because the battery must function under real-world loading conditions, a more thorough investigation must be conducted for testing involving elevation and external forces like compressive, tensile and torsion caused by crashes as design optimisation parameters [6]. Optimising the thermal management of composites in battery casings is another challenge. To avoid overheating and potential thermal runaway, heat produced during the charging and discharging of batteries needs to be efficiently dispersed. At the battery cell's contact area, the energy lost will be converted to heat [13]. According to [14], battery damage can come from thermal runaway or quicker cell ageing when the battery is heated. Even safety hazards like fire or explosions are likely in excessive temperature situations. The safety and endurance of battery systems depend on the design of composites with superior thermal conductivity and the creation of effective cooling systems [15].

Furthermore, overcoming concerns with electrical conductivity is necessary for the integration of composites in energy storage applications. For effective energy storage and distribution, adequate electrical paths must be established inside the battery system because composites are often insulating materials. Potential solutions to this problem include adding conductive fillers or creating composite constructions with inherent conductivity. By using cutting-edge and biodegradable materials that can be employed in these energy storage devices, researchers have concentrated on making these devices inexpensive as well as sustainable during the past ten years [16].

Despite the difficulties, composites have a bright future in battery casings and energy storage technologies. Due to its excellent strength-to-weight ratio and design flexibility, it is possible to create thin, light battery casings that have higher energy densities and better performance. Composites can also provide resistance to corrosion and chemical deterioration, extending the longevity and durability of energy storage devices. Additionally, continuing research and development projects concentrate on enhancing composite material manufacturing procedures to make them more affordable and scalable for large-scale battery production. Innovative battery casing designs and cutting-edge energy storage solutions may become possible because of advances in nanotechnology and the use of sophisticated fibres and resins. The main goal of current research is to increase energy density of composites. In comparison to other carbon materials, graphene has a large specific surface area, high conductivity and strong chemical stability, making it an excellent choice for the electrode material of a flexible supercapacitor made of nanocellulose and graphene. Cellulose can also be utilised as a foundation material or an internal electrolyte reservoir because it has a high aspect ratio, good mechanical characteristics, flexibility and hydrophilicity. Materials made of graphene-cellulose are suitable as electrode materials for supercapacitors with high energy and power densities. In addition, more functional and performance enhancements, as well as a decrease in production costs and environmental effects, are needed for the practical deployment of hybrid composite as next-generation materials. Although mechanical strength, thermal management and electrical conductivity present difficulties, composites have a lot of potential for use in battery casings and other energy storage applications. Composites are anticipated to play a crucial role in determining the future of high-performance and sustainable energy storage systems with sustained research and technical development [17, 18].

1.5 Summary

Selection of right material for battery enclosure is crucial part of the battery. Polymer composites are best suited because of their durability, efficiency, productivity and its capability to act as energy storage. Polymer composites are a good alternative because of high strength, lightweight, corrosive-resistant and ease processability. Polymer composites are used as energy storage and has replaced liquid electrolytes because of improved safety, enhanced stability and reduced cost. Many research have concluded that polymer composites are a good replacement as a material for battery casing.

Acknowledgments: The authors would like to thank the editors Mohd Sapuan Salit, Mohd Roshdi Hassan, Eris Elianddy Supeni and Azizan As'arry for their guidance and review of this article before its publication.

References

1. Du X, Zhang Z, Liu W, Deng Y. Nanocellulose-based conductive materials and their emerging applications in energy devices – a review. Nano Energy 2017;35:299–320.
2. Yusuf J, Firdaus AHM, Sapuan SM, Rashid U, Ilyas RA, Hassan MR, et al. Nanocellulose-graphene hybrid composites: fabrication, characterization, applications and environmental impact. Int J Biol Macromol 2024;282:137244.
3. Zhang Y, Park SJ. Imidazolium-optimized conductive interfaces in multilayer graphene nanoplatelet/epoxy composites for thermal management applications and electroactive devices. Polymer 2019;168:53–60.
4. Huang Y, Kormakov S, He X, Gao X, Zheng X, Liu Y, et al. Conductive polymer composites from renewable resources: an overview of preparation, properties, and applications. Polymers 2019;11:1–32.
5. Yusuf J, Sapuan SM, Ansari MA, Siddiqui VU, Jamal T, Ilyas RA, et al. Exploring nanocellulose frontiers: a comprehensive review of its extraction, properties, and pioneering applications in the automotive and biomedical industries. Int J Biol Macromol 2024;255:121–28.
6. Dhoke A, Dalavi A. A critical review on lightweight design of battery pack enclosure for electric vehicles. Int J Sustain Transp Technol 2021;4:53–62.
7. Huang YH, Cheng WL, Zhao R. Thermal management of Li-ion battery pack with the application of flexible form-stable composite phase change materials. Energy Convers Manag 2019;182:9–20.
8. Lee J, Abidi A, Mohammad Sajadi S, El-Shafay AS, Degani M, Sharifpur M. Study of the effect of the aspect ratio of a cylindrical lithium-ion battery enclosure in an air-cooled thermal management system. J Energy Storage 2022;45:103684.
9. Ladpli P, Nardari R, Kopsaftopoulos F, Chang FK. Multifunctional energy storage composite structures with embedded lithium-ion batteries. J Power Sources 2019;414:517–29.
10. Wan J, Xie J, Kong X, Liu Z, Liu K, Shi F, et al. Ultrathin, flexible, solid polymer composite electrolyte enabled with aligned nanoporous host for lithium batteries. Nat Nanotechnol 2019;14:705–11.
11. Yu X, Manthiram A. A long cycle life, all-solid-state lithium battery with a ceramic-polymer composite electrolyte. ACS Appl Mater Interfaces 2020;3:2916–24.
12. Yusuf J, Sapuan SM, Rashid U, Ilyas RA, Hassan MR. Thermal, mechanical, thermo-mechanical and morphological properties of graphene nanoplatelets reinforced green epoxy nanocomposites. Polym Compos 2024;45:1998–2011.
13. Li W, Garg A, Xiao M, Peng X, Le Phung ML, Tran VM, et al. Intelligent optimization methodology of battery pack for electric vehicles: a multidisciplinary perspective. Int J Energy Res 2020;44:9686–706.
14. Yang N, Zhang X, Shang B, Li G. Unbalanced discharging and aging due to temperature differences among the cells in a lithium-ion battery pack with parallel combination. J Power Sources 2016;306:733–41.
15. Zwicker MFR, Moghadam M, Zhang W, Nielsen CV. Journal of Advanced Joining Processes Automotive battery pack manufacturing – a review of battery to tab. J Adv Joining Processes 2020;1:100017.
16. Lasrado D, Ahankari S, Kar K. Nanocellulose-based polymer composites for energy applications—a review. J Appl Polym Sci 2020;137:48959.
17. Xing J, Tao P, Wu Z, Xing C, Liao X, Nie S. Nanocellulose-graphene composites: a promising nanomaterial for fl exible supercapacitors. Carbohydr Polym 2019;207:447–59.
18. Trache D, Thakur VK, Boukherroub R. Cellulose nanocrystals/graphene hybrids—a promising new class of materials for advanced applications. Nanomaterials 2020;10:1–34.

Eris Elianddy Supeni*, Azizan As'arry and Haider Jaafar Chilabi

2 Polymer composite-based turbine in wind energy harvesting

Abstract: This study explores the development and optimization of polymer composite-based wind turbine blades, integrating glass fiber reinforced plastic (GFRP) with shape memory alloy (SMA) to enhance performance in wind energy harvesting. Advances in materials science, aerodynamics, computational modelling, and structural analysis have been leveraged to improve blade efficiency, durability, and self-adaptive capabilities. The research employs finite element analysis (FEA) and artificial neural networks (ANN) to evaluate the mechanical behaviour of composite blades under varying loads. A graded beam model was developed to assess the effects of ply drop-off and material distribution on structural integrity. Experimental validation confirmed that SMA integration enhances blade deformation recovery, mitigating stress accumulation and improving aerodynamic stability. The results demonstrate that GFRP-SMA blades achieve a performance coefficient approaching the Betz limit (0.5923), reducing deflections and improving load response. Despite these advancements, challenges remain in optimizing SMA wire placement, adhesion, and actuation efficiency. Future work should focus on refining material interfaces, developing adaptive control mechanisms, and validating the model in full-scale wind turbine applications. This study contributes to the next-generation smart wind turbine blade design, addressing structural limitations while enhancing energy efficiency and operational resilience.

Keywords: glass fiber reinforced plastic (GFRP); shape memory alloy (SMA); artificial neural network (ANN); finite element analysis (FEA); smart wind turbine blade

2.1 Introduction

Wind turbine technology dormant for several years, experienced a revival at the end of the 20th century, opening up new possibilities. Advances in various technological fields were integrated into wind turbine development, significantly enhancing its resurgence. Among the key disciplines contributing to the advancement of modern wind turbines are

---

***Corresponding author: Eris Elianddy Supeni**, Department of Mechanical and Manufacturing Engineering, Faculty of Engineering, Universiti Putra Malaysia, Serdang, Selangor 43400, Malaysia, E-mail: eris@upm.edu.my
Azizan As'arry, Department of Mechanical and Manufacturing Engineering, Faculty of Engineering, Universiti Putra Malaysia, Serdang, Selangor 43400, Malaysia, E-mail: zizan@upm.edu.my
Haider Jaafar Chilabi, Department of Mechanical and Manufacturing Engineering, Faculty of Engineering, Universiti Putra Malaysia, Serdang, Selangor 43400, Malaysia; and Midland Refineries Company (MRC), Ministry of Oil, Republic of Iraq, Baghdad 10022, Iraq, E-mail: haider.chilabi@gmail.com

As per De Gruyter's policy this article has previously been published in the journal Physical Sciences Reviews. Please cite as:
E. E. Supeni, A. As'arry and H. J. Chilabi "Polymer composite-based turbine in wind energy harvesting". *Physical Sciences Reviews* [Online] 2025. DOI: 10.1515/psr-2024-0020 | https://doi.org/10.1515/9783111234762-002

material science, computer science, aerodynamics, analytical methods, testing and power control. Materials science has introduced advanced composites and lightweight materials, improving efficiency and durability.

Composites for the blades and alloys for the metal components [1, 2]. Developments in computer technologies facilitate design, analysis, monitoring and control. Aerodynamics design methods, originally developed for the aerospace industry, have now been adapted to wind turbines [3, 4]. Analytical methods have now developed to the point where it is possible to have a much clearer understanding of how the new design should perform than was previously possible. Testing using a vast array of commercially available sensors and data collection and analysis equipment allows designers to understand how the new turbines actually perform. Power electronics is a relatively new area which is just beginning to be used with wind turbines. Power electronic devices can control the turbine's generator smoothly which plays a crucial role in the electrical network: allowing the turbine to run at variable speed, producing more energy, reducing fatigue damage, and benefiting the utility in the process; facilitating operation in a small, isolated network; and transferring energy to and from storage [4].

Materials such as fibre-reinforced composites are becoming an attractive option in the wind turbine blade industry in recent years due to their several advantages such as their high strength and light weight. Furthermore, composite structures have been used widely in industry, such as in the aerospace, marine and automotive field, and they have now transitioned to become viable materials for other applications such as mechanical structures, particularly for lightweight structural applications. One major area for the lightweight structure is wind turbine blades [1].

Fibre composites are a family of advanced materials formed by combining at least two elements together to produce properties which are different from those of the base elements. Most composites consist of a matrix material and a reinforcement material, usually in a fibre form, added primarily to increase the strength and stiffness of the matrix. Composites like glass fiber reinforced plastics (GFRP) and carbon fiber reinforced plastics (CFRP) are also non-corroding and non-magnetic, and can be directionally tailored to achieve strength, stiffness, or flexibility where it is most needed [5]. Fibre composites are used to manufacture high-performing products and components.

Wind turbine blades are usually made in polymer matrix composites, often glued together. Such structures may lead to a number of possible failure modes. For example, cyclic loading that can ultimately results in wind turbine blades failure. Wind turbine are subjected to fluctuating wind and hence flapwise bending are acting on the blades. Components, which are subject to repeated axial bending, such as wind turbine blades, may eventually develop cracks, which ultimately may make the component, break. Therefore, failure of wind turbine blade could be a big catastrophic if there is no immediate solution for safety measures which has not been taken into consideration [5–7].

Modern wind turbine blades, made from glass fiber-reinforced plastics (GFRPs) due to their high strength-to-weight ratio, feature strategically oriented fibres in the spar caps and shear webs, combined with a reinforced truss framework, to efficiently withstand bending and shear loads while optimizing structural integrity and cost. The wind turbine blade's shell, designed for both aerodynamic efficiency and structural reinforcement, uses a sandwich construction with GFRP skins and a low-density core to enhance torsional resistance, manage shear loads, and provide bending stiffness, while also incorporating longitudinal and diagonal fibres to support flapwise and edgewise bending, particularly reinforcing the blade near the root where bending moments are most significant. The study proposes a conceptual model to enhance wind turbine blade efficiency by integrating glass fiber reinforced plastic (GFRP) with shape memory alloy (SMA), resulting in a thinner, lighter, and more cost-effective design that addresses deflections, eases stress on composites during bending, and improves blade performance and longevity through strategic reinforcement, particularly at critical areas like the spar box.

2.2 Materials and methods

2.2.1 FEA model and ANN model development

2.2.1.1 Finite element model development

Finite element analysis (FEA) is critical in developing GFRP wind turbine blades, enabling accurate prediction of global behaviours like stiffness, stress, and deflections. However, modelling localized deformations in smart composite blades presents challenges due to the non-linear and complex nature of these stresses, often requiring detailed and computationally expensive 3D models [8]. Calibration is essential to account for manufacturing tolerances and imperfections, ensuring that GFRP blades meet certification standards and provide reliable structural performance [9]. Studies such as Jensen et al. [10] highlight the importance of FEA in understanding both global and local blade behaviours [6, 8, 11].

In finite element analysis (FEA) of a wind turbine blade, a 3D structure under arbitrary loads was modelled using plate elements to simulate deflection. The blade, treated as a cantilevered box beam, was represented by a GFRP plate with a specified ply configuration (0°/90°/–45°/45°, etc.) and modelled in ABAQUS® using S4R elements, assuming perfect bonding between epoxy and glass fibre. The model incorporated a composite layup and included shape memory alloy (SMA) wires simulated with Flexinol 90 properties, allowing for the analysis of GFRP-SMA interaction under various configurations and conditions, with precise meshing for both materials.

A numerical model was developed in ABAQUS 6.12® to analyse the deflection behaviour of a graded GFRP beam using the SC8R element type. The model, based on a specific ply configuration and mechanical properties, assumed perfect bonding between epoxy and glass fibre. Material properties such as modulus of elasticity, shear modulus, and Poisson's ratio were used, ensuring consistent unit application throughout the

Table 2.1: Laminate configurations.

Configuration	Ply orientation (°)	Length (mm)
1	0	600
2	90	550
3	−45	520
4	45	480
5	−45	430
6	45	410
7	90	360
8	0	320
9	90	270
10	0	220
11	45	160
12	−45	110

Table 2.2: Mechanical properties of SMA, GFRP, Epoxy and Core.

Material	SMA Nitinol	Fibre E-glass UD	Epoxy Kinetix R246TX	Core Foam
E_1 (MPa)	32,000 (martensite) 83,000 (austenite)	34,412	10,500	20
E_2 (MPa)	–	6,531	–	–
ν_{12}	0.3	0.217	0.28	0.3
G_{12} (MPa)	–	2,433	–	–
G_{13} (MPa)	–	1,698	–	–
G_{23} (MPa)	–	2,433	–	–
ρ (kg/m^3)	6,450	2,000	–	–
Elastic	Isotropic	Lamina	Isotropic	Isotropic

analysis despite ABAQUS/CAE® being unit-independent. The ply configuration as shown in Table 2.1 and the mechanical properties as shown in Table 2.2, were used to simulate the model of a graded beam with the element type SC8R.

2.2.1.2 Types of interactions and boundary conditions

Efficient and accurate modelling of structural behaviour requires careful treatment of interactions between different materials in the structure. For the SMA wire and GFRP plate model, Constraint 2 was used, applying *MPC Constraints instead of *TIE Constraints. This approach models the composite plate and SMA wire with different elements, tying the slave and master faces using a *PIN MPC Type, ensuring that the slave surface's translational and rotational motions match those of the nearest master surface point throughout the simulation, as illustrated in Figure 2.1. The fabrication of the physical model was performed to facilitate the use of the model within the experimental work. The plate-like structure was made from fibreglass and the SMA wires were incorporated in the plate. The formulation of a mathematical model using discrete mathematical elements and their connections and interactions to capture the prototype behaviour is called discretization for this purpose:

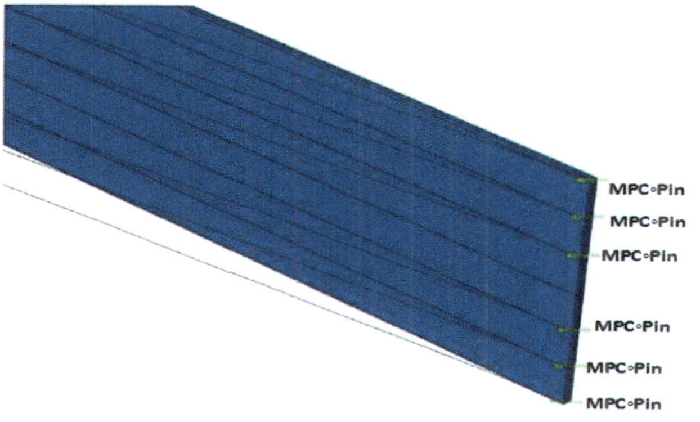

Figure 2.1: MPC implemented between the SMA wires and the GFRP plate (end section view).

(i) Joints/nodes were used to discretise elements and primary locations in the structure at the displacements which are of interest;
(ii) Elements were connected to each other at joints;
(iii) Masses, inertia, and loads were applied to elements and then transferred to joints.

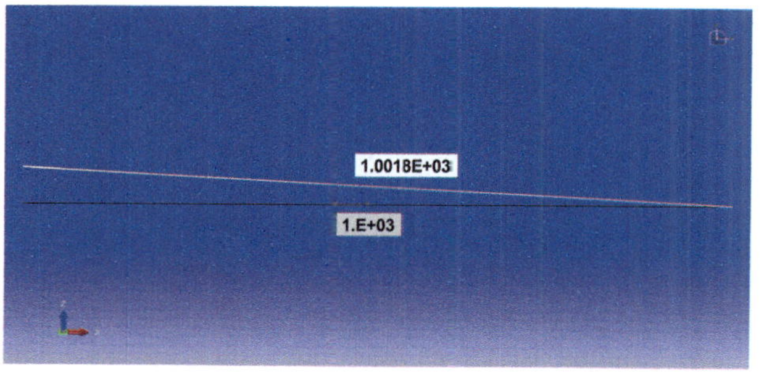

Figure 2.2: Sketching part of the wire.

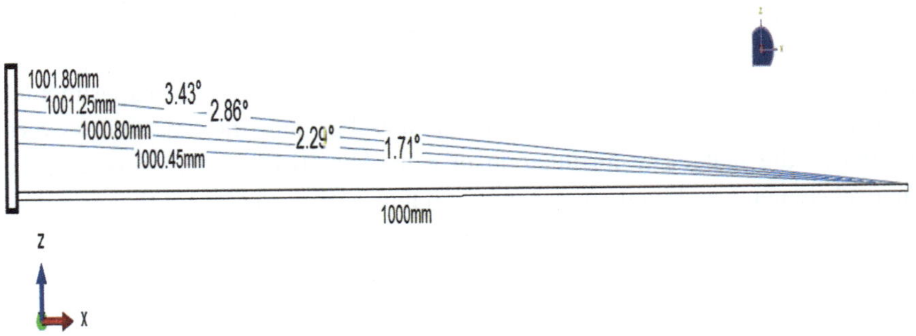

Figure 2.3: The orientation of the angle between the SMA wire and the GFRP plate in the side view angle.

Incorporating shape memory alloy (SMA) with Flexinol 90 properties into the finite element model, the T3D2 element type was used to simulate a 0.51 mm diameter wire, as shown in Figure 2.2. The mechanical properties, including Young's modulus, were analysed using stress-strain data from uniaxial tests. The wire, 1,001.80 mm in length, was optimized at a height with a 3.43° angle from the tip, as illustrated in Figure 2.3.

2.2.2 Experimental setup

2.2.2.1 Specimen fabrication

Unidirectional GFRP with ply configurations of 0°, 45°, −45° and 90° were measured using a protractor. The GFRP specimens were cut into the specified length and angle orientation as shown in Figure 2.4a–c which are illustrated respectively. The steel base plate was completely cleaned before carrying out the specimen preparation. It was cleaned using acetone solvent to remove any dirt and contamination. Then steel plate was wiped with a dry cloth until fully dry. Once it was cleaned, the steel plate was waxed with carnauba wax eight times. The wax agent was used as a release agent, which provided a smooth surface coating like a release film, enabled to remove the two layers between the fabricated specimen and the steel plate. This technique has been widely used in the fabrication of GFRP.

Specimens of fibreglass were cut from the roll with respect to their orientation into A3 paper size. Then, the cut GFRP was weighed to estimate the quantity of epoxy and hardener that would be needed for fabrication. In this fabrication, a hand lay-up method was used to control the specimen size. Bear in mind the following factors: the direction of the rollers must be in the direction of the ply orientation, a uniform distribution must be applied when rolling, and also any contamination must be avoided during preparation operations. Ensure that personal protective equipment (PPE) is worn for health and safety reasons because the preparation and handling of the specimen involves hazardous chemicals and caustic solvents.

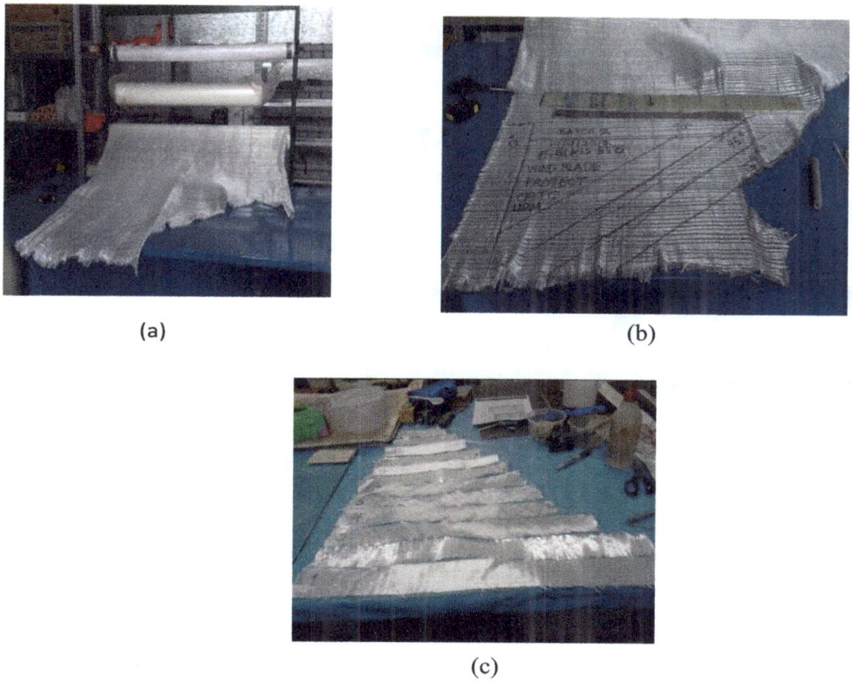

Figure 2.4: Photograph of specimen preparation and as follow: (a) roll of fibre; (b) angle orientation measurement; (c) material specimen preparation.

2.2.2.2 Preparation of the epoxy resin

GFRP with different orientations was weighed using a digital weighing scale. Let's say the total mass of the fibreglass in gram is m_{total}. The mass of the epoxy is m_{epoxy} and the mass of the hardener is m_{hardener}. These are determined by the weight fraction which is shown in Equation (2.1). The fraction mixtures of the epoxy and hardener are shown as follows:

$$m_{\text{total}} = 4 \times m_{\text{epoxy}} + \times m_{\text{hardener}} \tag{2.1}$$

The hardener solvent was stirred with the spatula until thoroughly mixed. The composition of the capacity of the resin and hardener mixture mainly depends on the total number of plies. For example, if there are four plies then the mixture of resin and hardener will be in a quarter ratio, respectively.

2.2.2.3 Calibration of SMA wires attached to a plate-like structure

A 300 mm length of the SMA wires is tested for calibration, which is shown in Figure 2.5 When the SMA wires are heated, they will contract by up to 10 % of their original length,

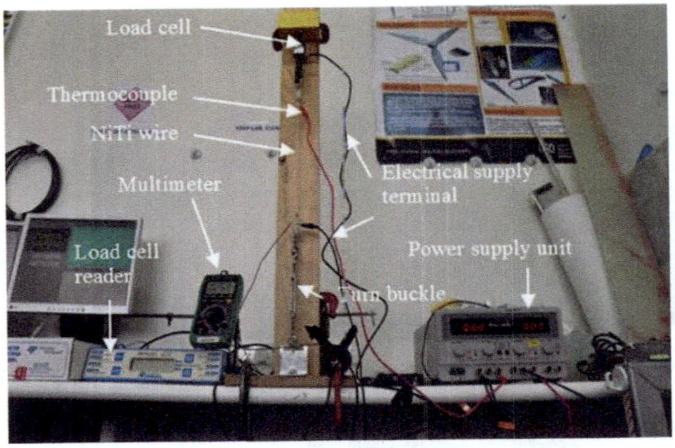

Figure 2.5: Photograph of the calibration setup for SMA.

but the total volume remains constant. A technique transducer, a load cell with a measuring limit up to 40 N, is used to determine the load during the test.

The wire is mounted vertically and stretched with a turn buckle at 4 % elongation. The top end of the wire is connected to the load cell while the bottom end has been connected with a turn buckle. A support block, which acts as a vertical pole, was used as a platform. The calibration was made with the aim of obtaining the readings for the tension force exerted with the amount of the electric current applied to the wire. The connection of the cable must be sturdy, and the coupling connection is inelastic. From these observations, the applied load can be estimated with respect to the electric current. As the SMAs were electrically heated, there was a temperature rise.

2.2.2.4 Deflection test for a GFRP plate

In order to demonstrate the deflection test for a GFRP plate, a test rig was custom designed and constructed at the Workshop of the Faculty of Engineering and Surveying, at the University of Southern Queensland. The test rig was made of square hollow steel with a dimension of 50(W) × 50(H) × 0.5(D) mm which is shown in Figure 2.5. The test rig is capable of a maximum hold load up to 700 N. The test rig is placed on a flat and roughly level surface in order to ensure body stability. To ensure that the plate is held tightly, the clamping zone has been fitted with two nuts and a gripper.

2.2.2.5 Experimental setup arrangement

In this section, an adequate power supply of GW GPC-3030DQ was used to generate a constant current flow through the six wires for a large scale model which is about 1/10

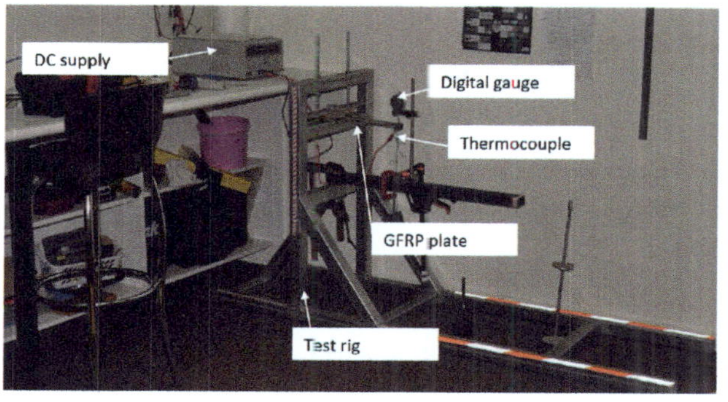

Figure 2.6: Experimental setup arrangement with the test rig.

scale to 2.5 m blade size. In order to supply a relatively high voltage (up to 21 V), the driven switches on the slave and master panels were turned into series mode. In the present work, SMA wires are deformed by straining them using a designated frame and then they are integrated with GFRP cantilevers that serve as bias springs. In the cold state, the GFRP plate cantilevers assume their flat shape, except for a slight residual bending that results from the static equilibrium with the wires in the martensitic phase. The detail of the experimental setup is shown in Figure 2.6.

Upon heating, the wires contract and bend the cantilevers out of the plane. In order to translate the axial forces generated upon heating into a bending moment and thus to allow out-of-plane displacement, the wires are placed eccentrically onto the GFRP plate. This was accomplished by anchoring one end of the wires to the fixed clamp of the GFRP cantilever (fixed anchor) and the other end of the wires to the flexible side of the cantilever (moving anchor).

The SMA wires on the GFRP plate assembly were mounted at the end plate and in parallel. The cable wire was connected to the tower installation height adjustable anchors, which is displayed in Figures 2.7 and 2.8.

The arrangement for ordering the wire, such as 2, 4 and 6, was quite easy, whereby, the wire electrical terminals were installed at the end of where the wire was clamped. For an odd order of wire arrangement, for examples 1, 3 and 5, the terminal wirings were made at both sides of the plate which were at the tip and the root side.

For safety considerations, all the wires have been insulated with the fibreglass sleeve cloth material. Similar to the deflection test in 300 mm, the SMA wires have been mounted parallel to the GFRP plate. The same arrangement of the SMA wires has been implemented, but the blade is longer as well as the angle being bigger (Figure 2.9). The size of the angle is determined by an angle of elevation to the plate. As the anchoring height increases, the angle of elevation increases as well.

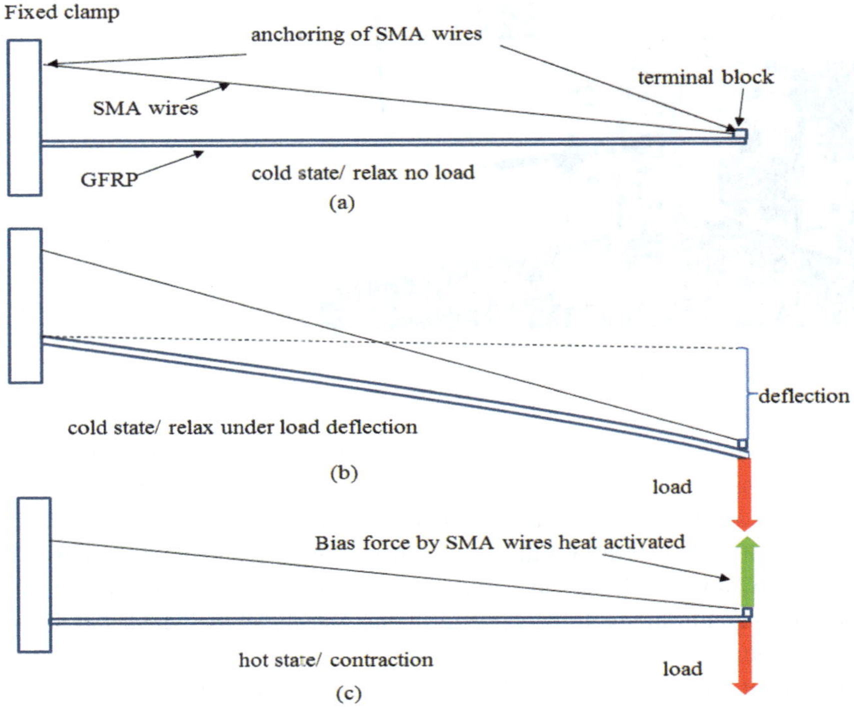

Figure 2.7: Schematic operating principle.

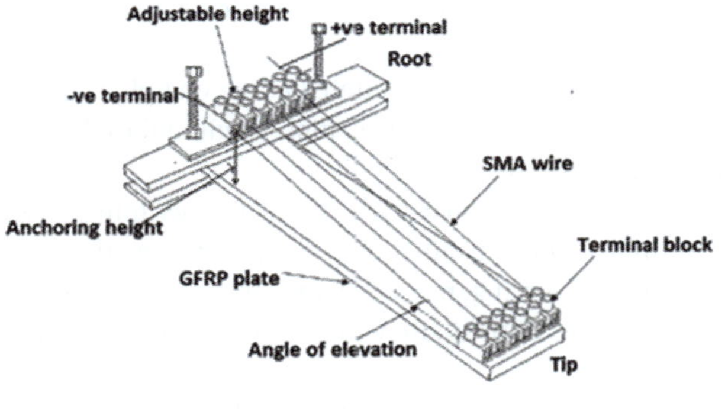

Figure 2.8: Schematic diagram of SMA wires arrangement.

A series of tests have been undertaken to determine the response of the deflection under heating condition as shown in Figure 2.10. A constant current has been generated to contract the SMA wires. The purpose of the experiment, which is shown in Figure 2.11, was to straighten the smart blade which was modelled by the plate. The straightening

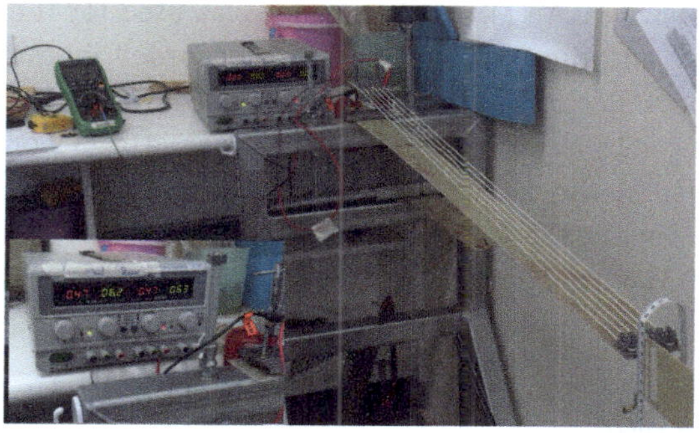

Figure 2.9: Photograph of experimental setup and values reading of the power supply in series mode driven when is current is activated (insert picture).

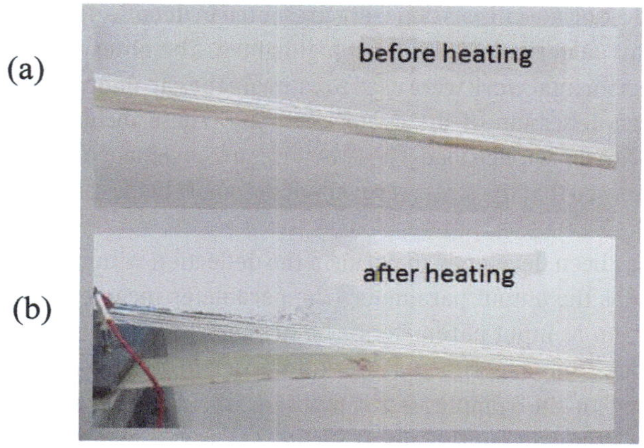

Figure 2.10: Photograph of the tip deflection (a) before heating, (b) after heating (section view).

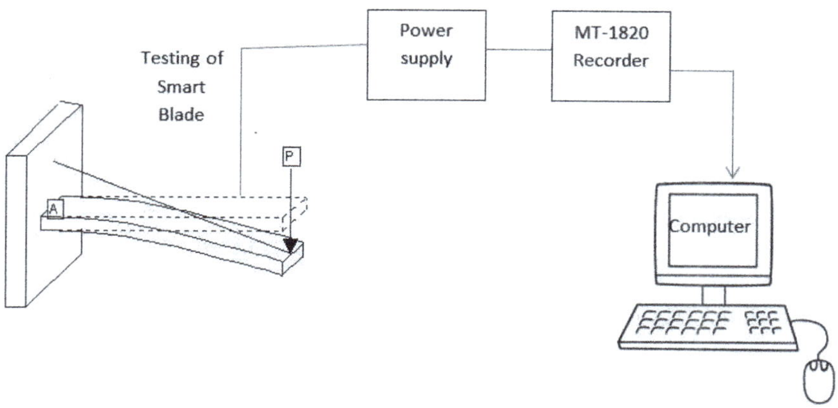

Figure 2.11: Schematic diagram of the experimental setup.

mechanism was actuated by the SMA wires. An MT-1820 recorder was used for the process of obtaining the temperature and time response and was recorded by the computer. The measurement of the tip deflection was precisely measured by a digital dial gauge. The SMA wires were powered by a current supplied in the range 0–25 V or 0–5 Amp: this was manually controlled by the dial knob which was provided on the constant power supply unit. The experiment was carried out at room temperature in the P2 Building, in the Structural Health Monitoring Laboratory, CEEFC, USQ Australia.

2.2.2.6 Summary of the experimental setup

In this chapter, the findings and associated discussion were presented in detail. A feasible study of the application of smart material has also been highlighted. The observations and data collected from the experimental work were used to estimate the capability of the SMA wires to recover the large deflection of the composite plate. These include the amount of the electric current with the applied load. The SMA wires arrangement also has been conducted. It has been observed that the stress recovery of the smart blades also has been successfully performed.

A neural network model has been developed to perform the deflection with respect to a number of wires required as the output parameter. The parameter includes load, current, time taken and deflection as input parameters. The network has been trained with experimental data obtained from experimental work. In order to remove all the technological difficulties related to the complicated composite fabrication, a simple proof-of-concept specimen was built by mounting pre-strained SMA wires on top of the laminated composite plate strip.

2.3 Results and discussion

The following section presents the results and discussion in alignment with the analyses outlined earlier. The content in this section is subdivided into five main sections, namely the characterization of the GFRP and the thermo-mechanical behaviour. Next, describes the calibration test, the deflection test and the deflection of the graded beam. Then, discusses the tuning FEA for large deflection and the tuning-up the ANN. After that, analyses the preliminary study of the mechanism, deflection and load relationship. Finally, discusses the SMA wires arrangement, stress recovery and performance analysis, followed by a concluding remark at the end of the chapter.

Figure 2.12: Tensile test setup with measurement of longitudinal and transverse strains by a contact extensometer.

2.3.1 Characterization of the GFRP

The tensile test of the composite materials glass fiber-reinforced plastic (GFRP) was performed using the well-established techniques and is governed by the standard test method. The test speed used is 50 mm/min for measuring strength and elongation and 1 mm/min for measuring modulus. The tensile test is performed in accordance with ASTM D3039 (D3039/3039M 2008). Figure 2.12 shows the tensile test setup with a contact extensometer used to measure elongation.

Table 2.3: Four parts of graded beam assembly.

Part	Material	No. of plies	Ply configuration
1	GFRP	4	[0°/90°/–45°/+45°]
2	GFRP	8	[0°/90°/–45°/+45°/–45°/+45°/90°/0°]
3	GFRP	12	[0°/90°/–45°/+45°/–45°/+45°/90°/0°/90°/0°/+45°/–45°]
4	EPS	N/A	N/A

The purpose of the test is to validate the mechanical properties of the E-glass unidirectional (UD), namely to determine the value of the mechanical properties such as elastic modulus and Poisson's ratio Table 2.3 shows the four parts of the coupon specimen which are described in the ply configurations.

2.3.2 Deflection of the graded beam

The full assembly of the graded beam was drawn precisely in two parts, which were made of GFRP and expanded polystyrene (EPS), as depicted in Figures 2.13–2.15,

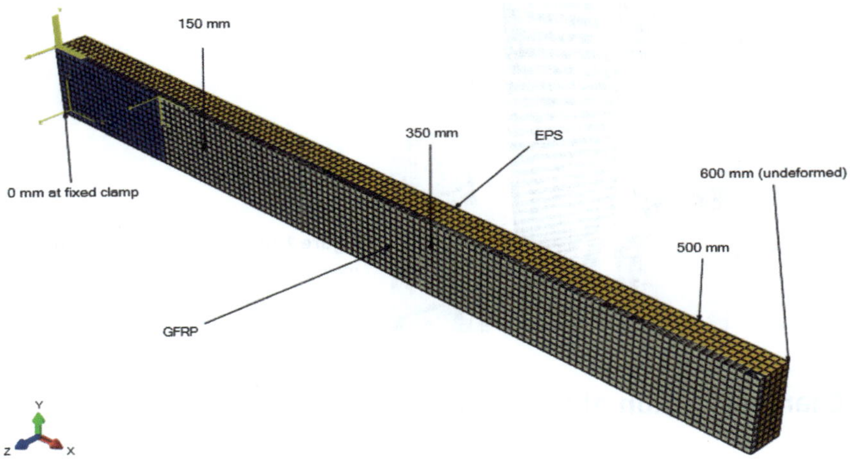

Figure 2.13: Graphical model representation with part of the assembly layout.

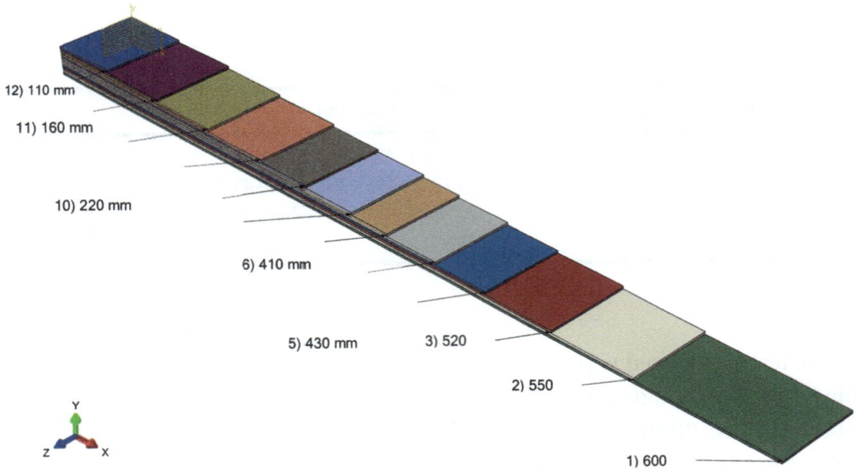

Figure 2.14: Graded beam specimen representing plies drop off imitation.

Figure 2.15 depicts the graphical model representation with part of the assembly layout. There are 12 sections which represent like a ply drop off in the turbine blade.

Figure 2.14 showcases a graded beam specimen designed to simulate ply drop-off behaviour, effectively replicating the gradual reduction of composite layers found in turbine blades. This representation provides insight into structural variations caused by ply terminations. Additionally, Figures 2.13 and 2.15 illustrate graphical model

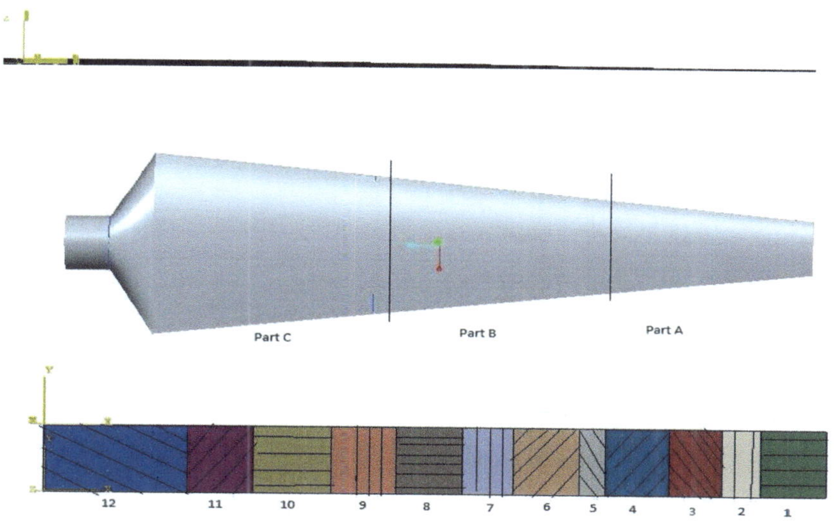

Figure 2.15: Graphical model representation with number of plies configuration (top view).

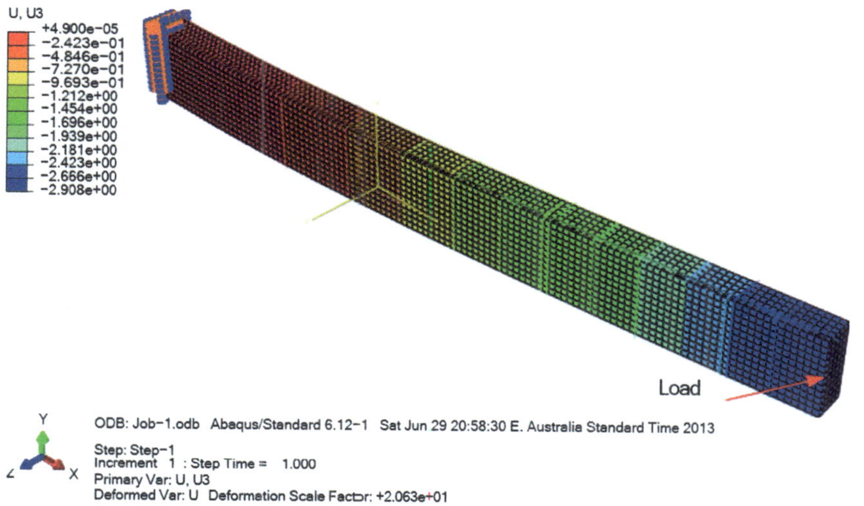

Figure 2.16: Deflection pattern profile for the whole graded beam.

representations, with Figure 2.13 depicting part of the assembly layout, while Figure 2.15 highlights 12 distinct sections that mimic ply drop-offs within the turbine blade structure.

Figure 2.16 illustrates the deflection pattern when the foam has been mounted to the beam. The deflection happened because the core of EPS has reduced the load bearing between the graded beam as there is less deflection when the load is applied.

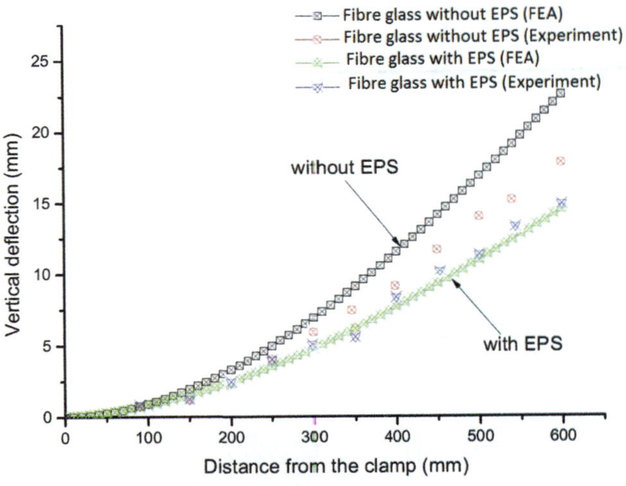

Figure 2.17: Comparison of vertical deflection of GFRP (with EPS/without EPS) between FEA and experiment.

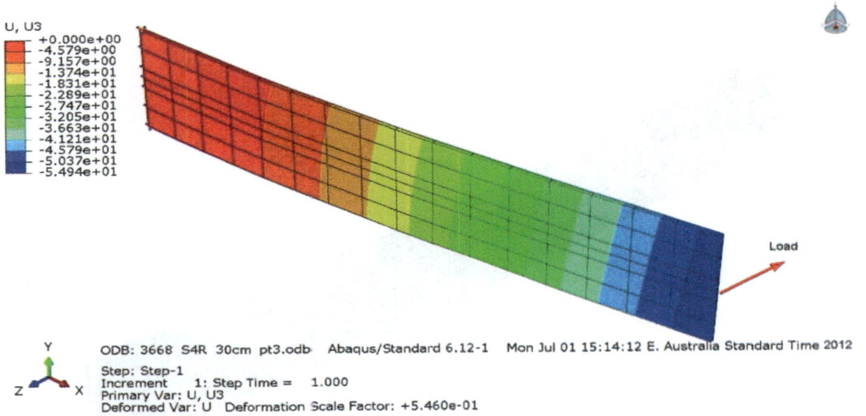

Figure 2.18: Deflection contour S4R of the GFRP plate.

Figure 2.17 shows the deflection is reduced remarkably with the EPS included in the sample. For comparison, this simulation was carried out using ABAQUS®, both models were tested without SMA.

2.3.3 Tuning FEA for large deflection of the model

A preliminary study was carried out without the SMAs incorporated, to simulate large deflections of the plate-like structure whose dimension was 300(W) × 60(H) × 4(D) mm.

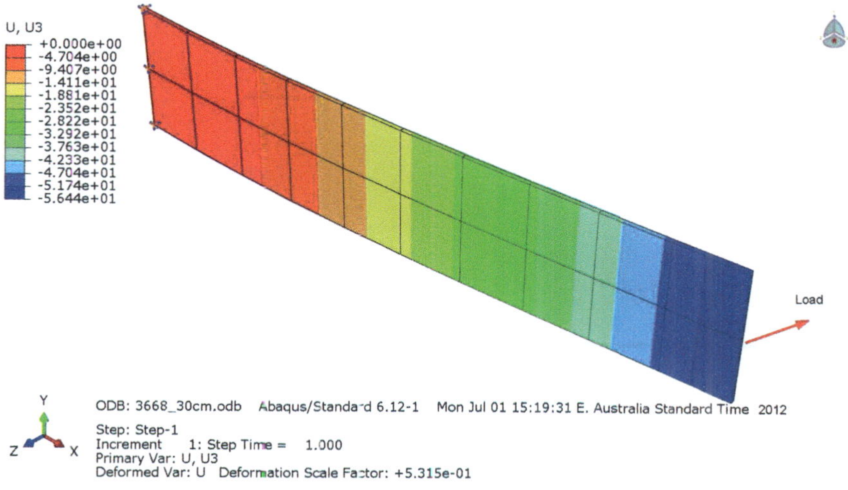

Figure 2.19: Deflection contour of SC8R of the GFRP plate.

Table 2.4: Comparison of FEA and experimental work without SMA for validation.

Method	FEA		Experiment
Element type	S4R conventional shell Reduced integration	SC8R continuum shell Reduced integration	
Tip deflection (mm)	54.78	56.44	55.40

This study was undertaken in order to visualize both models which were SC8R and S4R for deflection contours without SMA subjected to a concentrated load. Specifically, a sample loaded with 3.6 N was used to validate the model before the SMA wires were incorporated into the GFRP. The tip deflection curves were very similar along with the load applied to the specimen. This model is only accurate for the first portion of the load-deflection curve as it is not designed to predict the load-deflection at failure [11]. Both figures, Figures 2.18 and 2.19, confirm that the theoretical model is accurate. In both model specimens, the curve predicted by FEA was able to match this value in experimental condition very accurately. This value has been shown in Table 2.4. S4R recorded values close to the experimental values. So, it has been used for future element type modelling.

Either conventional shell elements or continuum shell elements might be used for the GFRP plates. However, using continuum shell elements would consume great deal of computation time. It would also cause many convergence problems due to the complicated geometric arrangement using eight nodes and very thin webs in the test specimen. Using conventional shell elements would be a more desirable option. Table 2.4 compares

Table 2.5: Plies configuration of the plate.

Configuration	Ply orientation (°)
1	0
2	90
3	−45
4	45
5	45
6	−45
7	90
8	0

the simulation results for using both solid and shell elements. The results are almost the same, giving a high confidence level in using conventional shell elements.

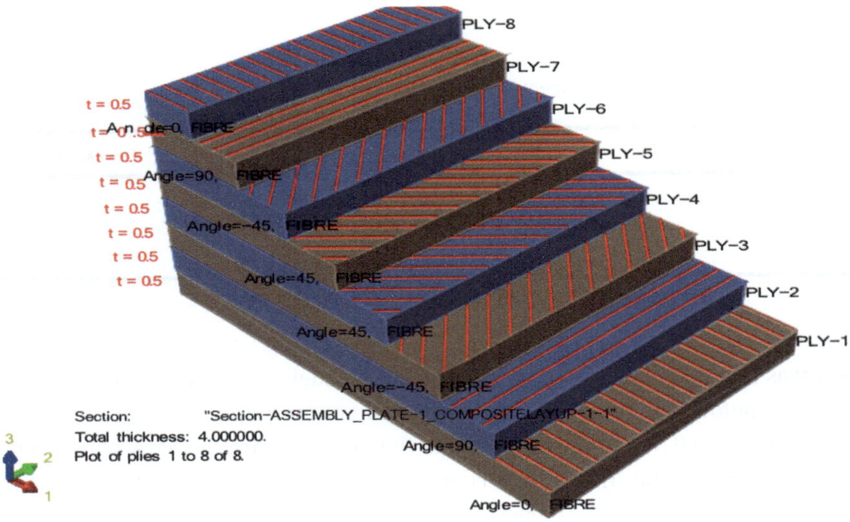

Figure 2.20: Schematic ply configuration of the actual blade.

2.3.4 Specification of specimen

The actual beam plate, with a configuration of eight plies, was fabricated in the facilities room at CEEFC, USQ Toowoomba, Australia. In this fabrication a large scale GFRP panel was used to fabricate the smaller plates [6]. The plate has dimensions of 1,000(W) × 60(H) × 4(D) mm. The plies configuration of the plate is shown in Table 2.5.

The fabrication of this specimen has the same mechanical properties used in Chapter 3. The ply configuration and the orientation is shown in Figure 2.20 below.

2.3.5 Summary of the results

In this chapter, the findings and associated discussion were presented in detail. A feasible study of the application of smart material has also been highlighted. The observations and data collected from the experimental work were used to estimate the capability of the SMA wires to recover the large deflection of the composite plate. These include the amount of the electric current with the applied load. The SMA wires arrangement also has been conducted. It has been observed that the stress recovery of the smart blades also has been successfully performed.

A neural network model has been developed to perform the deflection with respect to a number of wires required as the output parameter. The parameter includes load, current, time taken and deflection as input parameters. The network has been trained with experimental data obtained from experimental work. In order to remove all the technological difficulties related to the complicated composite fabrication, a simple proof-of-concept specimen was built by mounting pre-strained SMA wires on top of the laminated composite plate strip.

2.4 Conclusion and further work

2.4.1 Conclusions

This dissertation delivers a number of issues related to development of the pilot study of the smart blade composites. Their different constituents, namely fibre-glass and shape memory alloy elements, were first examined separately and the conclusion obtained extends beyond smart composites' blade applications.

In this study, the ability of the SMA wires to enable the plate-like structure to return to its original shape under load condition has been examined. This study shows that the proposed smart wind turbine blade, incorporating SMA as an active actuator, is feasible.

The results from the developed model show good agreement with the experimental result given the same system parameters. The model allows the performance parameters of the proposed smart turbine design to be analysed.

It has been proven that the stress recovery can be successfully achieved by straightening the deflection experienced by the wind turbine blade. In this work, a pilot study of the smart wind turbine blade was carried out in which a structural analysis based on the FEA was simulated and the results have been validated by the experimental results.

It has been shown that the prediction of performance can be analysed using ANN, which has been simulated by MATLAB. Throughout this analysis, the relationship between the number of SMA wires, the load, the current and the deflection have been successfully investigated. It was shown from ANN that all the models have predicted satisfactorily the performance behaviour of the SMA wires provided that the correlation factors (such as training, test and validation) which have been represented in the overall of data are interpreted properly.

The development of a smart blade using SMA wires has been efficiently proven not only to improve the performance coefficient close to 0.5923 (Betz limit) but also the stress recovery [12]. In this study, the large deflection has been reduced by the SMA reaction in straightening the blade into the original shape, which is an undeformed shape. SMA wires are materials that have the ability to undergo relatively large deformations and then recover the deformation when they are heat activated.

It is also felt that only time limitations and unprecedented experimental complications prevented further improvements in the control system as well as the development of the mechanism control. There is scope for practical innovation for the real scale application.

From the technological point of view, the choice of the composite constituents, the shape memory materials, and the manufacturing processes should be opti-mised according to the actual application, addressing the problems which were encountered in the present experimental work, such as the adherence of the shape memory elements to the fibre-reinforced composite.

2.4.2 Limitations of the study

The results obtained from this case study of SMA wires which were in (i) an embedded mode and (ii) a suspended mode are discussed and demonstrated to determine the feasibility of the SMA wires actuator concept. Significant operational inadequacies were, however, observed. Elimination of these inadequacies forms the basis for an improved actuator design, which is the subject of the current work.

The improvements required are detailed as follows:

(i) In the embedded mode, the temperature rise during contraction was not capable of actuating the SMA wires in large deflections of the wire since there is a restriction and uneven force acting on the wires. Increasing the number of wires is the best way to compensate, but it will cost more and be less efficient since there are lot of SMA wires involved.

(ii) Using the suspended SMA wires has been found that to be more practical compared to embedded ones. The temperature rise has been successful in actuating the SMA wires although for lower numbers of wires. However, the method of gripping the SMA wires was found to be ineffective and resulted in slippage of the wires which were connected between the GFRP and the SMA wires. Consequentially, uneven deflection angles in the pull up orientations were observed. This behaviour was notable as the loads on the system increased; Stress concentration at the points of attachment of the wires damaged the fibreglass hinge tube especially in the embedded mode [7]. This resulted in material degradation and failure was more likely to occur.

(iii) Due to the presence of multiple wires in both modes, which were the embedded and the suspended mode, difficulties were encountered in applying an equal amount of pre-strain to all the wires. It has been suggested that the uniformity of the pre-strain of the wire can be controlled by individual strain gauges.

(iv) The locking mechanism was found to be unreliable in taking back the deflection to the recovery position.

These limitations to be considered responsible for hindering the reliable tracking capability of the actuation system/self-straightening blade. The objective of this thesis is to eliminate these drawbacks to ensure a consistent tracking capability.

2.4.3 Further work

To enhance the practical application and efficiency of smart wind turbine blades, future work should focus on:

1. Improved SMA integration and actuation performance
   - Develop optimized SMA configurations that balance wire quantity, force distribution, and temperature control.
   - Investigate alternative pre-strain methods (e.g., individual strain gauges) to achieve uniform actuation.
2. Advanced material and adhesion techniques
   - Enhance the adhesion between SMA wires and fibre-reinforced composites to prevent slippage and stress-induced degradation.
   - Explore novel composite materials with improved heat transfer properties to facilitate efficient SMA activation.
3. Refinement of the locking and recovery mechanism
   - Design a more reliable locking system that effectively restores the blade's original position after actuation.
   - Implement smart control algorithms to adjust real-time blade deflection and recovery.
4. Experimental validation on a full-scale model
   - Extend the pilot study to real-scale wind turbine applications.
   - Conduct long-term durability testing under varied wind conditions to evaluate performance stability.

By addressing these areas, future research can significantly enhance the functionality, durability, and efficiency of smart wind turbine blades, making them viable for commercial deployment.

Acknowledgments: The authors would like to thank the editors S.M. Sapuan, Mohd Roshdi Hassan, Eris Elianddy Supeni and Azizan As'arry for their guidance and review of this article before its publication.

References

1. Eker B, Ovan A, Vardar A. Using of composite material in wind turbine blades. J Appl Sci 2006;6:2917–21.
2. Thomsen OT. Sandwich materials for wind turbine blades – present and future. J Sandw Struct Mater 2009; 11:7–26.
3. Eggleston D, Stoddard F. Wind turbine engineering design. New York: Van Nostrand Reinhold; 1987.
4. Burton T, Jenkins N, Sharpe D, Bossanyi E. Wind energy handbook, 2nd ed. Chichester: John Wiley & Sons, Ltd; 2011.
5. Sorensen B, Jorgensen E, Christian P, Jensen F. Improved design of large wind turbine blade of fibre composites based on studies of scale effects [PhD thesis]. Denmark: Ris National Laboratory Roskilde; 2004.
6. Mengal AN, Karuppanan S, Wahab AA. Structural analysis of basalt fiber reinforced plastic wind turbine blade. In: MATEC Web of Conferences. EDP Sciences, Les Ulis; 2014, vol 13:04019 p.
7. Kucher M, Johst P, Lizaranzu M, Lahuerta F, Böhm R. A micromechanical modeling approach for the estimation of the weathering-induced degradation of wind turbine blades. Mater Circ Econ 2023;5:17.
8. Liu JW, Liu PF, Leng JX, Wang CZ. Finite element analysis of damage mechanisms of composite wind turbine blade by considering fluid/solid interaction. Part I: full-scale structure. Compos Struct 2022;301:116212.
9. Hu W, Han I, Park SC, Choi DH. Multi-objective structural optimization of a HAWT composite blade based on ultimate limit state analysis. J Mech Sci Technol 2012;26:129–35.
10. Jensen F, Falzon B, Ankersen J, Stang H. Structural testing and numerical simulation of a 34 m composite wind turbine blade. Compos Struct 2006;76:52–61.
11. Pabut O, Allikas G, Herranen H, Talalaev R, Vene K. Model validation and structural analysis of a small wind turbine blade. In 8th International DAAAM Baltic Conference, Tallinn, Estonia; 2012. 19–21 pp.
12. Nallusamy S, Majumdar G. Testing of active vibration control on glass fiber reinforced plastic (GFRP) wind turbine blade using SMA and LabView. In: Hashmi MSJ, editor. Encyclopedia of materials: plastics and polymers. Amsterdam, The Netherlands: Elsevier; 2022: 745–54.

Eris E. Supeni*, Azizan As'arry and Haider Jaafar Chilabi

3 Advancement of energy harvesting using smart materials and composites

Abstract: This study explores advancements in energy harvesting using smart materials and composite structures, focusing on integrating shape memory alloys (SMAs) into wind turbine blades. The research investigates the structural and thermo-mechanical behavior of glass fiber-reinforced polymer (GFRP) composites and SMA wires, emphasizing their potential in improving blade adaptability and stress recovery. Finite Element Analysis (FEA) and experimental methods are utilized to evaluate deflection, load correlation, and actuation efficiency. Results demonstrate that suspended SMA wires enhance deflection recovery with lower energy loss compared to embedded configurations. A neural network model predicts deformation behavior based on load, current, and wire count. Additionally, SMA integration improves the power coefficient (Cp), bringing turbine efficiency closer to the Betz limit. Despite challenges such as nonuniform prestrain and material degradation at attachment points, this study confirms the feasibility of SMA-driven adaptive blades for wind energy applications. Future work includes optimizing SMA deployment in real-scale wind turbines through dynamic analysis and aerodynamic performance testing.

Keywords: shape memory alloy (SMA); glass fiber-reinforced polymer (GFRP); smart wind turbine blade; finite element analysis (FEA); deflection recovery

3.1 Introduction

To establish the foundation of this research, an extensive literature review has been conducted. This study examines various relevant studies that have influenced key decisions throughout the research It provides an overview of wind turbine operation, including aerodynamics, blade structure, and the evolutionary advancements and future potential of wind turbines. Additionally, it introduces the proposed conceptual framework and concludes with a summary outlining the research direction of this study. The Sandia Wind PACT Blade System Design Study (BSDS) introduced several advancements for large wind turbine blades [1]. The TPI-Dynamic Design-UC Davis team, led by Case van Dam,

***Corresponding author: Eris E. Supeni**, Department of Mechanical and Manufacturing, Faculty of Engineering, Universiti Putra Malaysia, Serdang, Selangor 43400, Malaysia, E-mail: eris@upm.edu.my
Azizan As'arry, Department of Mechanical and Manufacturing, Faculty of Engineering, Universiti Putra Malaysia, Serdang, Selangor 43400, Malaysia, E-mail: zizan@upm.edu.my
Haider Jaafar Chilabi, Department of Mechanical and Manufacturing, Faculty of Engineering, Universiti Putra Malaysia, Serdang, Selangor 43400, Malaysia; and Midland Refineries Company (MRC), Ministry of Oil, Baghdad 10022, Republic of Iraq, E-mail: haider.chilabi@gmail.com

As per De Gruyter's policy this article has previously been published in the journal Physical Sciences Reviews. Please cite as: E. E. Supeni, A. As'arry and H. J. Chilabi "Advancement of energy harvesting using smart materials and composites" *Physical Sciences Reviews* [Online] 2025. DOI: 10.1515/psr-2024-0021 | https://doi.org/10.1515/9783111234762-003

investigated thick airfoils with expanded trailing edges to improve structural integrity. They developed high-lift inboard airfoils with adjustable trailing-edge flaps, flatback airfoils, to optimize lift while considering drag increase. Additionally, aerodynamic control strategies, such as pitch regulation and smart-material-driven trailing-edge flaps, were explored [2, 3]. The ADASYS project designed a fully active full-scale rotor using piezoelectric-controlled trailing-edge flaps, effectively minimizing structural vibrations [4, 5]. Furthermore, research into morphing blade technologies, including camber control and microtabs, has been pursued to enhance aerodynamic efficiency [6, 7].

Metals have long been integral to engineering applications, and recent technological advancements have led to the development of smart materials, such as shape memory alloys (SMAs). The structural phase transitions in steel, first identified by Adolf Martens in the 1890s, laid the foundation for SMA research. The shape memory effect (SME) was discovered in 1932 by Arne Olander, while William Buehler and his team at the Naval Ordnance Laboratory developed NiTi-based SMAs in the 1960s [8, 9]. SMAs offer unique advantages, including high force-to-weight ratios and temperature-responsive actuation, making them valuable for aerospace and automotive applications. Their integration into wind turbine blades enables deflection recovery, stress mitigation, and enhanced operational efficiency. This study proposes an SMA-based smart blade system, utilizing NiTi alloys (e.g., Flexinol®) to improve blade adaptability and resilience. Experimental setups were designed to examine correlations between load, temperature, and SMA-induced force generation. The proposed composite blade-SMA integration aims to reduce large deflections, enhance stress recovery, and contribute to next-generation turbine efficiency.

Finite Element Analysis (FEA) is a crucial tool for evaluating the structural dynamics of wind turbine blades, focusing on natural frequencies, tip displacements, and stress–strain behavior. While shell models effectively capture overall rigidity, 3D solid models are essential for localized deformations, requiring calibration for geometric and manufacturing inconsistencies. Modern FEA integrates 3D CAD modeling and advanced computational algorithms to simulate dynamic wind loads, improving accuracy over traditional beam models. Studies confirm FEA's effectiveness in blade performance prediction, with research incorporating GFRP [10] and SMA elements in ABAQUS [11] to simulate mechanical interactions, advancing composite blade analysis.

In this study, ABAQUS® finite element analysis utilized SC8R continuum shell elements and S4R conventional shell elements to model the structural behavior of composite plates, while T3D2 truss elements simulated the anchoring of shape memory alloy (SMA) wires. SC8R elements, eight-node bricks with linear interpolation, offer high accuracy in 3D simulations but require significant computational resources. S4R elements, being four-node quadrilaterals with reduced integration, provide a balance between precision and computational efficiency, making them suitable for thin-walled structures. The primary distinction between these shell elements lies in thickness specification: S4R elements define thickness through section properties, whereas SC8R elements derive it from nodal geometry. T3D2 truss elements, which bear axial loads in tension and compression, restrict displacement to axial motion at each node and are commonly used in truss system evaluations.

Accurate representation of structural behavior requires precise treatment of material interfaces and boundary constraints. In composite layup configurations, ply-to-ply adhesion is ensured by implementing Constraint 1 using the TIE constraint, designating the primary ply as the master surface and subsequent plies as slave surfaces. For hybrid composites incorporating Shape Memory Alloy (SMA) wires and Glass Fiber-Reinforced Polymer (GFRP) plates, Constraint 2 employs multipoint constraints (MPC) to synchronize nodal displacements between interacting components. Discretization involves assigning key nodal points, linking elements, and applying mass, inertia, and loading conditions. The layered structure is established through sequential ply assignment, specifying material characteristics and orientations. Flexinol 90-based SMA wires, defined with T3D2 elements, are calibrated to evaluate stress–strain responses, including elastic modulus. An Artificial Neural Network (ANN) model assesses turbine blade efficiency by utilizing feedforward neural networks to analyze load response and deformation characteristics.

3.2 Experimental setup

This study presents the experimental arrangement including the specimen fabrication and the smart wind turbine blade fabrication. Next, explains the SMA characterization. Finally, describes the whole experiment for the deflection test, followed by a summary at the end of the study.

3.2.1 Investigation of SMA wires transition using DSC

This study details the experimental procedures for fabricating smart wind turbine blades, characterizing SMA, and conducting deflection tests. DSC analysis examined the SMA wire's solid–solid transition, revealing its reversible and hysteretic behavior. The study highlights how prestrain affects recovery force in NiTi alloys and the significance of austenite start and finish temperatures for actuation. Transformation temperatures were experimentally determined at CEEFC, with DSC tests conducted at a 10 °C/min rate to analyze phase transitions and thermal properties, in accordance with the Standard Test Method for the Transformation Temperature of Nickel–Titanium alloys by Thermal Analysis [12] as shown in Figure 3.1. Figure 3.2 shows graphically the DSC curve for SMA wire.

3.2.2 Calibration of SMA wires attached to a plate-like structure

A 300 mm SMA wire was calibrated using a 40 N load cell to measure force during heating (Figure 3.3). The wire, elongated to 4 % strain and mounted vertically, contracted by 10 % upon heating while maintaining volume. A support block provided stability, and a secure, inelastic connection ensured accuracy. The calibration established a

Figure 3.1: Photograph of DSC test equipment and the SMA specimen used; (a) photograph of DSC apparatus (b) roll of NiTi specimen.

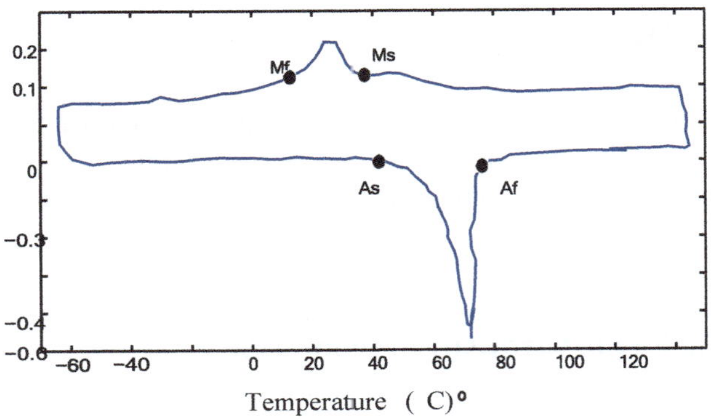

Figure 3.2: DSC curve for 0.50 mm diameter NiTi.

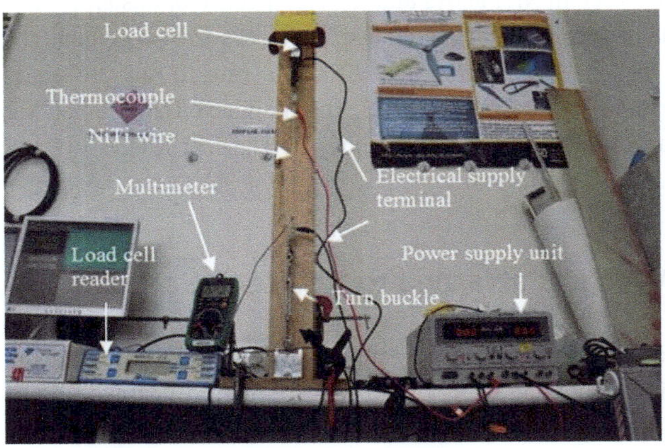

Figure 3.3: Photograph of the calibration setup for SMA.

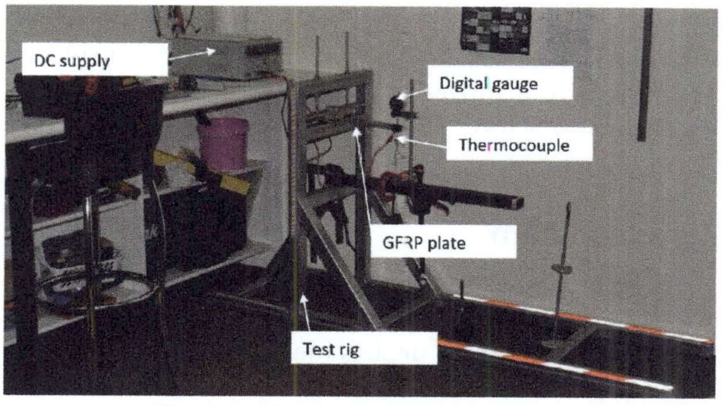

Experimental setup arrangement with the test rig

Figure 3.4: Experimental setup arrangement with the test rig.

relationship between tension force and electric current, with a temperature rise observed during electrical heating.

3.2.3 Experimental setup arrangement

In this section, a GW GPC-3030DQ power supply was used to maintain a constant current through six SMA wires for a large-scale model, approximately 1/10th the size of a 2.5 m wind turbine blade. The power supply operated in series mode to deliver up to 21 V. The SMA wires were prestrained and integrated with GFRP cantilevers serving as bias springs. In their cold state, the cantilevers remained flat with minimal residual bending from the martensitic phase equilibrium. Upon heating, the SMA wires contracted, inducing out-of-plane bending. Their eccentric positioning on the GFRP plate converted axial forces into bending moments. The wires were secured at both fixed and flexible ends and insulated with fiberglass cloth. Arrangements and wiring configurations varied based on the number of SMA wires used. Experimental data collection included deflection measurements before and after heating, recorded using a digital dial gauge. The SMA wires, powered by a manually controlled supply (0–25 V, 0–5 A), actuated the smart blade, and responses were recorded at room temperature using an MT-1820 recorder in the Structural Health Monitoring Laboratory, CEEFC, USQ, Australia. Figure 3.4 shows the experimental setup arrangement with the test rig.

3.3 Results and discussion

This study presents the findings and discussions of experiment, structured into five key sections. First, it covers the characterization of GFRP and its thermo-mechanical behavior. Next, it describes the calibration and deflection tests, including the graded beam deflection. Following this, the study examines FEA tuning for large deflections and ANN optimization. Subsequently, it analyzes the mechanism, deflection, and load correlation. Lastly, it discusses the SMA wire arrangement, stress recovery, and performance evaluation, concluding with final remarks.

3.3.1 Deflection of the graded beam

The full assembly of the graded beam was drawn precisely in two parts, which were made of GFRP and expanded polystyrene (EPS). Figure 3.5 depicts the graphical model representation with part of the assembly layout. There are 12 sections, which represent like a ply drop off in the turbine blade.

3.3.2 Tuning FEA for large deflection of the model

A preliminary study without SMA simulated large deflections in a 300 × 60 × 4 mm plate-like structure using SC8R and S4R models under a concentrated load. A 3.6 N load validated the model, showing close alignment of tip deflection curves. Figures 3.6 and 3.7 confirm FEA accuracy, with S4R matching experimental results making it the preferred model for future simulations.

Either conventional shell elements or continuum shell elements might be used for the GFRP plates. However, using continuum shell elements would consume a great deal of computation time. It would also cause many convergence problems due to the

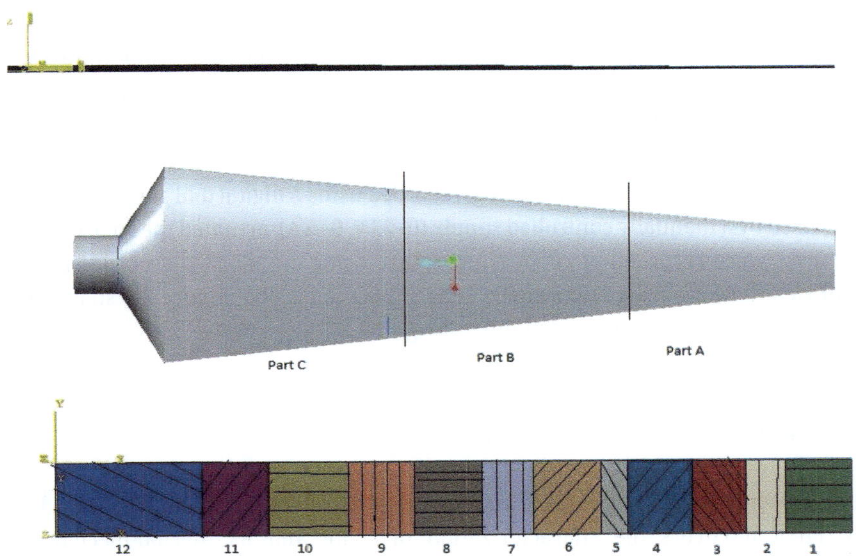

Figure 3.5: Graphical model representation with number of plies configuration (top view).

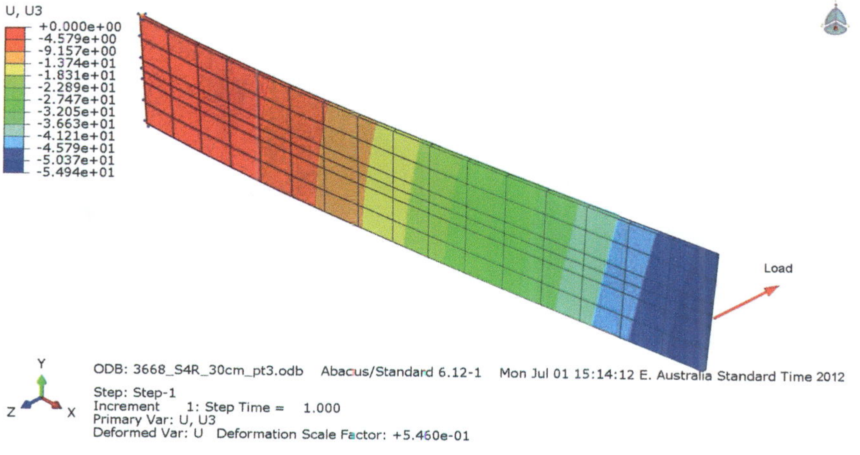

Figure 3.6: Deflection contour S4R of the GFRP plate.

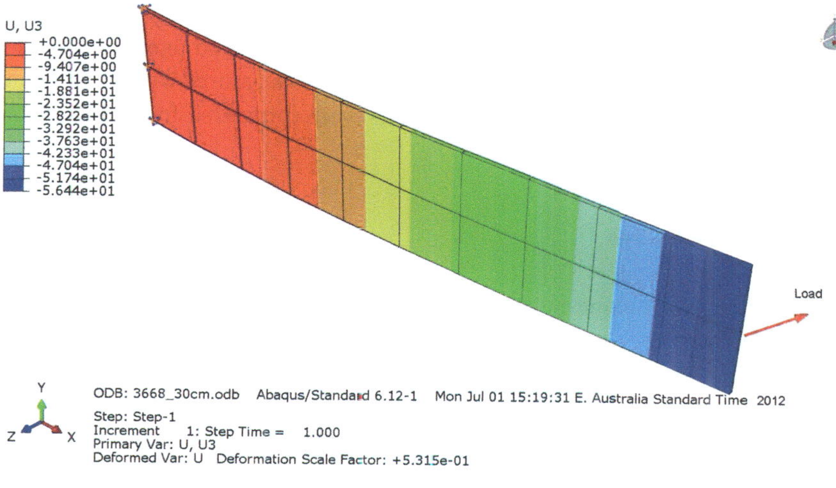

Figure 3.7: Deflection contour of SC8R of the GFRP plate.

complicated geometric arrangement using 8 nodes and very thin webs in the test specimen. Using conventional shell elements would be a more desirable option.

3.3.3 Preliminary study: use of the SMA mechanism

This study investigates SMA composite plate mechanisms through two approaches: embedded and suspended SMA wires. For embedded SMA, GFRP blades were reinforced

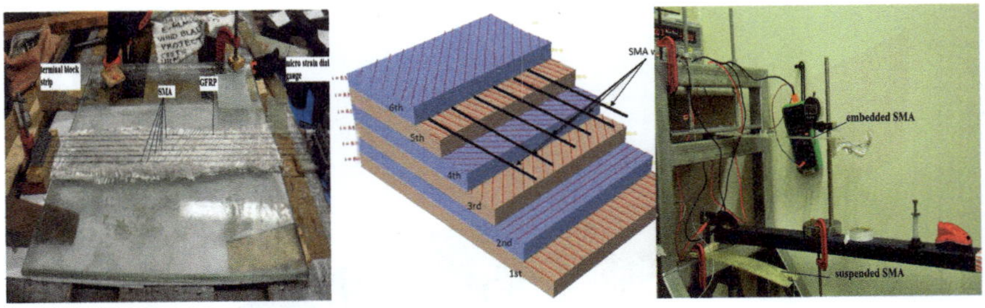

(a) (b) (c)

Figure 3.8: Integration of SMA wires into Composite Wind Blade Structures. (a) Embedded SMA fabrication; (b) schematic diagram of embedded SMA wires; (c) photograph preliminary study setup for embedded SMA and suspended SMA.

with SMA wires, transferring loads from the plate/skin. Two specimens (three-wire and six-wire configurations) were fabricated, with wires embedded beneath the final layer (Figure 3.8a). Wires were positioned away from the neutral axis to induce tension and were prestrained to 4 % using a micrometer strain gauge. A spirit level ensured horizontal alignment. Joule heating was applied via a 12 V DC power supply, and temperature/time response was recorded using MT-1820 software. Deflection was measured with a ± 0.01 mm accuracy micrometer gauge. Four key design aspects were considered for optimal SMA integration.

3.3.3.1 Embedded SMA wires

In wind turbine blades, maximum stress and strain occur at the skin, with complex distributions due to cross-sectional variations. SMA wires must be equidistantly aligned, clamped at 4 % prestrain, and tightly bonded for effective contraction. Figure 3.8b and c illustrates the SMA wire configuration and experimental setup, where embedded and suspended SMAs were tested under static conditions for deflection analysis.

Figure 3.8b and c illustrates the SMA experimental setup and temperature response over time. The actuation was repeatable with minimal hysteresis, influenced by SMA alloy type and fabrication process. Trained wires exhibited greater stability, and heating during manufacturing reduced loading history effects. However, Joule heating posed cooling challenges, affecting thermocouple response and data accuracy.

3.3.3.2 Suspended SMA wires

The suspended SMA configuration functions as an external reinforcement, with SMA wires anchored at both ends of the GFRP plate (Figure 3.9). This method was developed after the embedded model proved ineffective and caused damage. Suspended SMA wires

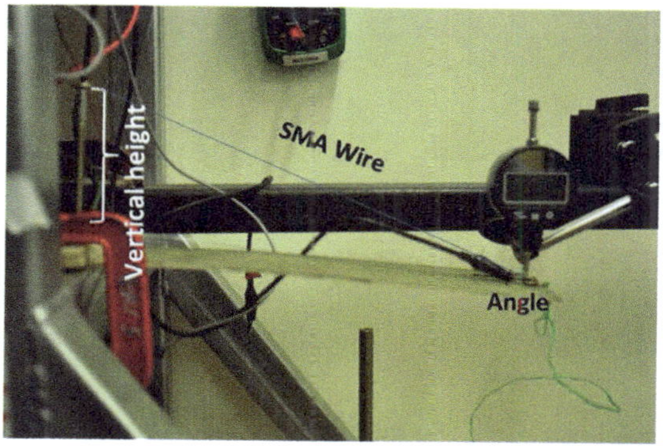

Figure 3.9: Photograph of preliminary suspended SMA wires deflection and load relationship.

efficiently restored the plate's original form, responded quickly even with a single wire, and generated less heat than embedded SMA [13].

The study confirms that suspended SMA wires outperform embedded ones in deflection recovery, with faster actuation, lower energy loss, and better structural integrity. Increasing SMA wire count and anchoring height enhances performance, while shorter wires improve heating efficiency. Embedded SMA wires suffer from heat dissipation and delamination, making them unreliable. Ohm's law governs current–voltage behavior, and conventional shell elements optimize simulations. Improved temperature monitoring and safety measures enhance system reliability.

The deflection simulation method models SMA wires and the composite matrix separately, allowing easy adjustments to material properties and geometry. Unlike application-specific composite models, this flexible approach enables design exploration.

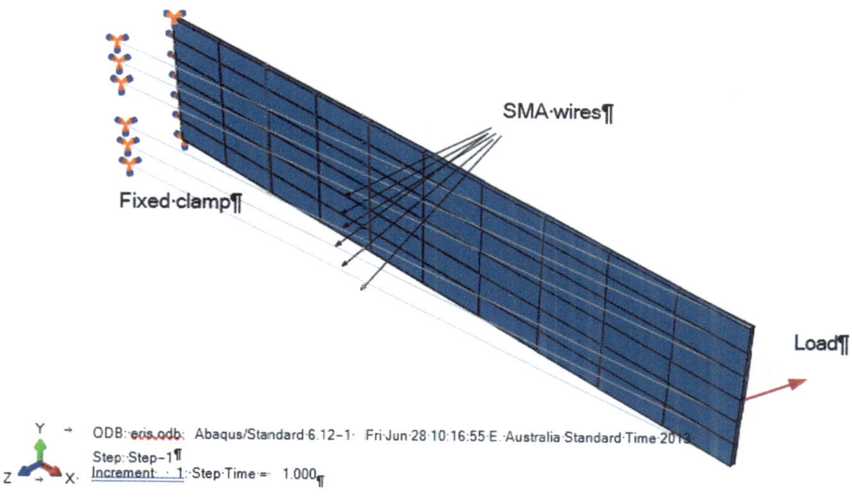

Figure 3.10: Simulation of smart wind blade model; detailed assembly of the complete structure.

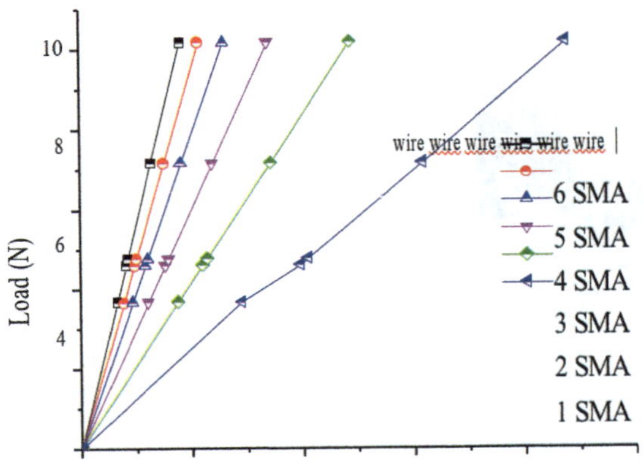

Figure 3.11: Calibration of the smart wind blade model under deflection.

Figure 3.10 demonstrates this capability through a 6-SMA wire deflection test, where a tip load was applied, and the blade root was clamped (6 DOF). Results confirm the numerical model's accuracy in capturing SMA's thermo-mechanical behavior.

Figure 3.11 presents the deflection–load relationship for different SMA wire configurations. FEA simulations confirm that increasing SMA wires reduces tip deflection under load. SMA activation via electrical current enhances lifting force and minimizes residual deflection.

3D structural analyses were conducted using plate elements to assess both in-plane and out-of-plane loads. The blade, modeled as a cantilevered box beam with its root fixed to the rotor hub, was simulated in ABAQUS. The model incorporated the geometry and ply configuration, utilizing 3D deformable shell and solid features for comprehensive analysis [11].

3.3.4 SMA wires arrangement

SMA wires, prestrained by 4 %, were mounted at designated anchoring heights. They were attached to the GFRP plate using a modified terminal block at the tip, permitting free movement, and secured at the root end with a screw terminal slot to prevent slippage. Continuous tension–relaxation cycles may eventually introduce slack in the wires.

3.3.5 Results of parametric studies

The study investigates the influence of anchoring height, SMA wire count, and heat sleeving on the deflection recovery of a smart wind blade system. Tests on a 300 mm plate demonstrate that increasing anchoring height enhances deflection recovery by directing SMA wire tension more vertically. Higher SMA wire counts further improve performance, as increased tension reduces deflection, with activation occurring at 1.5 A. Heat sleeving significantly improves heating efficiency by reducing heat loss. In a 1,000 mm plate, higher electrical currents counteract greater loads, aligning experimental results with FEA simulations. Stress analysis shows maximum strain near the blade root, and system constraints, including power requirements and response time, are identified. Increasing SMA wires reduces deflection across various loads, validating the smart blade's effectiveness in stress recovery and load alleviation.

3.3.6 Comparison of power performance

The power coefficient (Cp) measures a wind turbine's efficiency in converting wind energy into mechanical power, with the Betz limit setting the theoretical maximum at 0.5923 for three-bladed horizontal-axis wind turbines (HAWTs). Slootweg et al [14] developed a rotor model to approximate Cp, expressed as

$$\text{Pwt} = 0.5{*}\rho{*}\text{Cp}\,(\lambda, \beta){*}\text{A}{*}\text{vw}^3$$

where Pwt is the power extracted, ρ is air density, A is rotor area, vw is wind speed, λ is the tip speed ratio, and β is the blade pitch angle. For stall-controlled turbines, β is excluded, making Cp solely a function of λ. The study incorporated shape memory alloy (SMA) wires into the blades to enhance performance, adjusting the pitch angle (βSMA) based on SMA-induced deflection. Results indicated that blades with SMA wires achieved Cp values closer to the Betz limit compared to those without, suggesting improved efficiency. However, the SMA's response time was relatively slow, potentially affecting real-time performance adjustments.

3.3.7 Summary of the results

This work presented findings on SMA wire performance in deflection recovery. Experiments analyzed the effect of electrical current, applied load, and wire arrangement on stress recovery in smart blades. A neural network model was trained using experimental data to predict deflection based on load, current, time, and wire count. To simplify fabrication challenges, a proof-of-concept specimen was developed using prestrained SMA wires on a laminated composite plate.

In order to remove all the technological difficulties related to the complicated composite fabrication, a simple proof-of-concept specimen was built by mounting prestrained SMA wires on top of the laminated composite plate strip.

3.4 Conclusions

This study explored the development of smart wind turbine blades using SMA wires. The results confirm that SMA wires effectively restore blade deflection under load, proving their feasibility as active actuators. FEA simulations and experimental results showed strong agreement, validating the model's accuracy. A neural network model in MATLAB successfully predicted performance, analyzing the relationship between SMA wire count, load, current, and deflection. The smart blade design improved stress recovery and brought the performance coefficient closer to the Betz limit (0.5923). Despite time and experimental constraints, further optimization of composite materials and manufacturing processes can enhance real-scale applications. Addressing shape memory material adherence issues will be key for future innovations in smart wind turbine technology.

Acknowledgments: The authors would like to express their sincere gratitude to the editors – S.M. Sapuan, Mohd Roshdi Hassan, Eris Elianddy Supeni, and Azizan As'arry – for their valuable guidance and thorough review of this article prior to its publication.

References

1. Dam V, Mayda E, Chao D, Jackson K, Zuteck M, Berry D. Innovative structural and aerodynamic design approaches for large wind turbine blades. ASME/AIAA Wind Energy Symp 2005;0973.
2. Chopra I. Review of state of art of smart structures and integrated system. Am Inst Aeronaut Astronaut J 2002;40:2145–87.
3. Straub K. A feasibilty study of using smart materials for rotor control. Smart Mater Struct 1996;5:1–5.
4. Enekl, B, Klopper, V, Preibler, D. Full scale rotor with piezoelectric actuated blade flaps. In: Proceedings of the 28th annual forum of the European Rotorcraft Forum, Bristol, UK, 17–19 September 2002. Royal Aeronautical Society, London, UK; 2002.
5. Roth, D, Enenkl, B, Dietrich, O, Active rotor control by flaps for vibration reduction – full scale demonstrator and first flight test results In: Proceedings of the 32nd European Rotorcraft Forum. Maastricht, The Netherlands; 2006.
6. Barlas T, Kuik G. State of the art and prospectives of smart rotor control for wind turbines. J Phys Conf 2007; 75:1–20.
7. Nakafuji DY, van Dam CP, Michel J, Morrison P. Load control for wind turbines – a non-traditional microtab approach. In: Proceedings of the 40th AIAA/ASME aerospace sciences meeting and exhibit. Reston, Virginia, USA: American Institute of Aeronautics and Astronautics; 2002, pp. 321–30.
8. Hartl D, Lagoudas D. 'Aerospace applications of shape memory alloys proc. Imeche vol. 221 part g: J. aerospace engineering'. Proc IMechE Aero Eng 2007;221. PART g. https://doi.org/10.1243/09544100jaero211.
9. Mabe J, Cabell R, Butler G. Design and control of morphing chevron for take-off and cruise noise reduction. AIAA Aeroacoustics Conf 2005;26:1–15.
10. Jensen F, Falzon B, Ankersen J, Stang H. Structural testing and numerical simulation of a 34 m composite wind turbine blade. Compos Struct 2006;76:52–61.
11. Simulia. ABAQUS release note 6.12. USA: Dassault Systems; 2012.

12. ASTM International. 2010. Standard test method for transformation temperature of Nickel–Titanium alloys by thermal analysis ASTM F2004-05. West Conshohocken, PA: Committee F04, ASTM International.
13. Daniel A, Calin M, Ion V, Marius T, Cornel M. Contributions to design of systems actuated shape memory active elements. World Acad Sci Eng Technol 2010;7:222–7.
14. Slootweg J, Polinder H, Kling W. Representing wind turbine electrical generating systems in fundamental frequency simulations. IEEE Trans Energy Convers 2003;18:516–24.

Yusuf Jameel, Muhammad Firdaus Abdul Halim, Mohd Sapuan Salit*
and Abdul Habib

4 Advancements in composite materials for energy harvesting

Abstract: This chapter discusses mechanical properties, electrical conductivity and thermal stability of composite materials used in energy harvesting. It explores the use of composites in energy harvesters that are thermoelectric, electromagnetic, and piezoelectric. One of the main advantages of composite materials for energy harvesting is their high strength-to-weight ratio. Metal- and polymer-based composites are among the types of composites. The piece also explores the challenges that need to be solved for the technology to realize its full potential. If composite materials were less expensive and more durable, energy collecting systems might be more dependable and efficient. To ascertain the resilience and reliability of composite materials for application in energy harvesting, more research is necessary. Composite materials, however, have potential for energy harvesting. However, composite materials show promise for energy harvesting. Before they can reach their full potential, more research and development are needed. The major obstacles and opportunities for further research and development in composite materials for energy harvesting are covered in this chapter.

Keywords: composites; energy harvesting; magnetoelectric; piezoelectric

4.1 Introduction

Engineered materials with two or more constituent materials that have significant differences in their mechanical, chemical, and physical properties are called composites, or composite materials. The distinct qualities of each of its component pieces, the unique characteristics of these composites arise from their volume proportions and arrangements within the material system. Depending on their intended purpose, composites can be tailored to meet specific geometrical, structural, mechanical, chemical, and even aesthetic needs [1]. For electrically powered devices, composites that can harvest energy from their

***Corresponding author: Mohd Sapuan Salit**, Faculty of Engineering, Advanced Engineering Materials and Composites Research Centre (AEMC), Universiti Putra Malaysia, Serdang 43400, Selangor, Malaysia, E-mail: sapuan@upm.edu.my. https://orcid.org/0000-0002-4598-6358
Yusuf Jameel, Muhammad Firdaus Abdul Halim and Abdul Habib, Faculty of Engineering, Advanced Engineering Materials and Composites Research Centre (AEMC), Universiti Putra Malaysia, Serdang 43400, Selangor, Malaysia. https://orcid.org/0009-0004-4853-1449 (M.F. Abdul Halim). https://orcid.org/0009-0008-0648-548X (A. Habib)

As per De Gruyter's policy this article has previously been published in the journal Physical Sciences Reviews. Please cite as: Y. Jameel, M. F. Abdul Halim, M. S. Salit and A. Habib "Advancements in composite materials for energy harvesting" *Physical Sciences Reviews* [Online] 2024. DOI: 10.1515/psr-2024-0022 | https://doi.org/10.1515/9783111234762-004

surroundings are beneficial [2]. Composite materials are increasingly being used for energy harvesting. By converting ambient energy into electrical power, energy harvesting helps to increase overall energy efficiency while lowering dependency on traditional energy sources. Composite materials are increasingly being used in energy harvesting systems because of their unique mechanical and electrical properties, which make them appropriate for energy conversion [3]. Environmental energy sources such as heat, magnetic fields, and mechanical vibrations are used in energy harvesting [4].

4.1.1 Overview of energy harvesting

One approach to creating sustainable energy sources could be to harvest energy from various forms of energy that are wasted in the environment. But to achieve this, we must create intelligent materials and constructions that have the ability to transform a variety of energy-wasting sources into electrical energy. The technique of obtaining and transforming minute amounts of ambient energy from sources like solar, thermal, wind, or vibration into useful electrical energy is known as energy harvesting [5]. Energy harvesting is becoming a practical way to fuel a wide range of devices, particularly autonomous and low-power ones like wearable electronics, electric cars, and wireless sensors. Energy harvesting can increase a device's sustainability and dependability by lowering its reliance on external power sources or batteries [6]. Numerous energy sources offer opportunities for energy harvesting, including sunlight, mechanical vibration, electromagnetic radiofrequency, thermal gradient, and body motion as shown in Figure 4.1. Mechanical vibration is commonly generated in various mechanical systems, such as machinery, aeronautical components, wind-induced vibrations, automotive suspension systems, and flow-induced vibrations, making it a primary source of vibrational energy. This energy source is well-suited for low-power electronic devices, given its abundance and power density [8]. Table 4.1 gives an insight on current research on energy harvesting using composites.

4.1.2 Types of composite materials used in energy harvesting

There are a number of benefits to using composite materials rather than standard materials in energy harvesting systems [14]. In comparison to traditional materials, composites can offer greater mechanical strength and stiffness, enhanced thermal stability, and superior electrical conductivity [15]. Furthermore, composite materials can be engineered to possess particular mechanical and electrical characteristics to satisfy the demands of electric car energy harvesting systems.

- Polymer composites
- Piezoelectric composites
- Magnetoelectric (ME) composites
- Thermoelectric composites

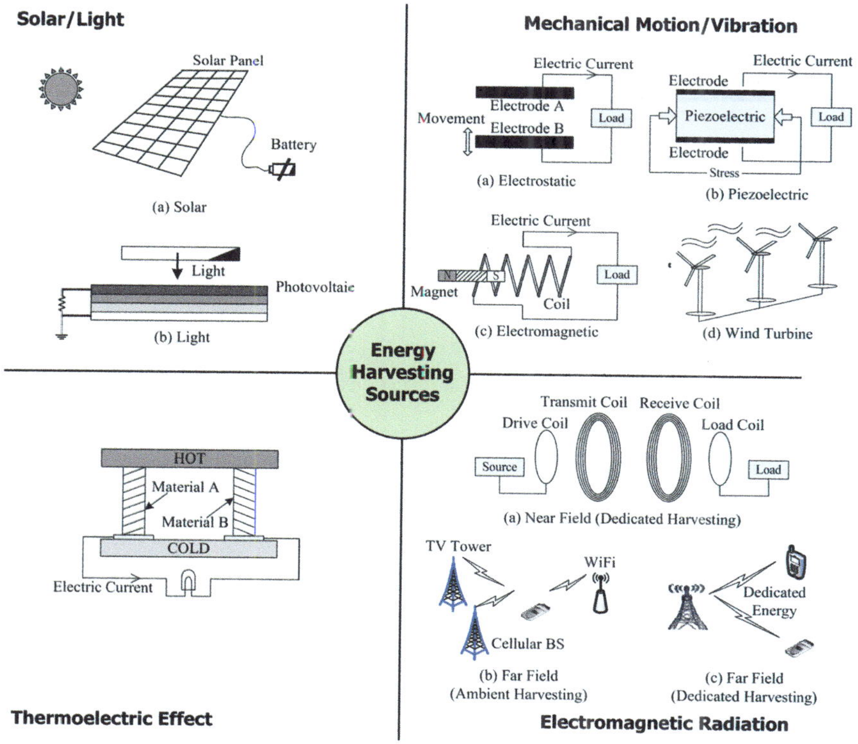

Figure 4.1: Types of energy sources [7].

Table 4.1: Current research on energy harvesting using composites.

No.	Source materials	Methods used	Properties	Applications	References
1.	Polycrystalline particles	Doping	Developed flexible piezoelectric nanogenerator with high output performance. Achieved maximal output power of 2 mW, 220 V, and 90 μA.	Enables energy harvesting for portable and wearable devices. Potential applications in biomedical energy solutions.	[9]
2.	Hexagonal boron nitride and molybdenum disulfide	Directional cross-linking during freezing process.	Outstanding effectiveness in synthesizing intrinsic piezoelectric composites based on 2D materials.	Development of sensitive wearable pressure and strain sensors. Creation of efficient ultrasound energy harvester.	[10]

Table 4.1: (continued)

No.	Source materials	Methods used	Properties	Applications	References
3.	Barium titanate/polylactic acid	Mathematical modeling of piezoelectric composites for energy prediction. Experimental validation with varying strains and compositions.	It explores various compositions, revealing that a 60 % $BaTiO_3$ composition can harvest up to 4.5 μW under specific strains.	Optimizes energy conversion efficiency in piezoelectric materials. Advances renewable energy applications through material composition exploration.	[11]
4.	PZT (lead zirconate titanate) nanofibers into PVDF (polyvinylidene fluoride) fibers	Modification of PZT nanofibers with dopamine hydrochloride. Hot-pressing method for multilayered film fabrication.	Peak piezoelectric coefficient (d33) of 29 pC/N achieved. Output voltage increased to 13.8 V from 3 V.	Wearable sensors and energy harvesting devices Human health monitoring system	[12]
5.	Polydimethylsiloxane open-cell foams and carbon nanotubes	Film fabrication	Generates stable electrical energy without thermoelectric devices, achieving a maximum output voltage of 1.18 mV and 50 % energy harvesting efficiency.	Thermoelectric energy harvesting technologies	[13]

4.1.3 Polymer composites

For energy harvesting applications, polymer-based composites are frequently preferred because of their great flexibility, low cost, and ease of processing. These composites' polymer matrix can be adjusted to enhance the material's overall energy harvesting performance as well as provide certain electrical and mechanical characteristics [16]. Carbon fiber–reinforced polymer composites, for example, are frequently utilized in energy harvesting systems because of their low weight, excellent electrical conductivity, and mechanical strength [17].

The mechanical properties of polymers and composites, known for their flexibility and compliance, hold great potential for future energy harvesting applications. Particularly in

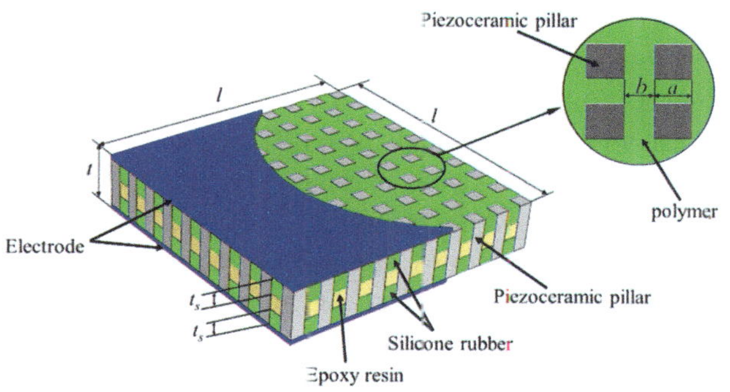

Figure 4.2: Structure of the 1–3 piezoelectric composite with 3-tier polymer structure [20].

scenarios involving low-frequency vibrations such as smart textiles or the recovery of micro-seismic or human-generated energy, electrostrictive polymer composites emerge as promising candidates [18]. Polymer-based composites are frequently preferred for energy harvesting applications because of their affordability, ease of manufacturing, and exceptional flexibility. The polymer matrix within these composites can be tailored to possess precise electrical and mechanical characteristics, and it can be fine-tuned to enhance the overall performance of energy harvesting systems [19].

4.1.4 Piezoelectric composites

Polymer-based piezoelectric composites are a common option for energy-harvesting applications. The benefits of piezoelectric materials in transforming mechanical energy into electrical energy have drawn a lot of attention to this area of study (refer Figure 4.2). Polymeric piezoelectric composites are more convenient than ceramic-based composites because they have faster processing times, a suitable voltage with enough power output, less expensive production, and mechanical flexibility [21].

With the help of wind or certain fluid flow–induced vibrations, these piezoelectric energy harvesters may capture energy from a variety of vibration sources, including translational and rotating motions, wind or some fluid flow–induced vibrations [22, 23].

4.1.5 Magnetoelectric (ME) composites

A ferromagnetic (FM) material and a piezoelectric material paired with different phase connectivity methods make up an ME composite. When a magnetic field is applied across a composite like this, the FM material experiences magnetostriction, which causes the material to undergo elastic strain. Through a mechanical coupling, this elastic strain is transmitted to the linked piezoelectric phase. An electric field is produced transversely across the piezoelectric phase because of this strain in the piezoelectric material. These

ME composites are used to extract the electric energy from this electric field. Put otherwise, an electric field is created when a magnetic field is applied across these composite materials. Magnetic energy is another common waste energy source that is present around electric current–carrying lines and appliances [24].

The direct magnetoelectric effect is the name given to this phenomenon. In certain ME composites, the opposite action also occurs, where the introduction of an electric field causes the creation of a magnetic field. Numerous applications, including energy harvesting, magnetoelectric transformers, phase shifters, AC and DC magnetic field sensors, and resonators, have investigated both effects [25].

Figure 4.3 illustrates a schematic of the magnetoelectric (ME) effect in composites. The direct ME effect is observed when a magnetic field is applied to the ME composite, resulting in a change in the shape of the ferromagnetic phase due to magnetostriction. This means that the ferromagnetic material experiences strain when subjected to a magnetic field. This strain is subsequently transmitted through mechanical coupling to the adjacent piezoelectric phase, leading to the generation of electric polarization in the piezoelectric material due to the piezoelectric effect [6].

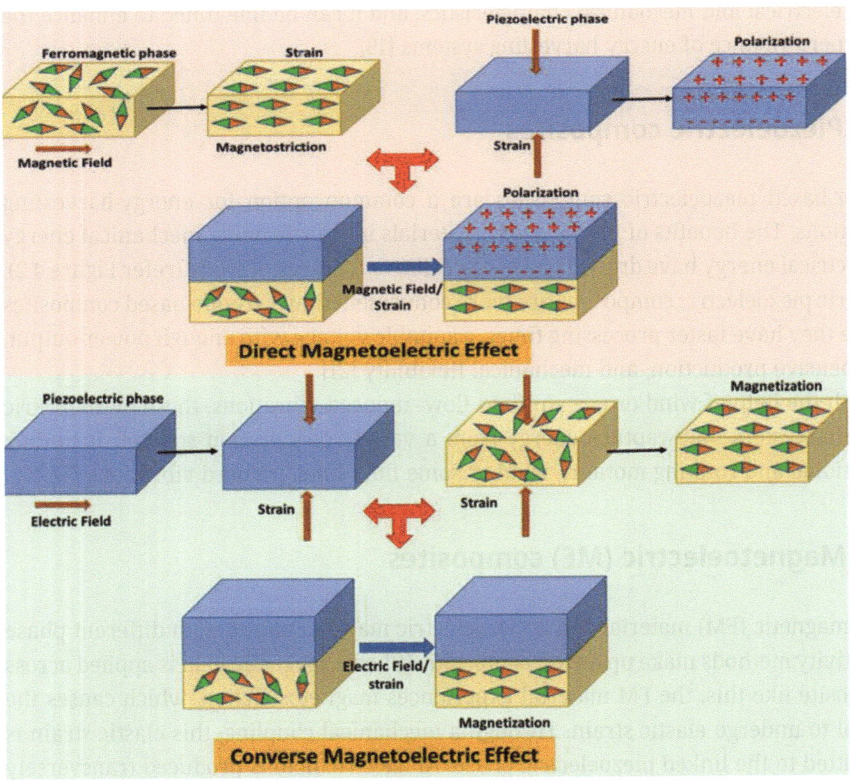

Figure 4.3: Schematic diagram of strain-mediated ME effect in an ME composite [6].

4.1.6 Thermoelectric composites

Using thermoelectric materials, thermoelectric energy harvesting entails transforming heat energy into electrical energy [26]. It has been demonstrated that composite materials have good thermal stability, which makes them ideal for thermoelectric energy harvesting devices. Numerous research works have looked into the use of composite materials for ambient heat and exhaust gas energy harvesting, among other thermoelectric energy harvesting applications. Thus, composite materials have demonstrated enormous promise for energy harvesting in electric cars. Excellent piezoelectric output, excellent electrical conductivity, and strong thermal stability are only a few of their special mechanical and electrical qualities that make them ideal for energy harvesting systems [27].

Thermoelectric (TE) functional materials are considered eco-friendly energy materials capable of converting heat into electricity using the movement of solid internal carriers, even when exposed to minimal temperature differences compared to the environment. In contrast to other emerging energy technologies, TE devices offer distinct benefits, including the absence of moving parts, a noise-free operation, and a long operational lifespan. Consequently, they are emerging as highly promising alternatives to conventional energy materials [28].

4.1.7 Characteristics of composite materials for energy harvesting

4.1.7.1 Mechanical properties

Composite materials have special qualities, especially in their mechanical qualities, that make them appropriate for energy harvesting applications. These materials combine mechanical robustness with greater energy efficiency, vital for products like wearable electronics and electric vehicles.

When choosing composite materials for energy harvesting, mechanical qualities are quite important. The ability of the materials to tolerate the mechanical stress and strain produced during energy conversion is essential to the performance of energy harvesting systems [29]. Depending on the matrix and reinforcing components utilized, composite materials have different mechanical properties. The excellent toughness and ductility of polymer-based composites, which are composed of a polymer matrix and reinforcing material, are well-known. Conversely, metal-based composites have great strength and stiffness because they are made of a metal matrix and reinforcing material like ceramic or polymer fibers. To guarantee the composite material's sturdiness, dependability, and efficiency in energy harvesting systems, its mechanical qualities must be optimized [30].

When compared to unfilled rubber, composites like those constructed of natural rubber and carbon nanotubes (CNTs) show notable gains in mechanical properties, with fracture toughness rising by about 484 % [31]. Kumar et al. [32] focused on piezoelectric energy harvesting with composites. Rubber composites were fabricated by incorporating titanium carbide (TiC) and molybdenum disulfide (MoS_2) into a silicone matrix. Excellent mechanical and electromechanical characteristics were achieved. The compressive modulus increased from 1.55 ± 0.08 MPa to 1.95 ± 0.07 MPa with TiC and to 2.02 ± 0.09 MPa with MoS_2. Likewise, stretchability enhanced from 133 ± 7 % to 153 ± 9 % with TiC and to 165 ± 12 % with MoS_2. Electromechanical evaluations indicated that at 30 % strain, the output voltage was 3.5 mV for TiC and 6.7 mV for MoS_2. Overall, the incorporation of TiC and MoS_2 into silicone rubber results in strong and adaptable composite materials. Such composites enhance energy generation, stretchability, and optimal stiffness, aligning with current theoretical frameworks.

When selecting composite materials for energy harvesting applications, it is of paramount importance to take into account their mechanical attributes. This includes factors like tensile strength, compressive strength, modulus of elasticity, and impact resistance. Additionally, it's essential to consider aspects such as durability, resistance to fatigue, and stability under repetitive loading. The mechanical properties of the composite material should be fine-tuned and optimized to guarantee the durability, reliability, and efficiency of energy harvesting systems [19].

4.1.7.2 Electrical conductivity

One of the most important properties of composite materials for energy harvesting is electrical conductivity. Energy harvesting is made feasible by these materials' efficient electrical conductivity. The mix of matrix and reinforcing components used in composite materials affects their electrical conductivity [33]. Composites based on metals typically have more electrical conductivity than composites based on polymers, which typically have lower electrical conductivity. Nonetheless, the incorporation of conductive fibers or fillers into the composite structure might enhance the electrical conductivity of polymer-based composites. It is crucial to remember that electrical conductivity might not be sufficient in and of itself to assess if a composite material is suitable for energy harvesting. It's also necessary to take into account other elements like thermal stability and mechanical qualities. In general, the efficacy and efficiency of energy harvesting are greatly dependent upon the electrical conductivity of composite materials [34].

Hexagonal boron nitride (h-BN) and molybdenum disulphide (MoS_2) few-layer nanosheets (NSs) exhibit remarkable piezoelectric characteristics, as shown in Figure 4.4. Kaihang et al. [10] presented a simple technique for creating intrinsic piezoelectric composites using NSs that don't require electric poling. In comparison to polyvinylidene difluoride (PVDF), the resulting PE composites have a maximum piezoelectric coefficient of up to 25.5–28.4 pC N^{-1}, and they are significantly more cost-effective, safe, and scalable than traditional materials. The group further develops ultrasonic energy harvesters and extremely sensitive wearable strain and pressure sensors using this combination. While the harvester produces around 1.18 V/2.31 μA under 1 W cm^{-2} ultrasonic input underwater, these sensors can detect finger bending and distinguish between walking and sprinting.

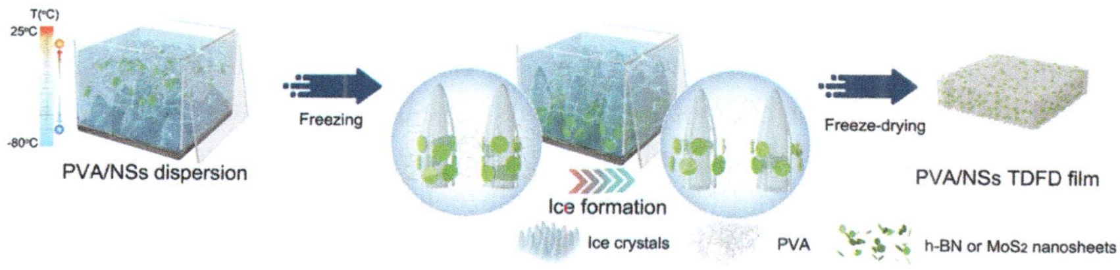

Figure 4.4: Schematic of the directional freezing and FD process for synthesizing polymer composites [10].

The electrical conductivity of composite materials plays a pivotal role in the context of energy harvesting for electric vehicles. It is this property that enables the effective transmission of electricity, thus making energy harvesting feasible. The electrical conductivity of composite materials can differ significantly based on the specific combination of reinforcing and matrix materials employed [19].

4.1.7.3 Thermal stability

One essential property of composite materials for energy harvesting is thermal stability. The reinforcing and matrix materials that are utilized also affect the thermal stability of composite materials. For example, composites based on metals have high thermal stability, but composites based on polymers often have lesser thermal stability. For the energy harvesting system to be reliable and long-lasting, thermal stability is crucial [35]. If the composite material cannot endure elevated temperatures, it becomes susceptible to degradation or potential failure, diminishing the effectiveness of the energy harvesting system. In extreme situations, this could pose safety risks. Therefore, when choosing composite materials for energy harvesting applications, it is crucial to consider their thermal resistance, it is critical to consider the material's thermal stability. The thermal stability of the energy harvesting system as a whole must be taken into account in addition to the thermal stability of the composite material. For example, to avoid excessive thermal stress impacting the energy harvesting system's performance and lifespan, the system must be built to adequately disperse heat. Accordingly, to guarantee the best possible performance and dependability of the energy harvesting system, a combination of a thermally stable composite material and an efficient thermal management system is required [36].

4.1.8 Applications of composite materials for energy harvesting

4.1.8.1 Piezoelectric energy harvesting

Using piezoelectric materials, the process of piezoelectric energy harvesting transforms mechanical energy into electrical energy. These materials are perfect for energy harvesting in electric vehicles because they produce an electrical potential when subjected to mechanical stress [37]. In piezoelectric energy harvesting systems for electric vehicles, composite materials derived from piezoelectric ceramics and polymers have shown promising results. These composite materials have better mechanical properties, like flexibility and resilience to mechanical stress, thanks to the polymers utilized in them. Furthermore, these composites' usage of piezoelectric ceramics improves the material's electrical conductivity, increasing its effectiveness in producing electrical energy. Piezoelectric energy harvesting is not without its restrictions, though. When compared to other energy harvesting systems, the energy generated is very little, which makes it more appropriate for low-power applications such as powering sensors or other low-power electronics [38].

4.1.8.2 Electromagnetic energy harvesting

Using electromagnetic generators, electromagnetic energy is transformed into electrical energy in the process of harvesting electromagnetic energy. Composite materials can be used to make electromagnetic generators; metal matrix composites, for example, have been used in electromagnetic energy harvesting systems for electric vehicles. For electromagnetic energy harvesting, metal matrix composite-based composite materials have a number of benefits. Because of their high electrical conductivity, metal matrix composites are a good choice for producing electricity from electromagnetic fields. Furthermore, because metal matrix composites are highly thermally stable, they are ideal for energy harvesting systems in electric vehicles, which produce high temperatures [39].

4.1.8.3 Thermoelectric energy harvesting

Using thermoelectric materials, thermoelectric energy harvesting is the process of transforming thermal energy into electrical energy. Thermoelectric materials are ideal for energy harvesting in electric vehicles because they produce an electrical potential when exposed to a temperature variation. Electric car thermoelectric energy harvesting systems have made use of composite materials derived from thermoelectric ceramics and polymers. There are various benefits to using thermoelectric energy harvesting in composite materials composed of thermoelectric ceramics and polymers. These composites' polymers can enhance their mechanical qualities, increasing their flexibility and ability to withstand mechanical stress. Furthermore, by increasing the composite's electrical conductivity, the thermoelectric ceramics utilized in these composites can increase their ability to produce electrical energy [40].

4.2 Challenges and opportunities

One of the biggest challenges is finding the right material combination that provides the desired electrical conductivity, thermal stability, and mechanical properties. These characteristics are often a compromise, and finding the right balance between them can be difficult. For example, increasing the electrical conductivity of a composite material often reduces its thermal stability, making it difficult to use in high-temperature applications. Another challenge is that the cost of composite materials used in energy production systems [41].

The absence of standards in composite materials for energy harvesting is another difficulty. To guarantee that energy harvesting devices are dependable, consistent, and economical, standards for materials, processing techniques, and testing protocols are required [42].

The mechanical characteristics of composite materials are frequently diverse, which may have an impact on how well they capture energy. For example, adding fillers to polymer nanocomposites can improve their mechanical and thermal durability, but choosing the right fillers is essential [43, 44]. For applications in electric vehicles, where weight reduction is crucial for efficiency, composites' mechanical properties – such as their strength and flexibility – are crucial. One of the main concerns is how long composite materials would last under operational loads. Damage to many composites can result in reduced energy harvesting system performance. Advanced composite materials can be too expensive to produce, which prevents their widespread use in energy harvesting technology. For these materials to be practical for commercial use, more reasonably priced substitutes are required [45].

In conclusion, there are several obstacles facing the development and application of composite materials for energy harvesting in electric vehicles. These include the need for the proper material combination, the exorbitant cost of these materials, the absence of industry standards, and the lack of knowledge regarding the durability and long-term performance of these materials. To effectively utilize composite materials' potential for energy harvesting in electric vehicles, several issues must be resolved. Research and development on hybrid polymer composites is a new field that aimed to solve issues with scalability and interfacial compatibility (refer Figure 4.5).

Figure 4.5: Environmental impact of hybrid composites [46].

4.3 Conclusions

Since composite materials are more advantageous than traditional materials in terms of mechanical properties, electrical conductivity, thermal stability, and cost-effectiveness, they have a lot of potential for energy harvesting in electric cars. Research on the creation and implementation of composite materials for energy harvesting in electric vehicles is expanding, with several studies investigating the various kinds of composite materials and their uses in the last few years. It is possible to develop more effective, dependable, and affordable energy harvesting systems for electric vehicles and to significantly contribute to the sustainable energy and transportation systems of the future by addressing the opportunities and challenges presented by these materials.

The chapter highlighted the restrictions faced by composite materials for energy harvesting due to their cost-effectiveness and durability. It is also mentioned that there is a lack of thorough study on the uses of composite materials in energy harvesting. In future, closing these gaps may result in more cost-effective, reliable, and efficient energy systems for environmentally friendly transportation. Conducting the research on hybrid composites will play a major role in the advancement in energy harvesting.

Acknowledgments: The authors would like to thank the editors S. M. Sapuan, Mohd Roshdi Hassan, Eris Elianddy Supeni, and Azizan As'arry for their guidance and review of this article before its publication.

References

1. Egbo MK. A fundamental review on composite materials and some of their applications in biomedical engineering. J King Saud Univ, Eng Sci 2021;33:557–68.
2. Harnden R, Carlstedt D, Zenkert D, Lindbergh G. Multifunctional carbon fiber composites: a structural, energy harvesting, strain-sensing material. ACS Appl Mater Interfaces 2022;14:33871–80.
3. Fan FR, Tang W, Wang ZL. Flexible nanogenerators for energy harvesting and self-powered electronics. Adv Mater 2016;4283–305. https://doi.org/10.1002/adma.201504299.
4. Mori T, Priya S. Materials for energy harvesting: at the forefront of a new wave. MRS Bull 2018;43:176–80.
5. Yusuf J, Firdaus AHM, Sapuan SM, Rashid U, Ilyas RA, Hassan MR, et al. Nanocellulose-graphene hybrid composites: fabrication, characterization, applications and environmental impact. Int J Biol Macromol 2024;282:137244.
6. Garg T, Goyal LM. Magnetoelectric composites-based energy harvesters. In: Vargas-Bernal DSR, Palma MR, editors. Rijeka: IntechOpen; 2023:Ch. 3 p.
7. Ku ML, Li W, Chen Y, Ray Liu KJ. Advances in energy harvesting communications: past, present, and future challenges. IEEE Commun Surv Tutorials 2016;18:1384–412.
8. Elsheikh A. Bistable morphing composites for energy-harvesting applications. Polym 2022;14. https://doi.org/10.3390/polym14091893.
9. Fan Y, Jia Z, Zhang Z, Gu S, Du W, Lin D. Flexible composites with rare-earth element doped polycrystalline particles for piezoelectric nanogenerators. Micromachines 2024;15. https://doi.org/10.3390/mi15111280.

10. Kaihang Z, Lu J, Cai X, Shah M, Wu J, Li J, et al. Nanosheet-doped polymer composites with high intrinsic piezoelectric properties for energy harvesting. Energy Environ Mater 2024;8:1–12.
11. Derraz M, Ennawaoui C, Mastouri H, Abouricha N, Rjafallah A, El Mehdi L, et al. Mathematical modeling for predicting electrical energy harvested using piezoelectric composite materials for smart system applications. Micro Nano Eng 2024;23:100253.
12. Yuan C, Li X, Yang X, Wu F, Gao G, Zhou H, et al. Application of multi-layered composite fiber films with enhanced piezoelectric performance in flexible energy harvesting devices. Ceram Int 2024;50. https://doi.org/10.1016/j.ceramint.2024.05.183.
13. Yu C, Li G. Electrical energy harvesting by connected form stable phase change material composites. Energy Convers Manag 2024;299:117851.
14. Bai S, Chunhua L. Overview of energy harvesting and emission reduction technologies in hybrid electric vehicles. Renew Sustain Energy Rev 2021;147:111188.
15. Shivamurthy B, Naik N, Thimappa BHS, Bhat R. Mechanical property evaluation of alkali-treated jute fiber reinforced bio-epoxy composite materials. Mater Today Proc 2020;28. https://doi.org/10.1016/j.matpr.2020.04.016.
16. Crossley S, Whiter RA, Kar-Narayan S. Polymer-based nanopiezoelectric generators for energy harvesting applications. Mater Sci Technol 2014;30:1613–24.
17. Wang Z, Kurita H, Nagaoka H, Narita F. Potassium sodium niobate lead-free piezoelectric nanocomposite generators based on carbon-fiber-reinforced polymer electrodes for energy-harvesting structures. Compos Sci Technol 2020;199:108331.
18. Cottinet PJ, Guyomar D, Guiffard B, Putson C, Lebrun L. Modeling and experimentation on an electrostrictive polymer composite for energy harvesting. IEEE Trans Ultrason Ferroelectr Freq Control 2010;57:774–84.
19. Naik N, Suresh P, Yadav S, Nisha MP, Arias-Gonzáles JL, Cotrina-Aliaga JC, et al. A review on composite materials for energy harvesting in electric vehicles. Energies 2023;16. https://doi.org/10.3390/en16083348.
20. Sun R, Wang L, Zhang Y, Zhong C. Characterization of 1-3 piezoelectric composite with a 3-tier polymer structure. Materials 2020;13. https://doi.org/10.3390/ma13020397.
21. Mishra S, Unnikrishnan L, Nayak SK, Mohanty S. Advances in piezoelectric polymer composites for energy harvesting applications: a systematic review. Macromol Mater Eng 2018;1800463. https://doi.org/10.1002/mame.201800463.
22. Song J, Dong FW, Guansong Shan TO, Itoh T. A black gauze cap-shaped bistable energy harvester with a movable design for broadening frequency bandwidth. Smart Mater Struct 2019;29:025015.
23. Lu Z, Chen J, Ding H, Chen L. Two-span piezoelectric beam energy harvesting Int J Mech Sci 2020;175. https://doi.org/10.1016/j.ijmecsci.2020 105532.
24. Zou H, Li M, Zhao L, Gao Q, Wei K, Zuo L, et al. A magnetically coupled bistable piezoelectric harvester for underwater energy harvesting. Energy 2020:119429. https://doi.org/10.1016/j.energy.2020.119429.
25. Pradhan DK, Kumari S, Rack PD, Kumar A. Applications of strain-coupled magnetoelectric composites. Encyclopedia Smart Mater 2021;2021. https://doi.org/10.1016/b978-0-12-815732-9.00048-6.
26. Kang TJ, Fang S, Kozlov ME, Haines CS, Li N, Kim YH, et al. Electrical power from nanotube and graphene electrochemical thermal energy harvesters. Adv Funct Mater 2012;477–89. https://doi.org/10.1002/adfm.201101639.
27. Wei J, Nie Z, He G, Hao L, Zhang Q. Energy harvesting from solar irradiation in cities using the thermoelectric behavior of carbon fiber. RSC Adv 2014:48128–34. https://doi.org/10.1039/c4ra07864k.
28. Gao C, Chen G. Conducting polymer/carbon particle thermoelectric composites: emerging green energy materials. Compos Sci Technol 2016;124:52–70.
29. Roscow JI, Pearce H, Khanbareh H, Kar-narayan S, Bowen CR. Modified energy harvesting figures of merit for stress- and strain-driven piezoelectric systems. Eur Phys J: Spec Top. 2019;1554:1537–54.
30. Borowiec M, Gawryluk J, Bochenski M. Influence of mechanical couplings on the dynamical behavior and energy harvesting of a composite structure. Polymers 2021;13:66.

31. Alam MN, Kumar V, Jung H-S, Park S-S. Fabrication of high-performance natural rubber composites with enhanced filler–rubber interactions by stearic acid-modified diatomaceous earth and carbon nanotubes for mechanical and energy harvesting applications. Polym 2023;15. https://doi.org/10.3390/polym15173612.
32. Kumar V, Alam MN, Yewale MA, Park S-S. Multifunctional aspects of mechanical and electromechanical properties of composites based on silicone rubber for piezoelectric energy harvesting systems. Polym 2024;16. https://doi.org/10.3390/polym16142058.
33. Ahmad S, Mujeebu MA, Farooqi M. Energy harvesting from pavements and roadways: a comprehensive review of technologies, materials, and challenges. Int J Energy Res 2019;43:1974–2015.
34. Afiqah N, Radzuan M, Bakar A, Sahari J. A review of electrical conductivity models for conductive polymer composite. Int J Hydrogen Energy 2016;42:9262–73.
35. Hu G, Ning X, Hussain M, Sajjad U, Sultan M, Ali H, et al. Potential evaluation of hybrid nanofluids for solar thermal energy harvesting: a review of recent advances. Sustain Energy Technol Assess 2021;48. https://doi.org/10.1016/j.seta.2021.101651.
36. Wu Z, Wang J, Liu Y, Hou S, Liu X, Zhang Q, et al. A review of spectral controlling for renewable energy harvesting and conserving. Mater Today Phys 2021;18:100388.
37. Qian W, Yang W, Zhang Y, Bowen CR, Yang Y. Piezoelectric materials for controlling electro - chemical processes. Nanomicro Lett 2020;12. https://doi.org/10.1007/s40820-020-00489-z.
38. Sodano HA, Inman DJ, Park G. Comparison of piezoelectric energy harvesting devices for recharging batteries. J Intell Mater Syst Struct 2005;16:799–807.
39. El-Sayed A-R, Tai K, Biglarbegian M, Mahmud S. A survey on recent energy harvesting mechanisms; 2016: 1–5 pp. https://doi.org/10.1109/CCECE.2016.7726698.
40. Anic M, Prodanovic M, Milenković S, Filipovic N, Grujović N, Zivic F. The review of materials for energy harvesting. 2021. https://doi.org/10.1109/BIBE52308.2021.9635169.
41. Jayamani E, Nair G, Soon K. Investigation of the dielectric properties of natural fibre and conductive filler reinforced polymer composites. Mater Today Proc 2020;22:162–71.
42. Ahmed O, Wang X, Tran M-V, Ismadi M-Z. Advancements in fiber-reinforced polymer composite materials damage detection methods: towards achieving energy-efficient SHM systems. Composites, Part B 2021; 223:109136.
43. Mrlik M, AlMaadeed MAS. 16 – fillers in advanced nanocomposites for energy harvesting. In: Dong Y, Umer R, Lau AK-T, editors. Woodhead Publ. Ser. Compos. Sci. Eng. Sawston, Cambridge: Woodhead Publishing; 2015:401–24 pp.
44. Yusuf J, Sapuan SM, Rashid U, Ilyas RA, Hassan MR. Thermal, mechanical, and morphological properties of oil palm cellulose nanofibril reinforced green epoxy nanocomposites. Int J Biol Macromol 2024;278: 134–421.
45. Ahmed S, Banerjee S, Sundar U, Ruiz H, Kumar S, Weerasinghe A. Energy harvesting: breakthrough technologies through polymer composites BT – smart polymer nanocomposites: energy harvesting, self-healing and shape memory applications. In: Ponnamma D, Sadasivuni KK, Cabibihan J-J, Al-Maadeed MA-A, editors. Cham: Springer International Publishing; 2017:1–42 pp.
46. Chen L, Zhang Y, Chen Z, Dong Y, Jiang Y, Hua J, et al. Biomaterials technology and policies in the building sector: a review. Environ Chem Lett 2024;22:715–50.

Yusuf Jameel, Abdul Halim Muhammad Firdaus, Mohd Sapuan Salit*
and Nazrin Asmawi

5 Mechanical and thermal properties of graphene reinforced poly (lactic acid) composites for battery casing in electric vehicles

Abstract: This study explores the development of graphene-reinforced poly (lactic acid) (PLA) composites as a sustainable material for electric vehicle (EV) battery casings. By incorporating graphene, a two-dimensional nanomaterial known for its exceptional mechanical and thermal properties, the PLA matrix exhibits significant improvements in performance. The optimized graphene content enhances the composite's tensile strength, Young's modulus, and ductility, ensuring robust mechanical protection for battery components. Additionally, graphene significantly increases the thermal conductivity of the composite, promoting effective heat dissipation and mitigating thermal degradation risks during battery operation. The research demonstrates the ability of graphene to act as a gas barrier, preventing the diffusion of oxygen and volatile by-products, which contributes to improved thermal stability and prolonged material durability. By tailoring the graphene loading and ensuring uniform dispersion, the composite achieves a balance between mechanical strength and thermal performance, making it ideal for lightweight, durable, and eco-friendly applications. This study underscores the potential of graphene-PLA composites in advancing sustainable technologies, particularly in the EV sector. It provides a foundation for future research aimed at optimizing fabrication methods, exploring synergistic filler combinations, and validating real-world performance to support the broader adoption of environmentally friendly materials in high-performance applications.

Keywords: Graphene; electric vehicle; poly(lactic acid) (PLA)

***Corresponding author: Mohd Sapuan Salit**, Advanced Engineering Materials and Composites Research Centre (AEMC), Faculty of Engineering, Universiti Putra Malaysia, Serdang 43400, Selangor, Malaysia, E-mail: sapuan@upm.edu.my. https://orcid.org/0000-0002-4598-6358
Yusuf Jameel and Abdul Halim Muhammad Firdaus, Advanced Engineering Materials and Composites Research Centre (AEMC), Faculty of Engineering, Universiti Putra Malaysia, Serdang 43400, Selangor, Malaysia
Nazrin Asmawi, Fakulti Kemanusiaan, Pengurusan Dan Sains, Universiti Putra Malaysia Kampus Bintulu Sarawak, Jalan Nyabau, 97008 Bintulu, Sarawak, Malaysia

As per De Gruyter's policy this article has previously been published in the journal Physical Sciences Reviews. Please cite as: Y. Jameel, A. H. M. Firdaus, M. S. Salit and N. Asmawi "Mechanical and thermal properties of graphene reinforced poly (lactic acid) composites for battery casing in electric vehicles" *Physical Sciences Reviews* [Online] 2024. DOI: 10.1515/psr-2024-0023 | https://doi.org/10.1515/9783111234762-005

5.1 Introduction

Graphene is defined as a two-dimensional (2D) hexagonal lattice structure comprise of single atom thick sheet of covalently bonded sp^2 carbon atoms [1]. Theoretically, it can measure up to 1 TPa of Young's modulus [2] and 130 GPa of intrinsic strength [3]. Even at higher densities of sp^3 defects, it can withstand higher breaking strength and stiffness. It is widely utilised as a reinforcing agent in polymer composites for its excellent mechanical, thermal, and electrical properties. The key aspect of developing high performance graphene-based polymer compositeslies on its high specific surface area to interact with polymer matrix. Its unique feature is sought-after for the fabrication of high-end products such as energy storage, sensors, automotive, electronics and biomedical devices [4]. Poly (lactic acid) (PLA) is a biopolymer synthesized through condensation polymerization of lactic acid monomers, a product from the fermentation of biomasses such as sugarcane, corn starch and other plant-based materials. As a biodegradable polymer, it gains massive attention as a potential alternative of conventional petrochemical plastics in producing sustainable, renewable and less environmental impact products. Initially, PLA is mainly utilised in medical service such as drug encapsulation, sutures, stents and tissue scaffold [5]. Nowadays, PLA had gained popularity as an additive manufacturing of 3D printing for it can easily melt, fabricated into complex design and release non-toxic fume [6].

Interestingly, only a small fraction of graphene is required to provide an enhancement effect to the polymeric material. While there are various studies had been conducted and variable results are recorded, the speculated values ranging from 0.1 to 3 %. Uniformly dispersed graphene in PLA provides significant improvement in gas barrier properties. The reliability of battery critically relies on the movement of electrons and ions to generate electrical energy. Enhanced gas barrier material prevents the invasion of any gases or liquids into the cell membrane, which reduce the efficiency of the device. Battery casing made of graphene reinforced PLA can act as supplementary barrier for oxygen and water vapour before reaching the battery's reactant. Additionally, thermal degradation can be suppressed while retain its mechanical robustness to sustain a protective structure to the battery from external impact.

This growth has brought advancement various improvements in battery systems due to the proliferation of electric vehicles, leading to maximize the search for state-of-the-art materials suitable for modem battery casings [7]. These casings help to ensure that battery cells are shielded from every form of environmental aggression, mechanical stress, high thermal conditions, and other safety requirements to prolong the lifespans of battery cells. To meet these requirements battery casings, need to exhibit a set of properties for which currently there are no ideal materials; it must be mechanically rigid, thermally stable, chemically resistant and it should be relatively lightweight [8]. Metals such as aluminum or other alloys have been traditionally used, however, due to vehicle lightweighting initiatives to increase energy efficiency of automobiles advanced composite materials have found novel applications as suitable replacements.

Overall, the polymer-based composites and specifically those containing fibers or fillers demonstrate high potential for practical application owning to their low density, tunable mechanical and thermal performance [9]. However, traditional composite systems have their challenges, including low thermal conduction and ultimate inadequacy concerning recycling methods [10]. Such issues have called for development of a new generation material which is the graphene-PLA material. The polymer PLA, which is derived from renewable resources, also meets the objectives of sustainable development of materials that are used to construct advanced composites [11]. Graphene improves PLA's strengths such as strength, thermal conductivity, and electricity conductivity and thus making it a perfect fit of EV battery casings that are light, strong, and eco-friendly [12]. This paper discusses analyzing the properties of EV battery casings, assessing the drawbacks of traditional materials and introducing new graphene-reinforced PLA composites into consideration.

5.2 Performance of battery casing

The performance of EV battery casings is critical for the safety, efficiency, and life of the battery system. Mechanical durability is one of the key performance metrics, as battery housings are subject to a myriad of operational conditions, vibrations, mechanical shocks and accidental impacts [13]. The stresses here can be from normal vehicle operation, road imperfections or collision events. Precise mechanical integrity is paramount to safeguarding the sensitive battery cells and avoiding structural deformities that might affect the vehicle's functionalities or create unsafe situations, such as thermal runaway [13]. The casing must be made of a material with high tensile strength, strong impact resistance, and good fatigue properties to withstand long-term and harsh working conditions [14].

Thermal management is, however, another key aspect of battery casing performance, which in turn, is not only about mechanical durability [14]. EV batteries usually need to work under certain temperature ranges in order to provide good efficiency and preventing overheating [15]. With poor thermal management, the battery life, functionality, or thermal runaway on a dramatic failure can decrease. Battery casings are critical in dissipating heat generated during charging and discharging cycles [15]. Thus, materials with reasonable thermal conductivity are crucial to disperse heat uniformly and avoid concentration of heat in the surroundings. At the same time, the casing needs to be an insulator to insulate the battery against outside temperature changes [16]. Moreover, emerging advanced materials capable of achieving core requirements such as high thermal conductivity without adding significant weight, such as graphene-reinforced composites, are becoming increasingly attractive [16]. Battery casings play a crucial role in ensuring the safety, reliability, and efficiency of electric vehicles (EVs) by balancing with mechanical robustness and thermal stability.

Research findings by Zhen Song et al. regarding effects of temperature on the performance on the battery power and FCS net power of hybrid electric vehicle [16]. These results

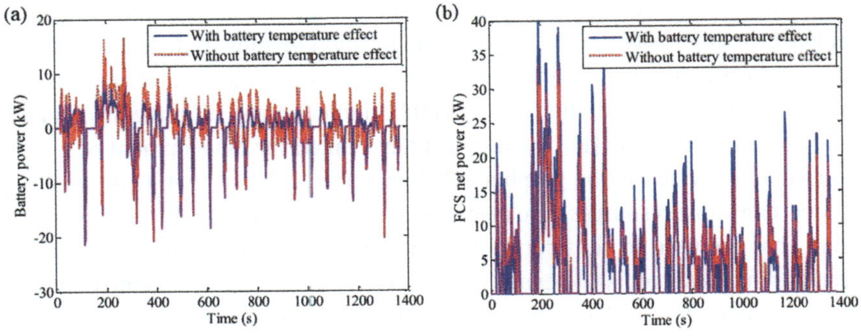

Figure 5.1: Simulation findings for the FTP72 urban driving cycle with and without account of the battery temperature effect: (a) battery power, (b) FCS net power [16].

revealed that with the consideration of the temperature effect, the dynamic amplitude of the total lithium-ion battery power during discharge was relatively small, and the fuel cell system could deliver the maximum power. From Figure 5.1(b) it is as well observed that the overall power capability from the fuel cell system is more than that is required for the FCHEVs which is consistent with the current top of the mind concept of FCHEVs where fuel cell is the main power supply while the lithium-ion battery is the backup power. Further, the energy reserve of the lithium-ion battery taking temperature into account can recover more of the braking energy during charging phase as shown in Figure 5.1(a). By incorporating temperature conditions of FTP72, NEDC 2000 and 1015, the fuel economy of the FCHEV improved by 2.33 %, 4.77 % and 3.02 % respectively. The characteristics of the extended PMP control strategy were also general for the simulation results of other control strategies when the optimal plane was defined as the extended PMP control strategy.

To enhance the PMP strategy they presided over four assumptions to control the PMP and to recognize the global optimal solution separately based on the temperature and SOC. Since the DP strategy can also yield the global optimal solution, comparisons of simulation results of the two-state variable PMP strategy and DP method were made. The results can be summarized as follows: FCHEV solves, in effect, four hypotheses and the strategy proposed in this article, based on PMP, ensures a reasonable difference in global optimality. Fuel consumption difference between the two strategies is less than 1.5 % for option (b). The PMP based power management strategy shows more significant time saving effect when the battery SOC is high, battery temperature range is high and when driving cycle time of the batteries are long. The process can also be used for other types of HEVs for example, engine/battery HEVs with high levels of electric driving.

5.3 Poly (lactic acid) (PLA)

5.3.1 Mechanical properties of poly (lactic acid) (PLA)

Figure 5.2 shows the flexural mechanical properties. The results as shown in Figure 5.2(a); indicate that all composite material's flexural strength (FS) value increased. With the

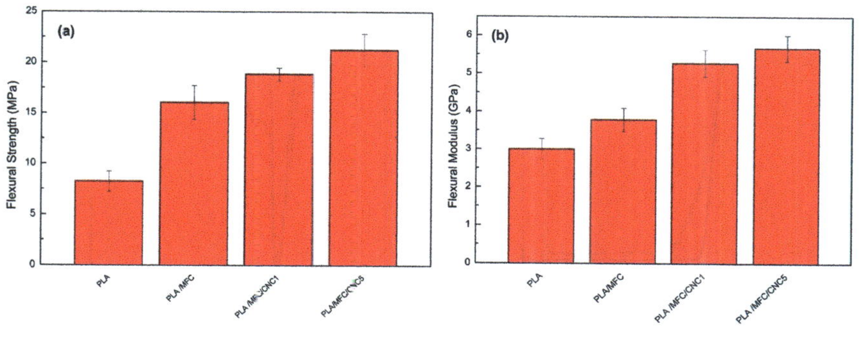

Figure 5.2: Flexural mechanical properties of PLA and the biocomposites. (a) Flexural strength and (b) modulus of the studied biocomposites [17].

confirmed introduction of 20 % w/w cellulose fibers, the PLA FS increased by approximately 80 %. Incorporating 1 % CNCs in place of 1 % cellulosic fibers increases the FS of the biocomposite to 130 % when compared with PLA. Moreover, the 5 % MFC-replacement level by CNCs also led to an increases of 175 % in resultant FS of the multiscale biocomposite. Such behavior reflects a better stress transfer respectively at a fiber-matrix interphase [18]. The function of the PLA-g-MA coupling agent cannot be responsible for the stress transfer efficiency, instead this may be due to the capacity of the CNCs as a form of strengthening in the interfacial area between the MFC and the PLA.

Moreover, the enhancement may be related to the entanglement between cellulose nanocrystals and cellulose microfibers in the composite that may hinder the fracture [19].

Figure 5.2(b) shows the flexural modulus of PLA and the biocomposites. Flexural modulus for pure PLA is 3 GPa, while the modulus of developed PLA/MFC is improved up to 25 %. But still, the elastic modulus of the PLA/MFC/CNC1 and PLA/MFC/CNC5 is 67 % and 82 % higher than that of the pure PLA. Again, due to CNCs acting as an interphase reinforcement between the PLA and MFC, this enhancement is observed [20].

5.3.2 Thermal properties of poly (lactic acid) (PLA)

Figure 5.3(a) presents the mass loss thermograms of PLA, PLA/MFC, PLA/MFC/CNC1 and PLA/MFC/CNC5. With their corresponding mass loss rate displayed in Figure 5.3(b). Figure 5.3(a) further proves that the degradation of PLA occurs in a single stage, without loss of mass until approximately 220 °C, followed by almost the same trend until 270 °C, at which point 3 % loss of mass is present. At 332 °C, 0 % of carbon residues indicating the 100 % degradation. It can also be seen In Figure 5.3(b), thermal degradation proceeds step by step, while the maximum degradation temperature of PLA is observed at 316 °C. Similar results were reported by Abubakar et al. [21] and Johari et al. [22].

As seen in Figure 5.3(a), mass loss in this temperature range (150–250 °C) is caused by depolymerization of non-cellulosic components of the fiber – hemicellulose (i.e., breaking

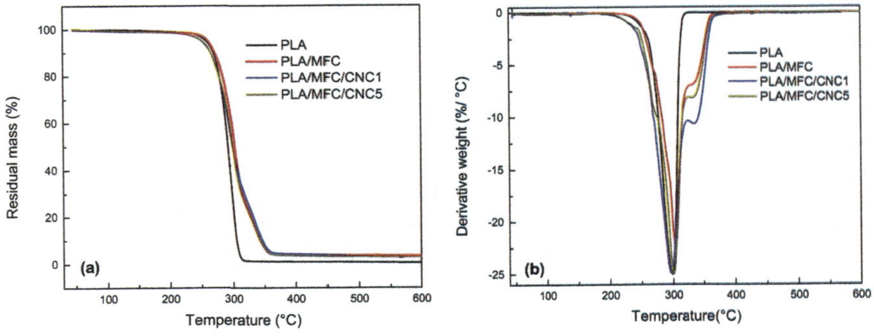

Figure 5.3: Thermal properties of biocomposites. (a) Residual mass thermograms and (b) their derivatives, for the studied biocomposites [17].

of glycoside bonds) and water adsorbed onto the fibers. As α-celluloses begin to disintegrate at around 300–350 °C, it requires considerable energy. This suggests that the sample degradation was limited to the amorphous structure of the sample and the crystalline fraction is thermally stable. This degradation of PLA/MFC involves the drying and dissipation of moisture, in addition to the breakdown of the glycosidic bonds of cellulose, releasing H_2O and CO_2, alkanes and other derivative hydrocarbons in a range of 230–370 °C, after which the decomposed products become facile to degrade [22]. Twenty six most of this process occurs between 150 °C and 700 °C where, as depicted in Figure 5.3(b), initially a loss of weight from moisture was observed and at 250 °C the observed decomposition of hemicelluloses and α-component. In contrast, the weight loss trend is not linear because this degradation related to the fibrous material of the cellulose was not achieved at the test temperatures and, additionally, this ash formation is directly concerned with the content of hemicellulose present in the material. The trend is also similar of mass losses for both materials until 240 °C but cellulose degradation happens at 330 °C. Above 370 °C, no further weight loss occurs, leaving a carbon residue level of 3.5 % attributable to cellulose.

5.4 PLA/graphene nanocomposites

5.4.1 Mechanical properties of poly (L-lactide) PLLA nanocomposites

Kim et al. [6] postulated that PLA/graphene nanocomposites can be applied in 3D printing filament to fabricate product with good electrical conductivity and mechanical flexibility. At 2 % loading, tensile strength increased by 44 % and maximum strain by 57 %. The strong interfacial adhesion between graphene and PLA matrix effectively distributes the strain to the neighbouring region so that crack propagation is hindered. Higher loading is prone to crack propagation due to slippery condition of graphene-to-graphene stack. On another note, it leads to loosening of bonding strength subsequently stress concentrated on exerted region cannot be transfer completely. Gu et al. [23] reported that the molecular

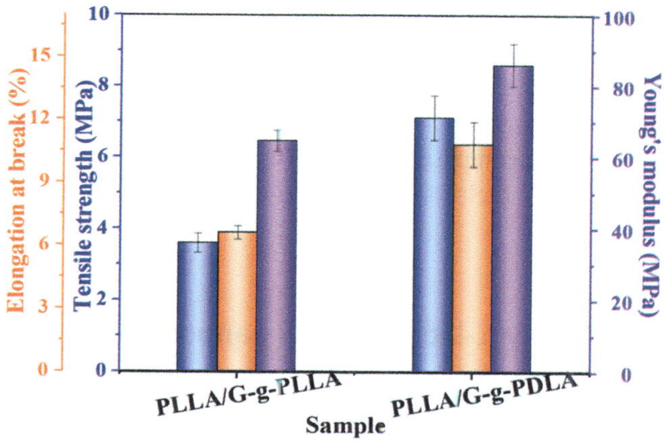

Figure 5.4: The mechanical properties of PLLA nanocomposites at 100 °C [23].

chain of poly (D-lactide) grafted graphene (G-g-PDLA) promoted the formation of stereocomplex crystallites (SC) at the interface region. At 100 °C, the intensified interfacial bonding of poly (L-lactide) (PLLA) and graphene catalysed by SC and its synergetic effect with graphene promoted a percolated network structure to toughen tensile strength. Tensile strength, elongation at break and young's modulus as demonstrated in Figure 5.4.

The analysis of the mechanical properties of PLLA/G-g-PLLA and PLLA/G-g-PDLA reveals significant differences between the two materials. Firstly, the tensile strength of PLLA/G-g-PDLA is approximately 6.5 MPa, which is considerably higher than the 4.5 MPa exhibited by PLLA/G-g-PLLA. This indicates that PLLA/G-g-PDLA is more resistant to breaking under tension, making it a more suitable choice for applications requiring high tensile strength.

In terms of elongation at break, PLLA/G-g-PDLA also outperforms PLLA/G-g-PLLA, with values of approximately 6 % and 5 %, respectively. This suggests that PLLA/G-g-PDLA has better flexibility and ductility, as it can stretch more before breaking. This property is particularly important in applications where materials need to withstand deformation without failing.

Furthermore, the Young's modulus of PLLA/G-g-PDLA is approximately 9 MPa, compared to 7 MPa for PLLA/G-g-PLLA. A higher Young's modulus indicates that PLLA/G-g-PDLA is stiffer and more rigid, which is beneficial for applications requiring materials that do not easily deform under stress.

According to Devinda et al. reseaech, the mechanical properties of GNP-reinforced virgin and recycled PC nanocomposites demonstrated notable improvements, particularly in Young's modulus, which increased by up to ~33 % for virgin PC and ~39.5 % for recycled PC at 10 wt% GNP loading. This enhancement highlights the reinforcing effect of GNP, which contributes to the stiffness of the nanocomposites. However, the failure strain decreased exponentially, indicating a transition from ductile to brittle behavior with higher GNP concentrations. This trade-off between stiffness and ductility is critical for determining optimal filler loading for specific applications. Among the key processing parameters, filler concentration was the most influential, while processing speed and barrel/die set temperature showed negligible effects. These findings suggest that the mechanical performance of PC/GNP nanocomposites can be effectively tailored by controlling the filler concentration, enabling their use in applications requiring enhanced stiffness with moderate ductility [24].

5.4.2 Thermal properties of poly (L-lactide) PLLA nanocomposites

Battery's temperature inevitably rose throughout manoeuvring EV and it contribute to the declination in overall performance and at worst, total shut down. Hence, thermal conductive material must be integrated to circulate the heat from exceeding maximum operating temperature, which might lead to fire and even explosion. Graphene is a highly conductive material (5,000 W m^{-1} K^{-1}) with high aspect ratio and large specific surface area [23]. Conductive network established within PLA nanocomposites is essential to dissipate heat generated by the battery into the surrounding. The heat dissipation maintains the operational condition of battery from exceeding the hazardous temperature. The introduction of graphene into PLA acts as a barrier to suppress the diffusion of oxygen and other gases, which beneficially inhibit the thermal degradation process. This effect is necessary to sustain to workability of the battery where prolonged exposure to high temperature is unavoidable. Gonçalves et al. [25] indicated that the high thermal conductivity of graphene facilitates heat transfer to overwhelm mass transfer barrier effect hence resulting in higher thermal degradation temperature of PLA/graphene nanocomposites. At 0.5 % loading, onset degradation temperature was elevated as diffusion of pyrolysis products were supressed effectively. Even so, trivial changes were recorded in glass transition and melting temperature. At higher loading of 5 %, Adesina et al. [26] determined that the presence of the honeycomb-like nanofiller brought forth similar shielding effect to restrain the diffusion of volatile decomposition products. The spark plasma sintering process used in the study yield highly dense nanocomposite sample, which left highest amount of residue above 400 °C. Gu et al. [23] stated that the formation of SC at the interface region is favourable for intermolecular interaction with PLLA. The study evaluated that PLLA/G-g-PDLA had higher thermal conductivity, superior heat distortion temperature and decent mechanical properties at high operating temperature compared to pristine PLLA and PLLA/G-g-PLLA. The interfacial thermal resistance is larger between filler and matrix rather than filler to filler, hence, improved dispersion of graphene through grafting method is an efficient approach to construct continuous conductive path in the nanocomposites. Figure 5.5 elucidates the thermal conductivity of pristine PLA and its nanocomposites at different loadings.

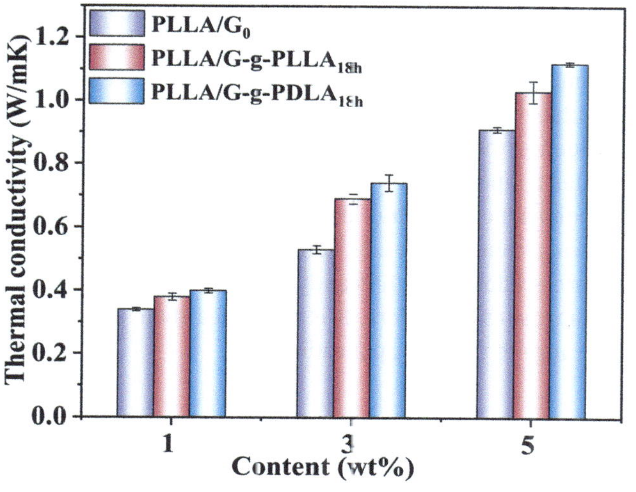

Figure 5.5: The thermal conductivity of PLLA nanocomposites at different graphene loadings [23].

The thermal conductivity analysis of PLLA/G_0, PLLA/G-g-PLLA$_{18}$h, and PLLA/G-g-PDLA$_{18}$h at different content levels (wt%) reveals distinct trends. The thermal conductivity of each material increases with rising content percentage. Specifically, for PLLA/G_0, the thermal conductivity is approximately 0.3 W/mK at 1 wt%, 0.5 W/mK at 3 wt%, and 0.6 W/mK at 5 wt%. For PLLA/G-g-PLLA$_{18}$h, the values are roughly 0.5 W/mK at 1 wt%, 0.7 W/mK at 3 wt%, and 0.8 W/mK at 5 wt%. The highest thermal conductivity is observed in PLLA/G-g-PDLA$_{18}$h, with around 0.7 W/mK at 1 wt%, 0.9 W/mK at 3 wt%, and close to 1.1 W/mK at 5 wt%.

The data indicates that PLLA/G-g-PDLA$_{18}$h exhibits the highest thermal conductivity among the three materials at each content level, making it the most efficient in conducting heat. PLLA/G-g-PLLA$_{18}$h also shows better thermal conductivity than PLLA/G_0 but is lower than PLLA/G-g-PDLA$_{18}$h. PLLA/G_0 has the lowest thermal conductivity, suggesting it may be more suitable for insulating applications where minimal heat transfer is required.

In summary, the analysis highlights the superior thermal conductivity of PLLA/G-g-PDLA$_{18}$h, making it the preferred material for applications demanding efficient heat transfer. Conversely, PLLA/G_0, with the lowest thermal conductivity, might be ideal for applications requiring insulation. The observed trends provide valuable insights for selecting materials based on their thermal conductivity properties for various engineering and structural applications.

A study by Giovanni et al. shows a clear increase in thermal conductivity with increasing graphene content is observed. In fact, at the temperature of 298.15 K, the thermal conductivity for the pure PLA is 0.173 W/mK, whereas for PLA with 6 wt% of graphene nanoplatelets (GNP) reaches the value of 0.470 W/mK, which corresponds to a significant improvement of about 171 % [27].

Therefore, since classical thermally insulating materials show a thermal conductivity of the order of 10^{-3} (W/m K), the proposed nanocomposites, in light of the measured values of conductivity and in combination with the other benefits typical of polymer materials, can be considered suitable for potential heat transfer applications. With reference to the temperature influence, the thermal conductivity increases slightly as it increases, at least in the investigated temperature range. This is because, with increasing temperature, the molecular vibrations increase, thus leading to a higher phonon propagation and hence to a higher thermal conductivity [28]. Carbon-based nanofillers, when dispersed within polymeric matrices allow to significantly improve the thermal conductivity of the resulting nanocomposites. Therefore, it is motivated their use as heat sinks to dissipate heat in a wide range of thermal applications [29].

5.5 Conclusion and future outlooks

This study highlights the potential of graphene-reinforced poly (lactic acid) (PLA) composites as a sustainable and high-performance material for electric vehicle (EV) battery casings. The integration of graphene into the PLA matrix significantly improves both mechanical and thermal properties, effectively addressing the inherent limitations of pure PLA. The enhancement in tensile strength, Young's modulus, and ductility demonstrates the material's ability to withstand mechanical stresses, ensuring robust protection for battery components. Optimal graphene loading and uniform dispersion within the matrix enhance interfacial bonding, allowing efficient stress transfer and minimizing crack propagation.

In terms of thermal properties, the incorporation of graphene greatly increases the thermal conductivity of the composite, ensuring effective heat dissipation. This improvement mitigates the risks associated with thermal degradation, thereby enhancing the safety and operational stability of EV batteries. Furthermore, graphene's role as a barrier to gas and pyrolysis product diffusion contributes to the thermal stability and longevity of the composite material. These characteristics make the material well-suited for EV applications, where lightweight yet durable components are critical.

The findings emphasize the importance of optimizing graphene content and dispersion techniques to achieve a balance between mechanical strength and thermal performance. This study establishes graphene-reinforced PLA composites as a promising alternative to conventional materials, supporting the development of sustainable and efficient technologies in the EV industry.

Future research should focus on scaling up production processes to meet industrial demands, exploring the use of additional fillers for synergistic property enhancements, and conducting real-world testing to validate the material's performance under operational conditions. These efforts will further solidify the role of graphene-PLA composites in advancing sustainable transportation solutions.

Acknowledgements: The authors would like to thank the editors S.M. Sapuan, Mohd Roshdi Hassan, Eris Elianddy Supeni and Azizan As'arry for their guidance and review of this article before its publication.

References

1. Wang D, You F, Hu GH. Graphene/polymer nanocomposites with high dielectric performance: interface engineering. Berlin, Europe: Elsevier Inc.; 2015.
2. Mohan VB, Tak Lau K, Hui D, Bhattacharyya D. Graphene-based materials and their composites: a review on production, applications and product limitations. Composites Part B 2018;142:200–20.
3. Remanan S, Das TK, Das NC. Graphene as a reinforcement in thermoset resins. Polym Nanocomposites Contain Graphene Prep Prop Appl 2022:317–41. https://doi.org/10.1016/B978-0-12-821639-2.00012-4
4. Silva M, Pinho IS, Covas JA, Alves NM, Paiva MC. 3D printing of graphene-based polymeric nanocomposites for biomedical applications. Funct Compos Mater 2021;2. https://doi.org/10.1186/s42252-021-00020-6.
5. Khammassi S, Tarfaoui M, Škrlová K, Měřínská D, Plachá D, Erchiqui F. Poly (lactic acid) (PLA)-Based nanocomposites: impact of vermiculite, silver, and graphene oxide on thermal stability, isotherma crystallization, and local mechanical behavior. J Compos Sci 2022;6. https://doi.org/10.3390/jcs6040112.
6. Kim M, Jeong JH, Lee JY, Capasso A, Bonaccorso F, Kang SH, et al. Electrically conducting and mechanically strong graphene-polylactic acid composites for 3D printing. ACS Appl Mater Interfaces 2019;11:11841–8.
7. Gao W, Li X, Ma M, Fu Y, Jiang J, Mi C. Case study of an electric vehicle battery thermal runaway and online internal short-circuit detection. IEEE Trans Power Electron 2021;36:2452–5.
8. Burd JTJ, Moore EA, Ezzat H, Kirchain R, Roth R. Improvements in electric vehicle battery technology influence vehicle lightweighting and material substitution decisions. Appl Energy 2021;283:116269.
9. Cywar RM, Rorrer NA, Hoyt CB, Beckham GT, Chen EY-X. Bio-based polymers with performance-advantaged properties. Nat Rev Mater 2022;7:83–103.
10. Wang Y, Desroches GJ, Macfarlane RJ. Ordered polymer composite materials: challenges and opportunities. Nanoscale 2021;13:426–43.
11. Cojocaru V, Frunzaverde D, Miclosina C-O, Marginean G. The influence of the process parameters on the mechanical properties of PLA specimens produced by fused filament fabrication – a review. Polymers 2022; 14:886.
12. Aumnate C, Potiyaraj P, Saengow C, Giacomin AJ. Reinforcing polypropylene with graphene-polylactic acid microcapsules for fused-filament fabrication. Mater Des 2021;198:109329.
13. Chen Y, Kang Y, Zhao Y, Wang L, Liu J, Li Y, et al. A review of lithium-ion battery safety concerns: the issues, strategies, and testing standards. J Energy Chem 2021;59:83–99.
14. Camargos PH, dos Santos PHJ, dos Santos IR, Ribeiro GS, Caetano RE. Perspectives on Li-ion battery categories for electric vehicle applications: a review of state of the art. Int J Energy Res 2022;46:19258–68.
15. He L, Jing H, Zhang Y, Li P, Gu Z. Review of thermal management system for battery electric vehicle. J Energy Storage 2023;59:106443.
16. Song Z, Pan Y, Chen H, Zhang T. Effects of temperature on the performance of fuel cell hybrid electric vehicles: a review. Appl Energy 2021;302:117572.
17. Ruz-Cruz MA, Herrera-Franco PJ, Flores-Johnson EA, Moreno-Chulim MV, Galera-Manzano LM, Valadez-González A. Thermal and mechanical properties of PLA-based multiscale cellulosic biocomposites. J Mater Res Technol 2022;18:485–95.
18. Fu W, Xu X, Wu H. Mechanical and biodegradable properties of l-lactide-grafted sisal fiber reinforced polylactide composites. J Reinforc Plast Compos 2014;33:2034–45.
19. Faludi G, Hári J, Renner K, Móczó J, Pukánszky B. Fiber association and network formation in PLA/lignocellulosic fiber composites. Compos Sci Technol 2013;77:67–73.

20. Huang L, Zhang X, Xu M, Chen J, Shi Y, Huang C, et al. Preparation and mechanical properties of modified nanocellulose/PLA composites from cassava residue. AIP Adv 2018;8:25116.
21. Birnin-Yauri AU, Ibrahim NA, Zainuddin N, Abdan K, Then YY, Chieng BW. Effect of maleic anhydride-modified poly(lactic acid) on the properties of its hybrid fiber biocomposites. Polymers 2017;9:165.
22. Johari AP, Mohanty S, Kurmvanshi SK, Nayak SK. Influence of different treated cellulose fibers on the mechanical and thermal properties of poly(lactic acid). ACS Sustainable Chem Eng 2016;4:1619–29.
23. Gu T, Sun D, Qi X, Yang J, Zhao C, Lei Y, et al. Synchronously enhanced thermal conductivity and heat resistance in poly(L-lactide)/graphene nanoplatelets composites via constructing stereocomplex crystallites at interface. Composites Part B 2021;224:109163.
24. Wijerathne D, Gong Y, Afroj S, Karim N, Abeykoon C. Mechanical and thermal properties of graphene nanoplatelets-reinforced recycled polycarbonate composites. Int J Light Mater Manuf 2023;6:117–28.
25. Gonçalves C, Pinto A, Machado AV, Moreira J, Gonçalves IC, Magalhães F. Biocompatible reinforcement of poly(Lactic acid) with graphene nanoplatelets. Polym Compos 2018;39:E308–20.
26. Adesina OT, Jamiru T, Sadiku ER, Ogunbiyi OF, Adegbola TA. Water absorption and thermal degradation behavior of graphene reinforced poly(lactic) acid nanocomposite. IOP Conf Ser Mater Sci Eng 2019;627. https://doi.org/10.1088/1757-899X/627/1/012015.
27. Spinelli G, Guarini R, Kotsilkova R, Batakliev T, Ivanov E, Romano V. Experimental and simulation studies of temperature effect on thermophysical properties of graphene-based polylactic acid. Materials 2022;15: 986.
28. Bird RB. Transport phenomena. Appl Mech Rev 2002;55:R1–4.
29. Kostagiannakopoulou C, Fiamegkou E, Sotiriadis G, Kostopoulos V. Thermal conductivity of carbon nanoreinforced epoxy composites. J Nanomater 2016;2016:1847325.

Moklis Muhammad Harussani*, Salit Mohd Sapuan* and Md Rubel

6 Limitations and future outlooks for char and its composites in energy harvesting application

Abstract: This book chapter explores the recent advancements and future prospects of carbon materials, specifically char reinforced composites, in the field of energy harvesting. Carbon materials have gained significant attention due to their unique properties, such as high electrical conductivity, thermal stability, and mechanical strength. The incorporation of char, a carbon-rich residue obtained from the incomplete combustion or pyrolysis of organic matter, as a reinforcing agent in composites, has shown promising results in energy harvesting applications. The chapter begins with an overview of chars and their key properties that make them suitable for energy conversion and storage, including their ability to facilitate electron transfer and withstand harsh environmental conditions. The focus then shifts to char, highlighting its abundance as a by-product of various industrial processes and its potential as a reinforcement material in composites. Furthermore, the chapter discusses recent development and limitations faced by char and its composites in energy harvesting applications, including solar cells, thermoelectric generators, and supercapacitors. Finally, the chapter outlines the future directions and the potential for exploring novel carbon sources and the integration of char composites in emerging energy harvesting technologies.

Keywords: carbonization; thermochemical; carbon nanostructures; fillers; biocomposite

6.1 Introduction

The increasing generation of waste materials coupled with the need for sustainable energy solutions has sparked growing interest in the development of advanced composite materials for energy harvesting applications. Carbon materials, including char,

***Corresponding authors: Moklis Muhammad Harussani**, Department of Transdisciplinary Science and Engineering, School of Environment and Society, Tokyo Institute of Technology, Meguro, Tokyo 152-8552, Japan; and Advanced Engineering Materials and Composites Research Centre (AEMC), Department of Mechanical and Manufacturing Engineering, Universiti Putra Malaysia, 43400 UPM Serdang, Selangor, Malaysia, E-mail: harussani.m.ab@m.titech.ac.jp. https://orcid.org/0000-0003-4475-8764; and **Salit Mohd Sapuan** Advanced Engineering Materials and Composites Research Centre (AEMC), Department of Mechanical and Manufacturing Engineering, Universiti Putra Malaysia, 43400 UPM Serdang, Selangor, Malaysia, E-mail: sapuan@upm.edu.my. https://orcid.org/0000-0002-4598-6358
Md Rubel, Department of Transdisciplinary Science and Engineering, School of Environment and Society, Tokyo Institute of Technology, Meguro, Tokyo 152-8552, Japan, E-mail: rubel.m.aa@m.titech.ac.jp

As per De Gruyter's policy this article has previously been published in the journal Physical Sciences Reviews. Please cite as: M. M. Harussani, S. M. Sapuan and M. Rubel "Limitations and future outlooks for char and its composites in energy harvesting application" *Physical Sciences Reviews* [Online] 2024. DOI: 10.1515/psr-2024-0024 | https://doi.org/10.1515/9783111234762-006

have emerged as promising reinforcements in composite systems due to their unique properties and potential for effective energy conversion [1–3]. With the continuous growth of population and industrialization, the generation of waste materials has become a pressing concern. These waste materials pose environmental challenges and require effective management strategies. However, waste materials, particularly biomass and carbon-rich waste, can be transformed into valuable resources through the process of upcycling. Upcycling involves several techniques–fast pyrolysis, carbonization, hydrothermal liquefaction, etc.–converting waste materials into higher-value products with improved properties and functionality [4, 5]. Char, a carbonaceous material obtained through the pyrolysis or carbonization of waste biomass or carbon-rich waste, is a prime example of upcycled material.

In recent years, there has been growing interest in the development of advanced composite materials for energy harvesting applications. Carbon materials, including char, have emerged as promising reinforcements in composite systems due to their unique properties and potential for effective energy conversion [6]. The utilization of carbon materials, such as char, in composite structures offers several advantages for energy harvesting. Carbon-based materials possess excellent electrical conductivity, thermal stability, and mechanical strength, making them ideal candidates for efficient energy conversion and storage [7–9]. By incorporating char as a reinforcement in composites, it becomes possible to harness and convert various forms of energy, including mechanical, thermal, and electromagnetic, into usable electrical energy.

By upcycling waste into char and utilizing it as a reinforcement in composite materials, we can not only address the environmental challenges associated with waste management but also contribute to sustainable energy solutions. The development and optimization of char composites for energy harvesting applications have the potential to revolutionize the way we harness and utilize various energy sources, leading to advancements in renewable energy technologies and the development of more sustainable and efficient energy systems. The development of char composites for energy harvesting applications has shown great promise. The incorporation of char into composite materials enhances their energy harvesting capabilities by effectively converting and transferring energy from the surrounding environment. Char composites can act as efficient transducers, enabling the conversion of mechanical, thermal, or electromagnetic energy into electrical energy through the unique properties of the carbon material [1, 3, 10–13]. These composites have demonstrated potential for applications such as self-powered sensors, wearable electronics, and energy harvesting from waste heat or vibrations.

The objective of this chapter is to provide a comprehensive review of the recent development and potential of carbon materials–char reinforced composites in energy harvesting applications. By synthesizing the findings from previous studies, we aim to highlight the importance of upcycling waste into char and the significant contributions that char composites offer in the field of energy harvesting. By bridging the gap between waste management and energy harvesting, char composites represent a promising pathway toward sustainable energy solutions and the utilization of waste materials as valuable resources. Thus, through the exploration of various studies, we will examine the advancements in char production techniques, the incorporation of char into composite materials, and the performance of char composites in energy conversion and storage.

6.2 Carbon material – Char and its key properties for energy harvesting

Char is a carbon-rich solid material produced through the incomplete combustion or thermal decomposition of organic materials at temperatures above 250 °C, in the absence of or with limited access to air. This process, known as carbonization or pyrolysis, removes volatile components like gases and liquids, leaving behind a solid residue. Due to its high carbon content and porous structure, char is increasingly studied for its potential applications in energy storage and conversion, as well as in environmental remediation.

6.2.1 Novel carbon sources

Char can be obtained from the pyrolysis of a wide range of feedstocks, including biomass, which includes wood, agricultural residues [14] products such as rice husks, corn stover, and sugarcane bagasse, plant matter [15] like bamboo and hemp, algae–seaweed and microalgae [16], pulp industry by-products [17], and various waste materials from municipal and industrial sources, including food waste and wood chips [18]. This approach aids in waste management and reduces landfill usage.

Moreover, char is also a major by-product of coal gasification, where coal is partially oxidized at high temperatures to produce syngas. This process leaves behind char, which can constitute up to 15–30 % of the initial coal feedstock [19]. Additionally, the production of activated carbon, which involves the carbonization and activation of carbon-rich materials like coconut shells, wood, or coal, generates char as a residue. This residue is often further processed but is still a testament to the large quantities of char produced.

6.2.2 Formation of char

One of the primary sources of char is biomass pyrolysis, a process extensively utilized for converting organic materials into bio-oil, syngas, and char. During pyrolysis, a significant portion of the biomass, typically around 10–40 % by weight, is converted into char, making it a readily available by-product [20]. The formation of char is closely associated with thermal decomposition processes such as incomplete combustion and pyrolysis, which involve the breakdown of organic matter under controlled conditions. A summary of each process is documented below:

6.2.2.1 Incomplete combustion

During incomplete combustion, organic matter is heated in the presence of limited oxygen. This process does not allow the material to fully oxidize into carbon dioxide and water, leading instead to the formation of carbonaceous char, volatile organic compounds, and gases like carbon monoxide. The primary reaction in incomplete combustion can be simplified as Equation (6.1):

$$\text{Organic matter} + O_2 \rightarrow \text{Char} + CO + \text{Volatile organic} \tag{6.1}$$

6.2.2.2 Pyrolysis

Pyrolysis is the thermal decomposition of organic material in the absence of oxygen, where the material undergoes a series of complex chemical reactions, breaking down large molecules into smaller, volatile compounds, and leaving behind a solid carbon-rich residue known as char [4, 21, 22]. The key steps in pyrolysis include dehydration, depolymerization, and the decomposition of organic compounds, ultimately yielding char, bio-oil, and syngas [8]. This process can be conducted under various conditions, resulting in different product distributions. Fast pyrolysis occurs at around 500 °C with a very short vapor residence time (~1 s), primarily producing liquid products. Intermediate pyrolysis also operates at around 500 °C but with a longer vapor residence time of 10–30 s, yielding a balanced mix of liquid, solid (char), and gas. Carbonization, a form of slow pyrolysis, takes place at around 400 °C with extended vapor residence times ranging from hours to days, leading to a higher production of solid (char) and gas. Each of these pyrolysis processes highlights the influence of temperature and residence time on the composition of the resulting products, with the formation of char during pyrolysis summarized as Equation (6.2).

$$\text{Organic matter} \xrightarrow{\Delta,\text{no } O_2} \text{Char} + \text{Bio oil} + \text{Syngas} \tag{6.2}$$

Overall, the amount and properties of char produced depend on several factors, including the type of feedstock, temperature, heating rate, and duration of the pyrolysis or combustion process [22, 23]. Additionally, the following are the key unique properties of char that make it suitable for applications in energy conversion and storage.

6.2.3 Key properties of char materials

6.2.3.1 Structural properties – ability to facilitate electron transfer

Char is predominantly composed of carbon, with a carbon content typically ranging from 70 % to 95 %, depending on the production process and feedstock. This high carbon content contributes to its stability and desirable properties. The structure of char varies based on its production method and starting material, potentially featuring amorphous structures or graphitic and turbostratic carbon arrangements. The presence of graphitic structures enhances electrical conductivity and mechanical strength. Additionally, char often possesses a porous structure, which significantly influences its specific surface area, adsorption capacity, and reactivity. These pores are categorized as micropores (less than 2 nm), mesopores (2–50 nm), and macropores (larger than 50 nm), each contributing to its suitability for energy applications [24].

6.2.3.2 Mechanical strength

Char exhibits varying mechanical properties depending on its production method and composition. It can possess moderate to high mechanical strength and can be engineered to have specific structural properties for applications requiring load-bearing capabilities [25]. Incorporating char into polymer matrices has been shown to significantly increase mechanical strength without compromising flexibility, as reported by Zhang et al. [26]. This characteristic is beneficial for developing structural materials in energy devices, such as electrodes that require both mechanical integrity and electrical conductivity. Additionally, the use of raw pyrolysis chars in epoxy composites has been found to enhance electrical conductivity cost-effectively compared to carbon fiber, with the conductivity of such composites increasing continuously up to 25 wt% char content [27].

6.2.3.3 Thermal stability

Char is known for its excellent thermal stability, capable of withstanding high temperatures without significant degradation or combustion. The thermal conductivity of char is generally low, and its thermal properties are influenced by factors such as carbon content, porosity, and structure, making it suitable for thermochemical energy conversion processes like gasification and pyrolysis [28]. Li et al. [29] reported that biochar derived from agricultural waste maintains its structural integrity at temperatures exceeding 700 °C, making it a suitable candidate for thermochemical processes such as gasification. The low thermal conductivity of char, noted by [27], further emphasizes its utility in preventing heat loss in energy storage systems.

6.2.3.4 High electrical conductivity

The electrical conductivity of char is a key property that enables its application in energy storage devices such as batteries and supercapacitors. This conductivity is influenced by its carbon content and structural characteristics, particularly the degree of graphitic arrangement, allowing efficient electron to transfer essential for quick charge and discharge cycles [25].

Char is predominantly a porous material rich in carbon, with content up to 50 %, often found in stable forms such as fullerene, graphene, and various amorphous phases [27]. The presence of sp^2 carbon structures, as noted by [30, 31], enhances the potential for electrical conductivity, particularly when char is produced at higher pyrolysis temperatures. For instance, [30] reported that char obtained by pyrolysis of truck tires at 490 °C had a lower electrical conductivity pyrolysis at temperatures; however, exceeding 800 °C can transition char from a semiconductor to a conductor material, significantly conductivity increased to values higher than commercial tire carbon black.

Additionally, Zhao et al. [32] found that high-temperature pyrolysis leads to more ordered graphitic structures in biochar, which further enhances its electrical conductivity, making it suitable for supercapacitors and batteries. Similarly, Özaytekin and Kar [33] showed a notable increase in the electrical conductivity of oligo(azomethine)/PET char composites, underscoring char's adaptability in conductive polymer materials. Various studies have also documented enhanced electrical conductivity in epoxy composites when carbon nanofillers are incorporated [9, 34, 35]. However, Ahmetli et al. [27] employed unprocessed pyrolysis chars in epoxy composite production, which proved to be a more cost-effective alternative compared to using carbon fibers. The current research examined how different reinforcement strategies, involving varying percentages of char (up to 30 %) in epoxy resin (ER)/char composites, impact electrical conductivity. Results showed a continuous rise in the conductivity of ER/char composites, reaching a peak at 25 wt%, suggesting the potential for adjusting the conductivity of epoxy composites to meet specific technological requirements.

6.2.3.5 High surface area and porosity

Char's high surface area and porous structure enhance its adsorption capacity and catalytic activity. These properties are particularly advantageous in energy storage devices like supercapacitors, where a high surface area facilitates greater charge storage [24].

Activated biochar, as studied by Wang et al. [36], has been found to support high dispersion of active catalytic sites in fuel cells, enhancing overall performance. This high surface area also plays a crucial role in supercapacitors, where it facilitates increased charge storage.

6.2.3.6 Chemical reactivity

The reactivity of char can vary depending on its structure, porosity, and surface properties. Char's surface can be chemically modified to enhance its reactivity, making it effective in various catalytic processes, including those in fuel cells. This tuneable reactivity allows char to participate in critical oxidation–reduction reactions for energy conversion [37, 38].

6.2.3.7 Low-cost and abundant resource

Derived from biomass waste, char is an inexpensive and abundant resource. Its availability, coupled with its desirable properties, positions char as a cost-effective material for large-scale energy storage and conversion applications [39]. Char can be combined with other materials to create composites with enhanced properties for specific applications. These composites can improve the mechanical strength, thermal stability, and electrical conductivity of energy storage devices, thereby enhancing their efficiency [40].

These processes contribute to the widespread availability of char, particularly in industries focused on energy production, waste management, and material recovery. The abundance of char from these various industrial processes underscores its potential as a low-cost feedstock for energy conversion and storage applications, offering a sustainable pathway for utilizing an otherwise underused material.

6.2.3.8 Environment stability

Char is generally resistant to chemical degradation and environmental factors. It is nontoxic and chemically inert, making it suitable for applications where stability and environmental compatibility are important. The specific properties of char can be tailored through different production techniques and the choice of feedstock. The versatile characteristics of char make it suitable for a wide range of applications, including energy storage, water treatment, soil amendment, catalysis, and filtration. Char, particularly biochar, contributes to carbon sequestration by storing carbon in a stable form for extended periods. This ability helps mitigate climate change, making char a sustainable choice for energy systems [15].

6.3 Recent development of char-reinforced composites in energy harvesting applications

Char is a versatile carbon-based material that can be used as a raw material or precursor for the production of a wide range of carbon-based products and materials, depending on the specific properties and processing requirements of the desired end-use application. In this section, we will discuss deeply on the current application of char, biochar, and other related carbon-based materials as fillers or matrix in composites within energy field as summarized in Table 6.1.

6.3.1 Phase change materials for energy storage

The development of efficient and cost-effective energy storage systems is crucial for reducing energy consumption, decreasing dependence on fossil fuels, and mitigating

Table 6.1: Recent development of char composites in energy fields.

Filler	Matrix	Fabrication method	Performance	Ref.
Phase change material (PCM)				
Biochar	Heptadecane		Thermal energy storage	[41]
Polyethylene glycol	Porous carbon (coconut shell, potato-white radish)	PCM	Energy storage performance of PEG as PCM	[42, 43]
n-alkanes	Biochar	PCM	Energy storage and conversion applications	[44, 45]
Stearic acid	Biochar (sunflower straw)	PCM	Thermal energy storage applications such as solar heat energy storage system and energy conservation buildings	[46]
Paraffin	Biochar (orange peel)	PCM		[47]
Lauric–stearic acid eutectic mixture	Biochar (corncob)	PCM	The thermal conductivity of LA-SA/CNCC was 87.5 % greater than that of LA-SA.	[48]
Polyethylene glycol	AC	Shape-stabilized PCM	Thermal energy storage and temperature control	[49]
H_2 storage capacity				
	AC (coffee bean waste, petroleum residue, anthracites, litchi wood, fungi)	H2 storage	Hydrogen adsorption on the AC increased with increasing pressure and 2.56 wt% and 0.41 wt% at 77 K 0.1 MPa and at 303 K 5.7 MPa, respectively	[50–54]
	Ultra-microporous carbon	Hydrogen storage capacity	High levels of hydrogen uptake at near-ambient temperatures and pressures (e.g., up to 0.8 wt% at 25 °C and 2 MPa).	[55]
Electrode in battery				
Sulfur	Mesoporous carbon	Li/S batteries	High surface-area cathode for rechargeable lithium batteries	[56]
Sulfur	KOH-activated carbon (pig bone)	Li/S batteries		[57]
Nanofiber	Mesoporous biochar	Electrode	Electrochemical energy storage	[58]
Catalytic applications				

Table 6.1: (continued)

Filler	Matrix	Fabrication method	Performance	Ref.
SO_3H	Biochar (pine chip)	Solid acids catalysts	Heterogenous catalysts for biodiesel production	[59]
SO_3H	Biochar (vegetable oil asphalt)	Solid acids catalysts	High catalytic activity for the esterification of free fatty acid and methanol	[60]
SO_3H	Biochar (fruit seed cake, woody biomass, peanut shell)	Solid acids catalysts	Catalytic transesterification of triolein and methanol	[61–63]
SO_3H	Biochar (bamboo, cotton, starch)	Solid acids catalysts	Catalysts for the hydrolysis of biomass	[64–66]
Biochar (cellulose)	Pt nanoparticles	Cathode electrocatalysts	Catalysis of the oxygen reduction reaction (ORR)	[67]
Iron oxide	Graphitized biochar (waste cornstalks-pomelo skins)	Cathode electrocatalysts for MFC	Exhibit a high oxygen reduction reactivity	[68]
Pd nanoalloys-biochar	Coconut-shell-derived carbon	Electrocatalyst	Highly selective catalysts for hydrogenation of nitroaromatic derivatives	[69]
Ru	Biochar	Catalysts	Catalytic hydrogenation of CO and CO_2 to produce CH4	[70]
Pt NPs	Biochar	Catalysts	Catalyst for the dehydrogenation of methylcyclohexane to form toluene, with a conversion of over 95 % and a selectivity for H2 and toluene of almost 100 % at 300 °C	[71, 72]
Metal NPs (Ni, Fe)	Biochar	Catalysts	Catalytic conversion of tar with a conversion efficiency of 96.5 % via copyrolysis with biomass	[73, 74]
TiO2 NPs	Coconut shell biochar	Photocatalyst	Photocatalytic degradation of methylene blue	[75, 76]
Supercapacitors				
	Activated hydrochar (cellulose, potato starch, and eucalyptus wood saw dust)	Electrical double layer capacitors (EDLCs)/supercapacitors	High-power electrochemical energy storage devices	[77]
	Activated carbon (wheat straw)	Supercapacitor	The specific capacitance of the activated carbon reached 251.1 F/g at 2 mV/s scan rate in the electrolyte of MeEt3NBF4/AN	[78]
	Nanoporous carbons (animal bone, waste paper, SSS, banana fiber)	Supercapacitor	Carbon–carbon composite electrode material for supercapacitors	[79–82]

negative environmental impacts. Thermal energy storage (TES) systems can play a significant role in achieving these goals, especially when integrated with renewable energy technologies.

Furthermore, biochar can be produced in a more sustainable and economical manner. Derived from biological waste, biochar offers a promising alternative to the carbon fillers mentioned earlier [83]. A one-step impregnation process was successfully used to generate a shape-stabilized n-octadecane/activated carbon nanocomposite (SPCM), according to research by Khadiran et al. [83]. As depicted in Figure 6.1, n-octadecane served as the phase change material (PCM) for thermal energy storage,

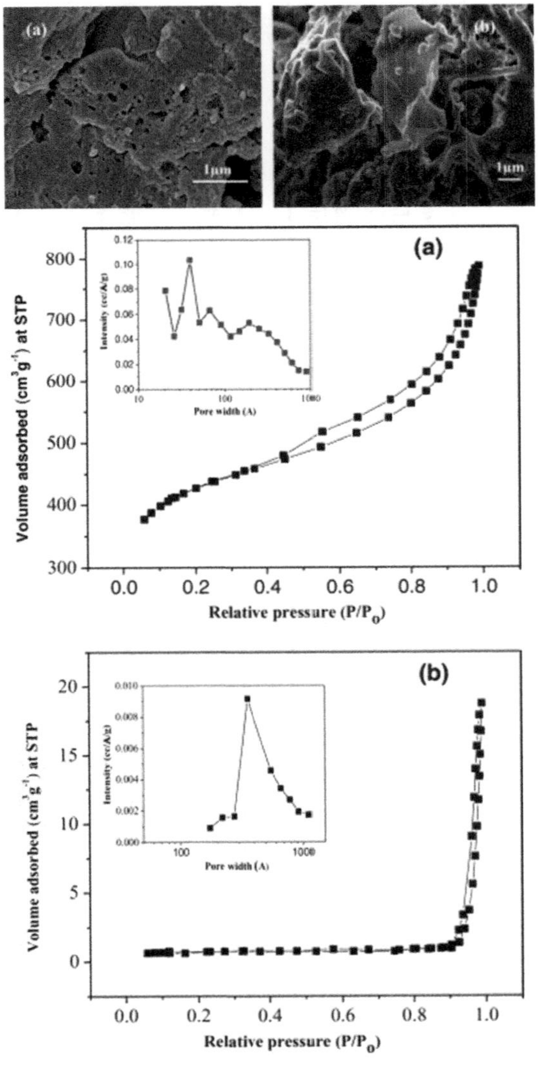

Figure 6.1: (i) FESEM micrographs of (a) AC and (b) SPCM5. (ii) N2 adsorption–desorption isotherms for (a) AC and (b) SPCM5, with insets displaying the pore size distribution for both AC and SPCM5 [83].

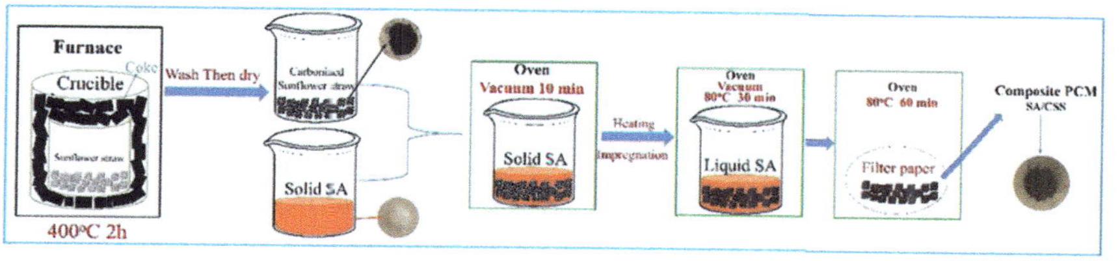

Figure 6.2: The experimental procedure for preparing the SA/CSS composite PCM [46].

while activated carbon (AC) was employed as an inorganic supporting material. Due to their superior surface texture properties, these materials have been used to address leakage problems and enhance the thermal conductivity of PCM [84, 85].

Wang et al. [42, 86] investigated the impact of expanded graphite and activated carbon porosity on PEG's ability to store energy as phase change material. By combining PEG with bio-carbons from fresh potatoes and white radish, Zhao et al. [43] created composite PCM. They also observed that the resulting composite PCM maintained its LHS features even after 200 melt/freeze cycles. Three different types of biochars were made by Atinafu et al. [44] in order to stabilize the form of n-dodecane and 1-dodecanol. The resultant composites showed an LHS capacity of 90 J/g and 73 J/g, respectively. Feng et al. [49] created a shape-stable PEG/mesoporous carbon composite using a vacuum impregnating process. Nicholas et al. [45] used biochar made from oil palm waste shell to impregnate n-octadecane for investigations on thermal control. Zhang et al. [48] used maize cob-derived carbon to impregnate lauric acid–stearic acid eutectic PCM, which demonstrated an 87 % increase in thermal conductivity above the eutectic PCM. When paraffin and orange peel–based biochar were combined, Xu et al. [47] produced a composite PCM that was leak-proof and had good thermal reliability. Wen et al. [46] in a different study synthesized carbon from sunflower straw, following the process detailed in Figure 6.2, and incorporated molten stearic acid into it. The results showed that the composite PCM exhibited excellent thermal stability at its working temperature. The SA/CSS composite had a melting point of 66.4 °C with a latent heat of 186.1 J/g and a freezing point of 65.9 °C with a latent heat of 186.7 J/g. Additionally, the thermal conductivity of SA/CSS reached 0.33 W/m/K, representing a 106.3 % increase compared to stearic acid alone. Overall, these findings suggest that SA/CSS possesses suitable thermal properties for applications in thermal energy storage systems, such as solar heat storage units and energy-efficient buildings.

6.3.2 Catalytic applications for energy harvesting and conversion

The use of functional materials based on biochar in fields including energy storage, environmental protection, and catalysis has received a lot of interest. New opportunities are opening up in areas such as improved chemical and biological analysis and sustainable chemistry, although these industries are often more hesitant to accept novel materials. This section addresses fresh prospects for using biochar-based materials in various applications and provides a quick summary of the current developments in these fields.

One example of biochar-based materials is biochar-based solid acids, which contain SO_3H groups and are commonly used as metal-free catalysts in a variety of chemical processes. As discussed in this study, biochar can be treated with sulfuric acid to create porous solid Brønsted acids. These biochar-derived solid acids have proven to be effective catalysts for a wide range of acid-catalyzed reactions, such as the hydrolysis of biomass, the esterification of organic acids in water, and the acylation of alcohols and amines, as well as the alkylation of aromatic compounds [87]. Due to the high density of hydrophilic functional groups on the flexible carbon matrix of biochar, biochar-based solid acids can integrate large amounts of hydrophilic molecules, such as H_2O, during catalysis [88].

Esterification is a widely catalyzed reaction by solid acids to produce biodiesel. Kastner et al. [59] demonstrated that biochar containing SO_3H groups can catalyze the esterification of free fatty acids and methanol, achieving a biodiesel yield of 97 % in 6 h. By sulfonating biochar made from vegetable oil asphalt, a solid acid catalyst with a flexible carbon framework and SO_3H groups was created [60]. This catalyst showed strong activity in esterifying methanol and free fatty acids, achieving a 94.8 % conversion in just 4.5 h. This high catalytic performance is attributed to the high acid strength (2.21 mmol g–1), the hydrophobic nature of the carbon structure (which prevents methanol hydration), and the hydrophilic functional groups (e.g., SO_3H, COOH, and phenolic OH), which allow methanol to access the protonated carboxylic groups of the free fatty acids more easily.

Biochar-based solid acid catalysts also show high performance in the catalytic transesterification of triolein and methanol [61–63], resulting in higher biodiesel yields. In addition, biochar-based solid acids act as efficient catalysts for the hydrolysis of biomass [64, 65]. Biochar sulfonic acid (BC–SO_3H) catalysts(BC–SO_3H) catalysts, prepared from bamboo, cotton, and starch, demonstrated higher turnover numbers (TON = 1.33–1.73) for cellulose hydrolysis compared to mineral acids (e.g., dilute H_2SO_4, TON = 0.02) and Amberlyst-15. Jiang and colleagues [89] reported that lignocellulosic biomass from corn crops could be effectively hydrolyzed into soluble sugars using biochar-based solid acids, with yields of xylose and arabinose reaching 78.1 %, much higher than previous reports. Importantly, these solid acids can be prepared from biomass residue.

Biochar materials have also been studied as supports for stabilizing metal nanoparticles in various catalytic applications due to their large surface areas and functionalities. Johnson et al. [67] reported that Pt nanoparticles supported on carbon produced from cellulose were promising catalysts for oxygen reduction reactions (ORR). Ma et al. [68] found that Fe_3O_4 nanoparticles on partially graphitized carbon, derived from cornstalks and pomelo skin, performed well as cathode electrocatalysts in microbial fuel cells (MFCs). MFCs with these synthesized cathode electrocatalysts achieved a high-power density of 1,502 mW m^{-2}, around 7 % lower after 18 cycles compared to Pt/C

(1,192 mW m^{-2}). These composites show potential as alternatives to Pt-based cathode electrocatalysts in MFCs due to their low charge transfer resistance, long-term stability, and high oxygen reduction activity.

In addition to their electrocatalytic properties, biochar-assisted metal nanoparticles have been studied for their catalytic activities in hydrogenation reactions. Chen et al. [69] found that Pd nanoalloys, such as Pd–B and Pd–Ni–B, supported on carbon derived from coconut shells, were highly effective and selective catalysts for the hydrogenation of nitroaromatic compounds. The high turnover frequencies in these reactions were attributed to the large surface area of the functional carbon and the well-dispersed Pd nanoparticles. Ru nanoparticles supported on activated biochar were also used as catalysts for the hydrogenation of CO and CO_2 to CH_4 [70]. Additionally, Zhang et al. [71] demonstrated that Pt nanoparticles supported on biochar were efficient catalysts for the dehydrogenation of methylcyclohexane to toluene, achieving over 95 % conversion and nearly 100 % selectivity for H_2 and toluene at 300 °C.

Metal nanoparticles can also be synthesized *in situ* on biochar supports. A simple one-pot method was recently developed to produce Ni/biochar composites with well-dispersed Ni nanoparticles using the pyrolytic gasification of rice husk biomass preloaded with Ni(II) [73, 74]. This in-situ-formed catalyst exhibited high catalytic performance for the conversion of tar, achieving 96.5 % efficiency through copyrolysis with biomass. Another straightforward method was reported for creating Fe nanoparticles embedded in biochar by thermally treating $Fe(NO_3)3$-impregnated pine wood biochar [90]. Some amorphous biochar materials have also been found to possess semiconductor characteristics [91–93], enhancing the reactivity of supported nanoparticles through electron transfer. This hybrid architecture holds potential for semiconductor photocatalysis based on the mechanism described by [75].

6.3.3 Carbon-based supercapacitor electrodes

Supercapacitors (SCs) are typically divided into three main types based on their energy storage mechanisms: electrical double-layer capacitors (EDLCs), pseudocapacitors (PCs), and hybrid supercapacitors (HSCs), also known as asymmetric supercapacitors (ASCs), where HSCs represent a subset of ASCs. EDLCs primarily use carbon-based materials such as activated carbon (AC), graphene, carbon nanotubes (CNTs), carbon aerogel (CA), carbide-derived carbon (CDC), and carbon fibers [94]. Similar to traditional dielectric capacitors, EDLCs store energy through electrostatic means by forming an electrical double layer (EDL) at the interface between the electrode and electrolyte. Here, the electrolyte is polarized by the applied voltage and functions as a dielectric (Figure 6.3). This process is purely physical and non-Faradaic. The incorporation of carbon materials in these systems enhances charge storage capacity and improves rate performance due to their porous structure and excellent conductivity. Recent innovations have shown that carbon-based composites can be effectively used in the negative electrodes (anodes) of

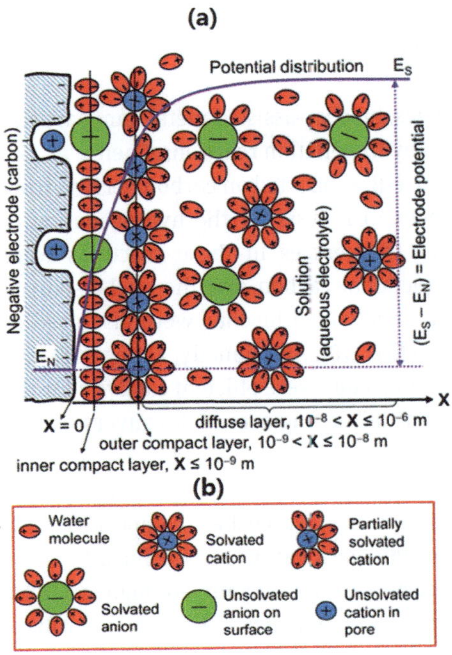

Figure 6.3: Diagrammatic illustrations of (a) the cross-sectional view of the electrical double-layer (EDL) structure at the interface between a porous carbon negative electrode and an aqueous electrolyte and (b) definitions of the symbols used in (a) [95].

supercapacitors or batteries, such as graphite and ACs, resulting in improved performance characteristics.

The SC electrodes are best suited for ACs. Because of their highly developed microstructure, superb conductivity, outstanding stability, high SSA, affordability, and ease of access, porous carbons are the subject of the most research. In addition to being largely proportional to the SSA of porous carbons [96], the capacitance of carbon-based SCs can also be increased by adding heteroatoms (such as oxygen, nitrogen, etc.) to the carbon framework, which creates a pseudo-capacitive effect [97]. Depending on the activation conditions and carbon sources employed, KOH activation of carbonaceous materials can produce ACs with uniform PSD, controlled surface functional groups, and drastically enhanced SSA and PV/MV.

In energy storage applications, one of the primary uses of KOH-activated carbons is in SCs, where charge accumulation occurs at the electrode/electrolyte interface [86, 98]. While SCs offer a higher specific power compared to other energy storage devices, they tend to have a lower specific energy. He et al. [99, 100] synthesized a series of ACs from coke using rapid microwave heating, studying the impact of microwave power, the KOH-to-coke mass ratio, and activation time (ranging from 27 to 37 min) on the pore structure of the ACs [99, 100]. With a surface area of up to 2,312 m^2/g, the ACs used as SC electrodes achieved a specific capacitance of 343 F/g, with an energy density of 8.0 W h/kg, and maintained 246 F/g even after 800 charge–discharge cycles.

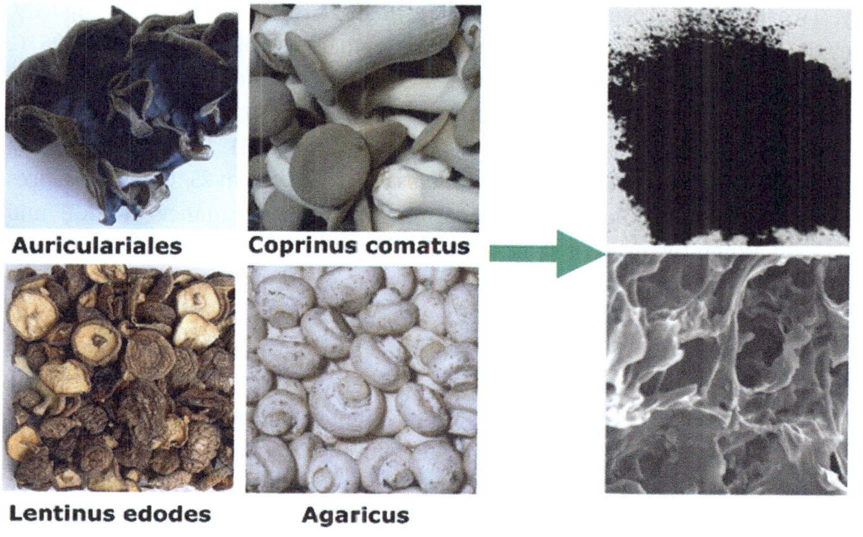

Figure 6.4: Conversion of different fungi into highly porous carbon materials through a precarbonization procedure, followed by chemical activation using KOH. (Left) images of four distinct types of fungi; (Right) photograph and SEM image of the porous carbon derived from fungi [50].

A variety of waste materials have been utilized to create the ACs, including cherry stones [101], fish scales [102], pig bone [79], waste paper [80], banana fibers [81], sunflower seed shells (SSS) [82], wheat straw [78], and wood sawdust [77]. Through KOH activation, natural fungi have been transformed into activated carbons (ACs) with surface areas as high as 2,264 m^2/g [50], see Figure 6.4. In contrast, porous carbons derived from the direct pyrolysis of starch-based gels resulted in lower surface areas (<600 m^2/g), but KOH activation of starch-based chars produced ACs with surface areas exceeding 2,000 m^2/g. Potato starch–based activated carbon spheres (PACS) were developed via stabilization and carbonization, followed by KOH activation. These PACS exhibited excellent specific capacitances, reaching up to 335 F/g, along with high surface areas (ranging from 1,605 to 2,342 m^2/g) and pore volumes of 0.74–1.24 cm^3/g.

6.3.4 Carbon materials for H_2 storage

Because of its ease of use, affordability, speed of kinetics, high specific surface area (SSA), and light weight, porous carbons have garnered a lot of interest in the physiadsorbent field for the storage of H_2. In general, the surface area and pore volume of porous carbons are directly correlated with their H_2 storage capacity. Therefore, KOH-activated carbons employed as the physisorbents have increased SSA and porosity as well as a well-defined micropore size distribution, which are extremely favorable for boosting H_2 adsorption capacity. Two physical processes are available to increase the adsorption capacity: (i) raising the storage pressure and (ii) lowering the temperature of adsorption to −196 °C.

In recent years, there has been a significant increase in studies on KOH-activated carbons for hydrogen adsorption. Researchers have created activated carbons (ACs) with varying porosities by using KOH to activate anthracite [51, 103] and pitch [52]. These ACs exhibited maximum hydrogen uptakes of 1.1–1.2 wt% at 25 °C and pressures between 10 and 20 MPa, which are comparable to the performance of AX21 and Maxsorb-3000. When the pressure was increased to 50 MPa, hydrogen uptake rose to 3 wt% [103]. Recently, natural materials like coffee bean waste [53], wood [54, 55], and corncobs [104] have been utilized as affordable precursors to produce ACs with specific surface areas ranging from 780 to 3,012 m^2/g and micropore volumes of 0.25–1.70 cm^3/g by varying the KOH dosage.

The spillover effect has led to a marked improvement in hydrogen storage capacity at room temperature for various metal-decorated porous carbon materials [105]. For instance, Pd-decorated carbon materials derived from sepiolite demonstrated hydrogen storage capacities four times greater than the unmodified carbon at room temperature [106]. Moreover, the incorporation of palladium nanoparticles into wood-derived ultra-microporous carbon not only significantly enhanced hydrogen storage capacity but also increased the rate of hydrogen uptake by lowering the activation energy for the adsorption process [55].

The hydrogen storage capabilities of porous carbon materials with microporous structures, derived from coffee bean waste via KOH activation, were examined at temperatures of 25 and −196 °C, focusing on pore structure. The process involved pyrolyzing the coffee beans at 500 °C for 1 h, followed by activation using steam at 800 °C for 2 h or KOH at 850 °C for 2 h [53]. The porous carbon achieved a maximum hydrogen storage of 0.6 wt% with a specific surface area of 2070 m^2/g. At 77 K, hydrogen storage capacity increased to 4.0 wt% for the same material. Notably, hydrogen storage capacity was found to correlate linearly with micropore volume.

Additionally, wood biomass underwent pyrolysis with phosphoric acid (H3PO4), followed by KOH activation [55]. The heat of adsorption for this slow process, which facilitates high uptake (17–20 kJ/mol), exceeded typical values for carbon materials, indicating slow adsorption kinetics and notable hysteresis in isotherms. Separately, Sun and Webley [104] explored activation methods for corncob biomass-based char using various activators. The resulting carbons from a two-step KOH activation achieved a BET specific surface area of 3,012 m^2/g and a pore volume of 1.7 cm^3/g. The microporous carbon with the highest surface area demonstrated hydrogen adsorption capacities of up to 2.0 wt% at 77 K under 1 atm and 0.44 wt% at 298 K under 5 MPa [104], as illustrated in Figure 6.5.

Zhu et al. [107] explored the potential of using pine needles as a carbon source for electrocatalytic hydrogen evolution. The carbon derived from pine needles, which was activated at 800 °C, proved to be an effective electrocatalyst for the hydrogen evolution

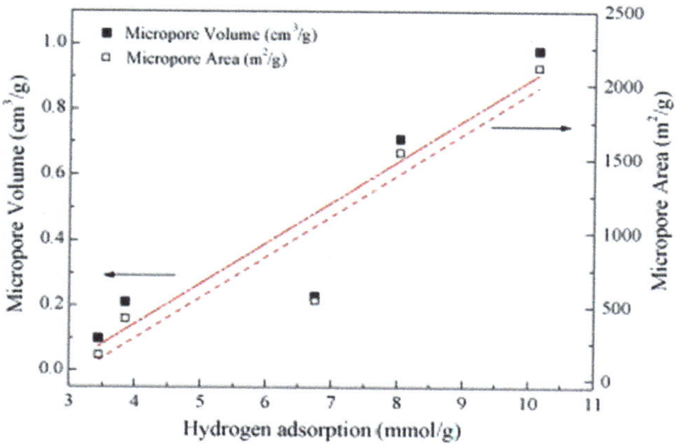

Figure 6.5: The correlation between micropore surface area and volume and hydrogen adsorption capabilities (at 77 K and 1 bar) [104].

reaction (HER), demonstrating a low onset potential of around 4 mV and a Tafel slope of approximately 45.9 mV/dec, alongside remarkable stability during extended cycling. When this carbon, activated at 900 °C, was employed as an electrode in supercapacitors, it achieved a high specific capacitance of 236 F/g at a current density of 0.1 A/g, excellent rate performance of 183 F/g at 20 A/g, and impressive long-term stability [107]. Notably, after 5,000 cycles at 2.0 A/g, the specific capacitance improved from approximately 205 to 227 F/g, attributed to enhanced activation and improved wetting of the electrodes.

Wang et al. [108] focused on willow catkins as a carbon source. They pyrolyzed the biomass at 600 °C for 1 h, then added cobalt(II) acetylacetonate and subjected the mixture to further pyrolysis at temperatures ranging from 700 to 900 °C to produce Co_3O_4-biochar nanocomposites [108]. The development of efficient, cost-effective, durable, and sustainable electrocatalysts is crucial for three important electrochemical reactions. In this context, they successfully synthesized a trifunctional electrocatalyst, Co_3O_4 nanoparticle-modified nitrogen-doped hollow hierarchical porous carbon microtubes (Co_3O_4/NCMTs), by pyrolyzing cobalt(II) complex willow catkin biomass in an argon-ammonia atmosphere. The resulting carbon materials preserved the original tubular structure of the willow catkins and featured a hollow micro/mesoporous architecture. The introduction of ammonia notably increased the nitrogen content within the carbon structure. The incorporation of doped nitrogen and Co_3O_4 nanoparticles generated more active sites for electrochemical reactions. When compared to the Pt/C catalyst, the Co_3O_4/NCMT-800 electrocatalyst exhibited superior catalytic stability and methanol tolerance during the oxygen reduction reaction (ORR), achieving a positive onset potential (E0 = 0.906 V) and half-wave potential (E1/2 = 0.778 V).

6.3.5 H_2 production

Biochar-based materials are significant not only for hydrogen storage but also for hydrogen production applications, including biological hydrogen generation, electrocatalysis, and photocatalytic hydrogen evolution. According to [109], incorporating biochar into anaerobic digestion systems resulted in a maximum hydrogen yield of 96.63 ± 2.8 mL H_2 per gram of carbon, compared to just 22.55 ± 2.2 mL H_2 per gram of carbon without biochar (see Figure 6.6). This study utilized biochar to improve hydrogen production from the organic fraction of municipal solid waste (OFMSW) through a co-culture of *Enterobacter aerogenes* and *E. coli*. The presence of biochar promotes biofilm formation and allows for effective microbial colonization due to its large surface area. Additionally, Zhang et al. [110] observed that adding biochar along with an appropriate concentration of Fe^{2+} enhanced hydrogen production from glucose, increasing the yield from 158.0 to 234.4 mL g^{-1} glucose, while reducing the lag phase from 4.25 to 2.12 h.

Beyond biological hydrogen production, electrocatalytic water splitting serves as an alternative to conventional fossil fuel–based catalytic systems for producing high-purity hydrogen. The application of biochar-based materials for the efficient catalytic hydrogen evolution reaction (HER) from water has become an essential method for generating clean and renewable hydrogen fuel. A promising strategy for creating high-performance HER electrocatalysts involves the incorporation of catalytically active species into biochar [111–114]. For instance, Zhou et al. [115] found that biochar enriched with sulfur and nitrogen, obtained from peanut root nodule biomass, effectively serves as an electrocatalyst for HER. This sulfur- and nitrogen-doped carbon possesses a porous multilayer structure, as depicted in Figure 6.7, with a specific surface area of 513.3 m^2/g and a high electrochemical area of 27.4 mF/cm^2. Electrochemical tests reveal that this material demonstrates considerable electrocatalytic activity for HER in 0.5 M H_2SO_4, exhibiting a low overpotential of only −0.027 V, a Tafel slope of 67.8 mV/dec, and remarkable catalytic stability.

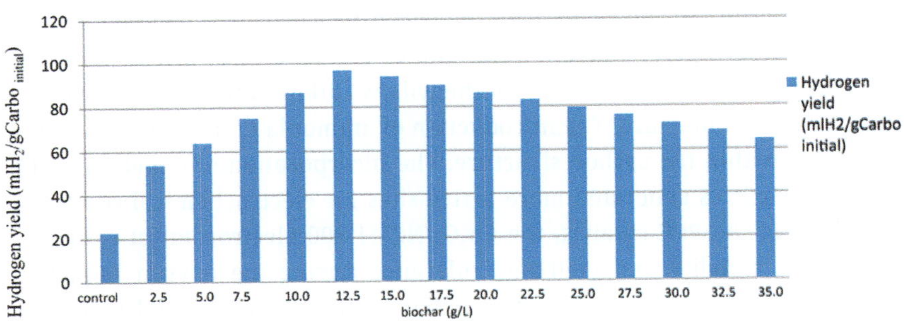

Figure 6.6: Impact of biochar addition on hydrogen yield [109].

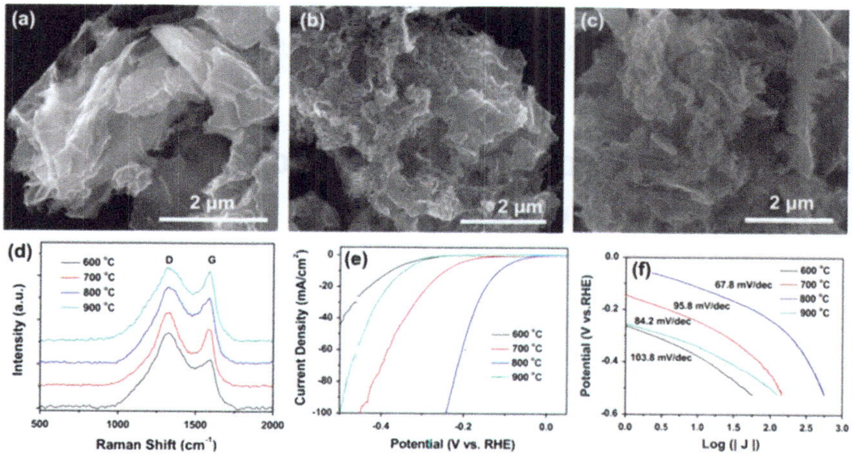

Figure 6.7: SEM images of RN produced at various temperatures: (a) 600 °C, (b) 700 °C, (c) 900 °C, along with (d) Raman spectroscopy results, (e) polarization curves, and (f) the corresponding Tafel plots derived from (e) for RN obtained at these different temperatures [115].

Building on the earlier work by Zhou et al. [115], researchers developed a biochar-nanostructured composite by combining sulfur- and nitrogen-doped biochar from sunflower seed shell biomass with Mo_2C nanoparticles. This composite requires only a 60 mV overpotential to achieve a current density of 10 mA cm^{-2}. Additionally, its unique structural characteristics contribute to its excellent durability and a faradaic efficiency close to 100 % [116]. In a different study, a carbon fiber aerogel produced from cotton wool biomass was used as a substrate for $MoSe_2$ nanosheets ($MoSe_2$/CFA). This synthesized material demonstrated remarkable electrochemical performance as a HER electrocatalyst, exhibiting a low onset potential of 104 mV versus RHE and a Tafel slope of 62 mV/dec [117].

6.3.6 Perovskite solar cells

Char composites have demonstrated promise in enhancing the performance of perovskite solar cells by improving charge transport and stability. Recent research indicates that adding char-derived materials to the perovskite layer or as additives in the electron or hole transport layers can boost both the efficiency and durability of these solar cells. The high surface area and electrical conductivity of char facilitate better charge collection and minimize recombination losses [118]. Furthermore, emerging technologies like perovskite solar cells and flexible energy harvesters can gain advantages from the use of char composites. Studies suggest that carbon-based materials, including char, may serve as alternative electrodes in perovskite solar cells, potentially reducing costs and enhancing sustainability [119]. Additionally, the flexibility, conductivity, and stability of char make it well-suited for applications in wearable and flexible electronics, which are becoming increasingly vital in the contemporary energy landscape [120].

6.4 Limitations faced by char-filled composites in energy harvesting applications

Carbon-based composites have attracted considerable interest in recent years due to their potential for energy harvesting applications, which involves converting ambient energy into usable electrical energy. Research indicates that polymer-based carbon composites can effectively combine the benefits of both carbon particles and polymers [121, 122]. These composites leverage the high conductivity of carbon along with the advantageous properties of polymers, such as low thermal conductivity, flexibility, and processability in solution or melt forms [123–125]. Furthermore, the interface between the polymers and carbon particles is strategically utilized to achieve an energy-filtering effect, enhancing thermoelectric performance, electrical conductivity, and overall capacity [12, 126]. Among various carbon materials, carbon nanotubes (CNTs) and graphene have become particularly prominent in energy harvesting applications, often being combined with conductive or insulating polymers due to their unique properties.

CNTs and graphene are noted for their exceptional electrical and thermal conductivity, as well as impressive optical and electronic characteristics, such as high light absorption and outstanding carrier mobility. These traits enable effective charge transport, enhancing the performance of energy harvesting devices like solar cells, thermoelectric generators, and piezoelectric systems. Both materials offer a high surface-to-volume ratio, which increases the active area for energy conversion and storage, making them advantageous for applications such as supercapacitors where larger surface areas can lead to higher charge storage capacity and faster charge/discharge rates. Additionally, they possess remarkable mechanical properties, including high tensile strength and flexibility, making them suitable for applications requiring durability, such as wearable energy harvesting devices or flexible electronics.

While activated carbon (AC) and carbon black (CB) can be utilized in certain energy harvesting contexts, their effectiveness varies based on specific applications. Activated carbon is commonly used in energy storage and conversion systems like supercapacitors and batteries due to its high surface area, porosity, and favorable electrochemical properties, which facilitate efficient charge storage. AC electrodes in supercapacitors can achieve high capacitance and rapid charge/discharge rates; however, their relatively low electrical conductivity may limit their use in high-power energy harvesting devices. In contrast, CB is primarily employed as a conductive filler in materials like rubber or plastics to enhance their electrical properties. Although CB has a large surface area and good conductivity, it is not usually employed directly in energy harvesting devices.

Instead, its main function is to improve the conductivity of composites or support charge transfer in electrode materials. Research has explored CB as a conductive component in composite battery cathodes for energy storage through electrochemical processes [127–129]. Initial studies have also examined CB as an active electrode material in supercapacitors [130, 131], while activated carbon microspheres have been used as active materials [132–135], with graphene being proposed for efficient energy conversion [10]. Other studies highlight CB as an effective component in composite electrodes for solar cells and photocatalysts [136–138]. In photocatalytic applications, the excellent electrical conductivity of CB facilitates rapid electron transfer and reduces electron–hole recombination in semiconductors like TiO_2 [139]. Moreover, CB additives can enhance the thermal stability of the anatase phase, preventing its transformation into rutile.

Char materials, typically produced through incomplete combustion or pyrolysis, often have a disordered structure, low electrical conductivity, and limited surface area, rendering them less suitable for energy harvesting applications. Their lack of essential properties – such as high conductivity, extensive surface area, and adequate charge storage capacity – restricts their direct applicability in energy conversion and storage. To address these limitations, char often undergoes modifications, such as densification, pelletization, or purification, to improve its characteristics. For instance, chemical activation can enhance the surface area and porosity of char, or it can be combined with other materials to boost electrical conductivity or structural integrity. Although these enhancements increase the suitability of char for energy applications, they also introduce additional complexity and potential costs.

Typically, char is refined to produce carbon black for further use. The production process involves slow pyrolysis at lower temperatures, utilizing biomass and municipal waste. The resulting char generally contains around 80–85 % carbon and 15–20 % inert ash, primarily silica and zinc. While char shares similarities with CB, it lacks the consistency and quality needed to serve as a direct substitute due to metal contaminants and variable particle size. Therefore, refining char is essential for obtaining high-quality carbon black, with its quality depending on the feedstock type and pyrolysis temperature. During industrial production, the char is assessed for quality before refinement to ensure it meets standards. Recent research has focused on optimizing char content, particle size, and morphology to improve the specific capacitance, cycling stability, and overall energy storage performance of the resulting composites.

To harness char in energy harvesting, various approaches can be explored. Although char materials have limitations for direct use, they can still be modified or integrated into hybrid systems to enhance their applicability in energy harvesting. The development of char-based composites holds significant potential for advancing renewable energy technologies, fostering energy sustainability, and enabling the integration of energy harvesters into diverse applications, including sensors, wireless devices, and Internet of Things (IoT) systems.

6.5 Future outlooks and opportunities

The outlook for char-filled composites in energy harvesting is promising, driven by advances in material properties and manufacturing techniques. Opportunities include optimizing novel carbon sources and integrating these composites into emerging technologies, which fosters sustainability and enhances performance across diverse energy applications, as described below. The research and development of char-filled composites for energy harvesting hold promising future outlooks. Here are some potential areas of focus and advancements:

1. Optimization of pyrolysis condition: Research is needed to optimize pyrolysis conditions (temperature, pressure, residence time) for different carbon sources to maximize char yield and quality [4, 22].
2. Enhanced conductivity: Researchers can explore methods to enhance the electrical conductivity of char-filled composites. This can involve optimizing the composition, structure, and processing techniques to improve the interfacial contact and percolation network of the char particles within the composite. Higher conductivity would enable more efficient energy harvesting and conversion.
3. Tailored porosity: Manipulating the porosity of char-filled composites can offer advantages in energy harvesting applications. Researchers can investigate techniques to control the pore size, distribution, and connectivity within the composite. Tailored porosity can enhance the mass transport of reactants or ions, improving the performance of energy storage or conversion devices. Detailed studies on the physical and chemical properties of char from novel sources are essential. This includes surface area, porosity, and adsorption capabilities, which influence its performance in energy systems [140] and [141].

   Char composites can also serve as catalysts in biofuel production processes. The high surface area and catalytic properties of char can enhance the efficiency of converting biomass into biofuels, thus providing a renewable source of energy that can replace conventional fossil fuels [142]. This not only promotes the use of renewable resources but also reduces the carbon footprint associated with energy production.
4. Surface functionalization: Surface functionalization of char particles can enhance the compatibility and interaction between the char and the composite matrix. Functionalization techniques, such as chemical modifications or coating treatments, can improve the interfacial adhesion, dispersion, and stability of the char particles within the composite. This can lead to better mechanical strength, electrical conductivity, and overall performance of the composite.
5. Hybrid composites: Combining char with other functional fillers or nanomaterials can create hybrid composites with synergistic properties. Researchers can explore incorporating char with materials like carbon nanotubes, graphene, or metal nanoparticles to enhance conductivity, catalytic activity, or optical properties. These

hybrid composites can provide multifunctionality and improved energy harvesting capabilities.

6. Advanced manufacturing techniques: Developing advanced manufacturing techniques, such as additive manufacturing (3D printing) or scalable processing methods, can enable the fabrication of complex-shaped char-filled composite structures. This would allow for the design and customization of energy harvesting devices with specific geometries and architectures, enhancing their performance and integration into various applications.
7. Multiscale modeling and simulation: Utilizing computational modeling and simulation techniques can provide insights into the behavior and performance of char-filled composites at different length scales. Multiscale modeling can help optimize the composite's microstructure, predict its electrical, thermal, and mechanical properties, and guide the design of efficient energy harvesting systems.
8. Application-specific optimization: Research efforts can be focused on tailoring char-filled composites for specific energy harvesting applications. For example, optimizing the composite properties for thermoelectric generators, piezoelectric devices, or supercapacitors. Application-specific optimization can lead to improved efficiency, durability, and cost-effectiveness in energy harvesting technologies.

   Assessing the cost-effectiveness of using novel carbon sources compared to traditional materials is crucial. This includes evaluating the supply chain, processing costs, and potential market value [143]. Life cycle assessments should be conducted to understand the environmental benefits and potential drawbacks of using alternative carbon sources. This includes evaluating carbon footprint reductions and waste management improvements [18].

Overall, the development of char-filled composites for energy harvesting applications is an active and evolving research area. Continued efforts in material synthesis, characterization, processing techniques, and device integration will contribute to the advancement of efficient and scalable energy harvesting systems. The development of char-filled composites for energy harvesting presents significant opportunities for advancement. Key improvements include enhanced conductivity, tailored porosity, surface functionalization, and the creation of hybrid composites. Innovative manufacturing techniques and multiscale modeling can further refine these materials, while application-specific optimizations boost their efficiency. Utilizing novel carbon sources like algae and waste products not only enhances sustainability but also reduces costs. Moreover, char composites, derived from biomass or waste materials, offer significant potential for contributing to a more sustainable and circular energy ecosystem. The utilization of char, particularly when sourced from waste products such as agricultural residues or municipal solid waste, aligns with the principles of circular economy by turning waste into valuable resources, thus reducing environmental burden and reliance on fossil fuels. Additionally, integrating char composites into emerging technologies such as perovskite solar cells, flexible harvesters, and hybrid storage systems supports a more sustainable energy ecosystem. Continued research is essential for driving progress in both performance and environmental benefits, in alignment with circular economy principles.

6.6 Conclusions

Carbon materials, with their exceptional properties, have proven to be ideal candidates for energy conversion and storage technologies. In this chapter, we explored the potential of biomass and waste-derived char-filled composites in energy harvesting applications. Char materials derived from biomass and waste sources have gained attention as a renewable and sustainable option for energy conversion and storage. The various fabrication techniques employed, including blending, impregnation, and *in situ* formation, have contributed to tailoring the composite properties to meet specific application requirements. However, their direct use in energy harvesting devices is limited due to their inherent properties such as limited electrical conductivity and surface area. Thus, the optimization of surface modifications, composite design, and carbonization processes can help overcome these limitations. Continued research and development efforts are needed to unlock the full potential of biomass and waste-derived char-filled composites in energy harvesting applications and pave the way for a greener and more sustainable energy future.

Acknowledgments: Authors thank to Universiti Putra Malaysia for providing the research facility and the Monbukagakusho (MEXT) scholarship for the financial support.

References

1. Chang BP, Rodriguez-Uribe A, Mohanty AK, Misra M. A comprehensive review of renewable and sustainable biosourced carbon through pyrolysis in biocomposites uses: current development and future opportunity. Renew Sustain Energy Rev 2021;152:111666.
2. Harussani M, Sapuan S, Nadeem G, Rafin T, Kirubaanand W. Recent applications of carbon-based composites in defence industry: a review. Def Technol 2022;18:1281–300.
3. Khodabakhshi S, Fulvio PF, Andreoli E. Carbon black reborn: structure and chemistry for renewable energy harnessing. Carbon 2020;162:604–49.
4. Harussani M, Sapuan S, Rashid U, Khalina A, Ilyas R. Pyrolysis of polypropylene plastic waste into carbonaceous char: priority of plastic waste management amidst COVID-19 pandemic. Sci Total Environ 2022;803:149911.
5. Mian MM, Zeng X, Nasry Aal NB, Al-Hamadani SM. Municipal solid waste management in China: a comparative analysis. J Mater Cycles Waste Manag 2017;19:1127–35.
6. Cha JS, Park SH, Jung SC, Ryu C, Jeon JK, Shin MC, et al. Production and utilization of biochar: a review. J Ind Eng Chem 2016;40:1–15.
7. Chen N, Pilla S. A comprehensive review on transforming lignocellulosic materials into biocarbon and its utilization for composites applications. Compos C: Open Access 2022;7:100225.

8. Harussani MM, Rashid U, Sapuan SM, Abdan K. Low-temperature thermal degradation of disinfected COVID-19 non-woven polypropylene–based isolation gown wastes into carbonaceous char. Polymers 2021;13:3980.
9. Norizan MN, Moklis MH, Demon SZN, Halim NA, Samsuri A, Mohamad IS, et al. Carbon nanotubes: functionalisation and their application in chemical sensors. RSC Adv 2020;10:43704–32.
10. Boota M, Hatzell K, Alhabeb M, Kumbur E, Gogotsi Y. Graphene-containing flowable electrodes for capacitive energy storage. Carbon 2015;92:142–9.
11. Feng Z, Adolfsson KH, Xu Y, Fang H, Hakkarainen M, Wu M. Carbon dot/polymer nanocomposites: from green synthesis to energy, environmental and biomedical applications. Sustain Mater Technol 2021;29: e00304.
12. Gao C, Chen G. Conducting polymer/carbon particle thermoelectric composites: emerging green energy materials. Compos Sci Technol 2016;124:52–70.
13. Zhai Y, Dou Y, Zhao D, Fulvio PF, Mayes RT, Dai S. Carbon materials for chemical capacitive energy storage. Adv Mater 2011;23:4828–50.
14. Yrjälä K, Ramakrishnan M, Salo E. Agricultural waste streams as resource in circular economy for biochar production towards carbon neutrality. Curr Opin Environ Sci Health 2022;26:100339.
15. Lehmann J, Joseph S. Biochar for environmental management: science, technology and implementation. New York: Taylor & Francis; 2024.
16. Alazaiza MY, Albahnasawi A, Eyvaz M, Al Maskari T, Nassani DE, Abu Amr SS, et al. An overview of green bioprocessing of algae-derived biochar and biopolymers: synthesis, preparation, and potential applications. Energies 2023;16:791.
17. Baldock JA, Smernik RJ. Chemical composition and bioavailability of thermally altered Pinus resinosa (Red pine) wood. Org Geochem 2002;33:1093–109.
18. Kumar A, Bhattacharya T. Biochar: a sustainable solution. Environ Dev Sustain 2021;23:6642–80.
19. Wu L, Liu J, Reddy BR, Zhou J. Preparation of coal-based carbon nanotubes using catalytical pyrolysis: a brief review. Fuel Process Technol 2022;229:107171.
20. Yang S, Wu M, Wu C. Application of biomass fast pyrolysis part I: pyrolysis characteristics and products. Energy (Calg) 2014;66:162–71.
21. Abdullah N, Sulaiman F, Safana AA. Pyrolysis of torrefied oil palm wastes for better biochar. Mal J Fund Appl Sci 2017;13:124–8.
22. Bridgwater AV. Review of fast pyrolysis of biomass and product upgrading. Biomass Bioenergy 2012;38: 68–94.
23. Antal MJ, Grønli M. The art, science, and technology of charcoal production. Ind Eng Chem Res 2003;42: 1619–40.
24. Liu X, Xu G, Xiao H, Wei X, Yang L. Free-standing hierarchical porous assemblies of commercial TiO_2 nanocrystals and multi-walled carbon nanotubes as high-performance anode materials for sodium ion batteries. Electrochim Acta 2017;236:33–42.
25. Bai P, Han X, He Y, Xiong P, Zhao Y, Sun J, et al. Solid electrolyte interphase manipulation towards highly stable hard carbon anodes for sodium ion batteries. Energy Storage Mater 2020;25:324–33.
26. Zhang J, Fernández-Blázquez JP, Li XL, Wang R, Zhang X, Wang DY. A facile technique to investigate the char strength and fire retardant performance towards intumescent epoxy nanocomposites containing different synergists. Polym Degrad Stabil 2022;202:110000.
27. Ahmetli G, Kocaman S, Ozaytekin I, Bozkurt P. Epoxy composites based on inexpensive char filler obtained from plastic wasteindex: waste and natural resources. Polym Compos 2013;34:500–9.
28. Chuah T, Wan Azlina A, Robiah Y, Omar R. Biomass as the renewable energy sources in Malaysia: an overview. Int J Green Energy 2006;3:323–46.
29. Li Y, Gupta R, Zhang Q, You S. Review of biochar production via crop residue pyrolysis: development and perspectives. Bioresour Technol 2023;369:128423.

30. Pantea D, Darmstadt H, Kaliaguine S, Roy C. Electrical conductivity of conductive carbon blacks: influence of surface chemistry and topology. Appl Surf Sci 2003;217:181–93.
31. Kennedy Z, Christ J, Evans K, Arey B, Sweet L, Warner M, et al. 3D-printed poly (vinylidene fluoride)/carbon nanotube composites as a tunable, low-cost chemical vapour sensing platform. Nanoscale 2017;9:5458–66.
32. Zhao Y, Feng D, Zhang Y, Huang Y, Sun S. Effect of pyrolysis temperature on char structure and chemical speciation of alkali and alkaline earth metallic species in biochar. Fuel Process Technol 2016;141:54–60.
33. Özaytekin İ, Kar Y. Synthesis and properties of composites of oligoazomethine with char. J Appl Polym Sci 2012;123:815–23.
34. Athanasopoulos N, Baltopoulos A, Matzakou M, Vavouliotis A, Kostopoulos V. Electrical conductivity of polyurethane/MWCNT nanocomposite foams. Polym Compos 2012;33:1302–12.
35. Haque MA, Mina MF, Moshiul Alam A, Rahman MJ, Bhuiyan MAH, Asano T. Multiwalled carbon nanotubes-reinforced isotactic polypropylene nanocomposites: enhancement of crystallization and mechanical, thermal, and electrical properties. Polym Compos 2012;33:1094–104.
36. Wang Y, Song Y, Li N, Liu W, Yan B, Yu Y, et al. Tunable active sites on biogas digestate derived biocharfor sulfanilamide degradation by peroxymonosulfate activation. J Hazard Mater 2022;421:126794.
37. Hamza MF, Salih KA, Adel A-H, Zayed YE, Wei Y, Liang J, et al. Sulfonic-functionalized algal/PEI beads for scandium, cerium and holmium sorption from aqueous solutions (synthetic and industrial samples). Chem Eng J 2021;403:126399.
38. Moklis MH, Shuo C, Boonyubol S, Cross JS. Electrochemical valorization of glycerol via electrocatalytic reduction into biofuels: a review. ChemSusChem 2023;17:e202300990.
39. Joseph S. Biochar for environmental management: an introduction. In: Biochar for environmental management: science and technology. New York: Routledge; 2015:1–13 pp.
40. Liu M, Almatrafi E, Zhang Y, Xu P, Song B, Zhou C, et al. A critical review of biochar-based materials for the remediation of heavy metal contaminated environment: applications and practical evaluations. Sci Total Environ 2022;806:150531.
41. Hekimoğlu G, Sarı A, Arunachalam S, Arslanoğlu H, Gencel O. Porous biochar/heptadecane composite phase change material with leak-proof, high thermal energy storage capacity and enhanced thermal conductivity. Powder Technol 2021;394:1017–25.
42. Wang C, Feng L, Li W, Zheng J, Tian W, Li X. Shape-stabilized phase change materials based on polyethylene glycol/porous carbon composite: the influence of the pore structure of the carbon materials. Sol Energy Mater Sol Cell 2012;105:21–6.
43. Zhao Y, Min X, Huang Z, Liu Y, Wu X, Fang M. Honeycomb-like structured biological porous carbon encapsulating PEG: a shape-stable phase change materialwith enhanced thermal conductivity for thermal energy storage. Energy Build 2018;158:1049–62.
44. Atinafu DG, Chang SJ, Kim S. Infiltration properties of n-alkanes in mesoporous biochar: the capacity of smokeless support for stability and energy storage. J Hazard Mater 2020;399:123041.
45. Nicholas AF, Hussein MZ, Zainal Z, Khadiran T. Palm kernel shell activated carbon as an inorganic framework for shape-stabilized phase change material. Nanomaterials 2018;8:689.
46. Wen R, Zhang W, Lv Z, Huang Z, Gao W. A novel composite Phase change material of Stearic Acid/Carbonized sunflower straw for thermal energy storage. Mater Lett 2018;215:42–5.
47. Xu Y, Zhang X, Wu B, Xu Y, Wen R, Liu Y, et al. Preparation and performance of shape-stable phase change materials based on carbonized-abandoned orange peel and paraffin. Fullerenes, Nanotub Carbon Nanostruct 2019;27:289–98.
48. Zhang W, Zhang X, Zhang X, Yin Z, Liu Y, Fang M, et al. Lauric-stearic acid eutectic mixture/carbonized biomass waste corn cob composite phase change materials: preparation and thermal characterization. Thermochim Acta 2019;674:21–7.
49. Feng L, Zheng J, Yang H, Guo Y, Li W, Li X. Preparation and characterization of polyethylene glycol/active carbon composites as shape-stabilized phase change materials. Sol Energy Mater Sol Cell 2011;95:644–50.

50. Wang J, Senkovska I, Kaskel S, Liu Q. Chemically activated fungi-based porous carbons for hydrogen storage. Carbon 2014;75:372–80.
51. Fierro V, Szczurek A, Zlotea C, Marêché J, Izquierdo M, Albiniak A, et al. Experimental evidence of an upper limit for hydrogen storage at 77 K on activated carbons. Carbon 2010;48:1902–11.
52. de Castro MM, Martínez-Escandell M, Molina-Sabio M, Rodríguez-Reinoso F. Hydrogen adsorption on KOH activated carbons from mesophase pitch containing Si, B, Ti or Fe. Carbon 2010;48:636–44.
53. Akasaka H, Takahata T, Toda I, Ono H, Ohshio S, Himeno S, et al. Hydrogen storage ability of porous carbon material fabricated from coffee bean wastes. Int J Hydrogen Energy 2011;36:580–5.
54. Huang CC, Chen HM, Chen CH, Huang JC. Effect of surface oxides on hydrogen storage of activated carbon. Sep Purif Technol 2010;70:291–5.
55. Bhat VV, Contescu CI, Gallego NC, Baker FS. Atypical hydrogen uptake on chemically-activated, ultramicroporous carbon. Carbon 2010;48:1331–40.
56. Liang C, Dudney NJ, Howe JY. Hierarchically structured sulfur/carbon nanocomposite material for high-energy lithium battery. Chem Mater 2009;21:4724–30.
57. Wei S, Zhang H, Huang Y, Wang W, Xia Y, Yu Z. Pig bone derived hierarchical porous carbon and its enhanced cycling performance of lithium–sulfur batteries. Energy Environ Sci 2011;4:736–40.
58. Liu WJ, Tian K, He YR, Jiang H, Yu HQ. High-yield harvest of nanofibers/mesoporous carbon composite by pyrolysis of waste biomass and its application for high durability electrochemical energy storage. Environ Sci Technol 2014;48:13951–9.
59. Kastner JR, Miller J, Geller DP, Locklin J, Keith LH, Johnson T. Catalytic esterification of fatty acids using solid acid catalysts generated from biochar and activated carbon. Catal Today 2012;190:122–32.
60. Shu Q, Gao J, Nawaz Z, Liao Y, Wang D, Wang J. Synthesis of biodiesel from waste vegetable oil with large amounts of free fatty acids using a carbon-based solid acid catalyst. Appl Energy 2010;87:2589–96.
61. Dawodu FA, Ayodele O, Xin J, Zhang S, Yan D. Effective conversion of non-edible oil with high free fatty acid into biodiesel by sulphonated carbon catalyst. Appl Energy 2014;114:819–26.
62. Dehkhoda AM, Ellis N. Biochar-based catalyst for simultaneous reactions of esterification and transesterification. Catal Today 2013;207:86–92.
63. Zeng D, Liu S, Gong W, Wang G, Qiu J, Chen H. Synthesis, characterization and acid catalysis of solid acid from peanut shell. Appl Catal Gen 2014;469:284–9.
64. Guo H, Lian Y, Yan L, Qi X, Smith RL. Cellulose-derived superparamagnetic carbonaceous solid acid catalyst for cellulose hydrolysis in an ionic liquid or aqueous reaction system. Green Chem 2013;15:2167–74.
65. Yamaguchi D, Kitano M, Suganuma S, Nakajima K, Kato H, Hara M. Hydrolysis of cellulose by a solid acid catalyst under optimal reaction conditions. J Phys Chem C 2009;113:3181–8.
66. Wu Y, Fu Z, Yin D, Xu Q, Liu F, Lu C, et al. Microwave-assisted hydrolysis of crystalline cellulose catalyzed by biomass char sulfonic acids. Green Chem 2010;12:696–700.
67. Johnson L, Thielemans W, Walsh DA. Synthesis of carbon-supported Pt nanoparticle electrocatalysts using nanocrystalline cellulose as reducing agent. Green Chem 2011;13:1686–93.
68. Ma M, Dai Y, Zou J, Wang L, Pan K, Fu H. Synthesis of iron oxide/partly graphitized carbon composites as a high-efficiency and low-cost cathode catalyst for microbial fuel cells. ACS Appl Mater Interfaces 2014;6 13438–47.
69. Chen T, Li D, Jiang H, Xiong C. High-performance Pd nanoalloy on functionalized activated carbon for the hydrogenation of nitroaromatic compounds. Chem Eng J 2015;259:161–9.
70. Wang S, Wang H, Yin Q, Zhu L, Yin S. Methanation of bio-syngas over a biochar supported catalyst. New J Chem 2014;38:4471–7.
71. Zhang C, Liang X, Liu S. Hydrogen production by catalytic dehydrogenation of methylcyclohexane over Pt catalysts supported on pyrolytic waste tire char. Int J Hydrogen Energy 2011;36:8902–7.
72. Trincado M, Banerjee D, Grützmacher H. Molecular catalysts for hydrogen production from alcohols. Energy Environ Sci 2014;7:2464–503.

73. Shen Y, Zhao P, Shao Q, Ma D, Takahashi F, Yoshikawa K. In-situ catalytic conversion of tar using rice husk char-supported nickel-iron catalysts for biomass pyrolysis/gasification. Appl Catal B Environ 2014;152: 140–51.
74. Shen Y, Yoshikawa K. Tar conversion and vapor upgrading via in situ catalysis using silica-based nickel nanoparticles embedded in rice husk char for biomass pyrolysis/gasification. Ind Eng Chem Res 2014;53: 10929–42.
75. Umeyama T, Imahori H. Photofunctional hybrid nanocarbon materials. J Phys Chem C 2013;117:3195–209.
76. Li Y, Zhang S, Yu Q, Yin W. The effects of activated carbon supports on the structure and properties of TiO_2 nanoparticles prepared by a sol–gel method. Appl Surf Sci 2007;253:9254–8.
77. Wei L, Sevilla M, Fuertes AB, Mokaya R, Yushin G. Hydrothermal carbonization of abundant renewable natural organic chemicals for high-performance supercapacitor electrodes. Adv Energy Mater 2011;1:356–61.
78. Li X, Han C, Chen X, Shi C. Preparation and performance of straw based activated carbon for supercapacitor in non-aqueous electrolytes. Microporous Mesoporous Mater 2010;131:303–9.
79. Huang W, Zhang H, Huang Y, Wang W, Wei S. Hierarchical porousindex: porous carbon obtained from animal bone and evaluation in electric double-layer capacitors. Carbon 2011;49:838–43.
80. Kalpana D, Cho S, Lee S, Lee Y, Misra R, Renganathan N. Recycled waste paper–a new source of raw material for electric double-layer capacitors. J Power Sources 2009;190:587–91.
81. Subramanian V, Luo C, Stephan AM, Nahm K, Thomas S, Wei B. Supercapacitors from activated carbon derived from banana fibers. J Phys Chem C 2007;111:7527–31.
82. Li X, Xing W, Zhuo S, Zhou J, Li F, Qiao SZ, et al. Preparation of capacitor's electrode from sunflower seed shell. Bioresour Technol 2011;102:1118–23.
83. Khadiran T, Hussein MZ, Zainal Z, Rusli R. Shape-stabilised n-octadecane/activated carbon nanocomposite phase change material for thermal energy storage. J Taiwan Inst Chem Eng 2015;55:189–97.
84. Gu X, Liu P, Liu C, Peng L, He H. A novel form-stable phase change material of palmitic acid-carbonized pepper straw for thermal energy storage. Mater Lett 2019;248:12–15.
85. Sarabandi D, Roudini G, Barahuie F. Activated carbon derived from pine cone as a framework for the preparation of n-heptadecane nanocomposite for thermal energy storage. J Energy Storage 2019;24: 100795.
86. Wang G, Zhang L, Zhang J. A review of electrode materials for electrochemical supercapacitors. Chem Soc Rev 2012;41:797–828.
87. Zhou CH, Xia X, Lin CX, Tong DS, Beltramini J. Catalytic conversion of lignocellulosic biomass to fine chemicals and fuels. Chem Soc Rev 2011;40:5588–617.
88. Toda M, Takagaki A, Okamura M, Kondo JN, Hayashi S, Domen K, et al. Biodiesel made with sugar catalyst. Nature 2005;438:178.
89. Jiang Y, Li X, Wang X, Meng L, Wang H, Peng G, et al. Effective saccharification of lignocellulosic biomass over hydrolysis residue derived solid acid under microwave irradiation. Green Chem 2012;14:2162–7.
90. Yan Q, Wan C, Liu J, Gao J, Yu F, Zhang J, et al. Iron nanoparticles in situ encapsulated in biochar-based carbon as an effective catalyst for the conversion of biomass-derived syngas to liquid hydrocarbons. Green Chem 2013;15:1631–40.
91. Chen S, Rotaru A-E, Shrestha PM, Malvankar NS, Liu F, Fan W, et al. Promoting interspecies electron transfer with biochar. Sci Rep 2014;4:5019.
92. Harussani M, Sapuan S. Development of kenaf biochar in engineering and agricultural applications. Chem Afr 2022:1–17. https://doi.org/10.1007/s42250-021-00293-1.
93. Kappler A, Wuestner ML, Ruecker A, Harter J, Halama M, Behrens S. Biochar as an electron shuttle between bacteria and Fe (III) minerals. Environ Sci Technol Lett 2014;1:339–44.
94. Rudra S, Seo HW, Sarker S, Kim DM. Supercapatteries as hybrid electrochemical energy storage devices: current status and future prospects. Molecules 2024;29:243.

95. Chen GZ. Supercapacitor and supercapattery as emerging electrochemical energy stores. Int Mater Rev 2017;62:173–202.
96. Frackowiak E. Carbon materials for supercapacitor application. Phys Chem Chem Phys 2007;9:1774–85.
97. Zhang LL, Zhao X. Carbon-based materials as supercapacitor electrodes. Chem Soc Rev 2009;38:2520–31.
98. Zhai Y, Dou Y, Zhao D, Fulvio PF, Mayes RT, Dai S. Carbon materials for chemical capacitive energy storage. Adv Mater 2011;23:4828–50.
99. He X, Geng Y, Qiu J, Zheng M, Long S, Zhang X. Effect of activation time on the properties of activated carbons prepared by microwave-assisted activation for electric double layer capacitors. Carbon 2010;48: 1662–9.
100. He XJ, Long SA, Zheng MD, Shui HF, Wang T. Optimization of activated carbon preparation by orthogonal experimental design for electrochemical capacitors. Sci Adv Mater 2010;2:545–51.
101. Olivares-Marín M, Fernández JA, Lázaro MJ, Fernández-González C, Macías-García A, Gómez-Serrano V, et al. Cherry stones as precursor of activated carbons for supercapacitors. Mater Chem Phys 2009;114: 323–7.
102. Chen W, Zhang H, Huang Y, Wang W. A fish scale based hierarchical lamellar porousindex: porous carbon material obtained using a natural template for high performance electrochemical capacitors. J Mater Chem 2010;20:4773–5.
103. Jorda-Beneyto M, Suárez-García F, Lozano-Castello D, Cazorla-Amorós D, Linares-Solano A. Hydrogen storage on chemically activated carbons and carbon nanomaterials at high pressures. Carbon 2007;45: 293–303.
104. Sun Y, Webley PA. Preparation of activated carbons from corncob with large specific surface area by a variety of chemical activators and their application in gas storage. Chem Eng J 2010;162:883–92.
105. Liu W-J, Jiang H, Yu HQ. Development of biochar-based functional materials: toward a sustainable platform carbon material. Chem Rev 2015;115:12251–85.
106. Back CK, Sandí G, Prakash J, Hranisavljevic J. Hydrogen sorption on palladium-doped sepiolite-derived carbon nanofibers. J Phys Chem B 2006;110:16225–31.
107. Zhu G, Ma L, Lv H, Hu Y, Chen T, Chen R, et al. Pine needle-derived microporous nitrogen-doped carbon frameworks exhibit high performances in electrocatalytic hydrogen evolution reaction and supercapacitors. Nanoscale 2017;9:1237–43.
108. Wang B, Xu L, Liu G, Zhang P, Zhu W, Xia J, et al. Biomass willow catkin-derived Co 3 O 4/N-doped hollow hierarchical porous carbon microtubes as an effective tri-functional electrocatalyst. J Mater Chem A 2017; 5:20170–9.
109. Sharma P, Melkania U. Biochar-enhanced hydrogen production from organic fraction of municipal solid waste using co-culture of Enterobacter aerogenes and *E. coli*. Int J Hydrogen Energy 2017;42:18865–74.
110. Zhang J, Fan C, Zang L. Improvement of hydrogen production from glucose by ferrous iron and biochar. Bioresour Technol 2017;245:98–105.
111. Chen WF, Iyer S, Iyer S, Sasaki K, Wang CH, Zhu Y, et al. Biomass-derived electrocatalytic composites for hydrogen evolution. Energy Environ Sci 2013;6:1818–26.
112. Cui W, Liu Q, Xing Z, Asiri AM, Alamry KA, Sun X. MoP nanosheets supported on biomass-derived carbon flake: one-step facile preparation and application as a novel high-active electrocatalyst toward hydrogen evolution reaction. Appl Catal B Environ 2015;164:144–50.
113. Lin Y, Pan Y, Zhang J. CoP nanorods decorated biomass derived N, P co-doped carbon flakes as an efficient hybrid catalyst for electrochemical hydrogen evolution. Electrochim Acta 2017;232:561–9.
114. Liu X, Zhang M, Yu D, Li T, Wan M, Zhu H, et al. Functional materials from nature: honeycomb-like carbon nanosheets derived from silk cocoon as excellent electrocatalysts for hydrogen evolution reaction. Electrochim Acta 2016;215:223–30.
115. Zhou Y, Leng Y, Zhou W, Huang J, Zhao M, Zhan J, et al. Sulfur and nitrogen self-doped carbon nanosheets derived from peanut root nodules as high-efficiency non-metal electrocatalyst for hydrogen revolution reaction. Nano Energy 2015;16:357–66.

116. An K, Xu X, Liu X. Mo_2C-based electrocatalyst with biomass-derived sulfur and nitrogen co-doped carbon as a matrix for hydrogen evolution and organic pollutant removal. ACS Sustainable Chem Eng 2018;6: 1446–55.
117. Zhang Y, Zuo L, Zhang L, Huang Y, Lu H, Fan W, et al. Cotton wool derived carbon fiber aerogel supported few-layered MoSe2 nanosheets as efficient electrocatalysts for hydrogen evolution. ACS Appl Mater Interfaces 2016;8:7077–85.
118. Liu H, Geng C, Wei P, Chen H, Zheng S, Wang H, et al. Improving the performance and stability of large-area carbon-based perovskite solar cells using N, O co-doped biomass porous carbon. J Alloys Compd 2022;912:165123.
119. Roy P, Ghosh A, Barclay F, Khare A, Cuce E. Perovskite solar cells: a review of the recent advances. Coatings (Oakv) 2022;12:1089.
120. Liu Y, Zhou J, Chen L, Zhang P, Fu W, Zhao H, et al. Highly flexible freestanding porous carbon nanofibers for electrodes materials of high-performance all-carbon supercapacitors. ACS Appl Mater Interfaces 2015; 7:23515–20.
121. Wang H, Yu C. Organic thermoelectrics: materials preparation, performance optimization, and device integration. Joule 2019;3:53–80.
122. Zhang YC, Pang H, Dai K, Huang YF, Ren PG, Chen C, et al. Conductive network formation during annealing of an oriented polyethylene-based composite. J Mater Sci 2012;47:3713–19.
123. Chen G, Xu W, Zhu D. Recent advances in organic polymer thermoelectric composites. J Mater Chem C 2017;5:4350–60.
124. Piao M, Kim G, Kennedy GP, Roth S, Dettlaff-Weglikowska U. Thermoelectric properties of single walled carbon nanotube networks in polycarbonate matrix. Phys Status Solidi (b) 2013;250:1468–73.
125. Zhang YC, Dai K, Tang JH, Ji X, Li ZM. Anisotropically conductive polymer composites with a selective distribution of carbon black in an in situ microfibrillar reinforced blend. Mater Lett 2010;64:1430–2.
126. Meng C, Liu C, Fan S. A promising approach to enhanced thermoelectric properties using carbon nanotube networks. Adv Mater 2010;22:535–9.
127. Lu S, Wang H, Zhou J, Wu X, Qin W. Atomic layer deposition of ZnO on carbon black as nanostructured anode materials for high-performance lithium-ion batteries. Nanoscale 2017;9:1184–92.
128. Reddy TB. Linden's handbook of batteries, 4th ed. New York: McGraw-Hill Education; 2011.
129. Zhang L, Zhang M, Wang Y, Zhang Z, Kan G, Wang C, et al. Graphitized porous carbon microspheres assembled with carbon black nanoparticles as improved anode materials in Li-ion batteries. J Mater Chem A 2014;2:10161–8.
130. Portet C, Yushin G, Gogotsi Y. Electrochemical performance of carbon onions, nanodiamonds, carbon black and multiwalled nanotubes in electrical double layer capacitors. Carbon 2007;45:2511–18.
131. Surovikin YV, Shaitanov A, Resanov I, Syr'Eva A. The properties of nanodispersed carbon black particles after thermal treatment. Procedia Eng 2015;113:519–24.
132. Boota M, Hatzell K, Beidaghi M, Dennison C, Kumbur E, Gogotsi Y. Activated carbon spheres as a flowable electrode in electrochemical flow capacitors. J Electrochem Soc 2014;161:A1078.
133. Campos JW, Beidaghi M, Hatzell KB, Dennison CR, Musci B, Presser V, et al. Investigation of carbon materials for use as a flowable electrode in electrochemical flow capacitors. Electrochim Acta 2013;98: 123–30.
134. Presser V, Dennison CR, Campos J, Knehr KW, Kumbur EC, Gogotsi Y. The electrochemical flow capacitor: a new concept for rapid energy storage and recovery. Adv Energy Mater 2012;2:895–902.
135. Zhang C, Hatzell KB, Boota M, Dyatkin B, Beidaghi M, Long D, et al. Highly porous carbon spheres for electrochemical capacitors and capacitive flowable suspension electrodes. Carbon 2014;77:155–64.
136. Cong Y, Li X, Dong Z, Yuan G, Cui Z, Zhang J. Core–shell structured carbon black@ TiO_2 microsphere with enhanced visible light photocatalytic activity. Mater Lett 2015;138:200–3.
137. Leary R, Westwood A. Carbonaceous nanomaterials for the enhancement of TiO_2 photocatalysis. Carbon 2011;49:741–72.

138. Li L, Zhu W, Zhang P, Chen Z, Han W. Photocatalytic oxidation and ozonation of catechol over carbon-black-modified nano-TiO_2 thin films supported on Al sheet. Water Res 2003;37:3646–51.
139. Linsebigler AL, Lu G, Yates Jr JT. Photocatalysis on TiO_2 surfaces: principles, mechanisms, and selected results. Chem Rev 1995;95:735–58.
140. Lima RMAP, Dos Reis GS, Thyrel M, Alcaraz-Espinoza JJ, Larsson SH, de Oliveira HP. Facile synthesis of sustainable biomass-derived porous biochars as promising electrode materials for high-performance supercapacitor applications. Nanomaterials 2022;12:866.
141. Zhang C, Liu L, Zhao M, Rong H, Xu Y. The environmental characteristics and applications of biochar. Environ Sci Pollut Control Ser 2018;25:21525–34.
142. Sohi SP, Krull E, Lopez-Capel E, Bol R. A review of biochar and its use and function in soil. Adv Agron 2010; 105:47–82.
143. Kung CC, McCarl BA, Cao X. Economics of pyrolysisindex: pyrolysis-based energy production and biochar utilization: a case study in Taiwan. Energy Policy 2013;60:317–23.

Mageswari Manimaran, Mohd Nurazzi Norizan*, Mohamad Haafiz Mohamad Kassim, Mohd Sapuan Salit, Mohd Nor Faiz Norrrahim, Mohd Ridhwan Adam and Ahmad Ilyas Rushdan

7 Functionalization and performance of hybrid nanocellulose from plant-based/metal oxide nanocomposites for sustainable energy applications

Abstract: Growing concerns over our dependence on finite, non-renewable resources like petroleum and metals have driven the development of eco-friendly technologies centered on advanced hybrid nanomaterials. Among these, the use of renewable nanocellulose – ranging in size from 1 to 100 nm – has gained significant attention in nanotechnology research. Derived from sustainable sources, nanocellulose offers notable advantages; however, challenges persist when integrating it with metal oxide nanoparticles (MONPs). These challenges include high reactivity in cellular environments, elevated production costs, and a tendency to aggregate, leading to instability in both liquid and dry states. Aggregation can impair uniform dispersion and result in sediment formation in certain applications. A promising solution to these challenges is hybridizing MONPs with functionalized nanocellulose, a method widely adopted by researchers. This approach is cost-effective, environmentally sustainable, and produces a renewable material with low density, excellent stability, superior mechanical properties, and

---

***Corresponding author: Mohd Nurazzi Norizan**, Bioresource Technology Division, School of Industrial Technology, Universiti Sains Malaysia, Minden, Penang 11800, Malaysia; and Green Biopolymer, Coatings & Packaging Cluster, School of Industrial Technology, Universiti Sains Malaysia, Minden, Penang 11800, Malaysia, E-mail: mohd.nurazzi@usm.my

Mageswari Manimaran and Mohamad Haafiz Mohamad Kassim, Bioresource Technology Division, School of Industrial Technology, Universiti Sains Malaysia, Minden, Penang 11800, Malaysia; and Green Biopolymer, Coatings & Packaging Cluster, School of Industrial Technology, Universiti Sains Malaysia, Minden, Penang 11800, Malaysia

Mohd Sapuan Salit, Advanced Engineering Materials and Composites Research Center (AEMC), Department of Mechanical and Manufacturing Engineering, Universiti Putra Malaysia, Serdang, Selangor, 43400, Malaysia. https://orcid.org/0000-0002-4598-6358

Mohd Nor Faiz Norrrahim, Research Centre for Chemical Defence, Universiti Pertahanan Nasional Malaysia, Kem Perdana Sungai Besi, Kuala Lumpur 57000, Malaysia

Mohd Ridhwan Adam, School of Chemical Sciences, Universiti Sains Malaysia, Minden, Penang 11800, Malaysia

Ahmad Ilyas Rushdan, School of Chemical and Energy Engineering, Faculty of Engineering, Universiti Teknologi Malaysia, 81310 UTM Johor Bahru, Johor, Malaysia; and Centre for Advanced Composite Materials (CACM), Universiti Teknologi Malaysia, 81310 UTM Johor Bahru, Johor, Malaysia

As per De Gruyter's policy this article has previously been published in the journal Physical Sciences Reviews. Please cite as: M. Manimaran, M. Nurazzi Norizan, M. H. M. Kassim, M. S. Salit, M. N. F. Norrrahim, M. R. Adam and A. I. Rushdan "Functionalization and performance of hybrid nanocellulose from plant-based/metal oxide nanocomposites for sustainable energy applications" *Physical Sciences Reviews* [Online] 2024. DOI: 10.1515/psr-2024-0025 | https://doi.org/10.1515/9783111234762-007

biocompatibility. However, several questions remain unresolved, such as the most commonly used functionalization techniques for MONPs hybridization, the underlying mechanisms, and the specific benefits of this hybridization. Based on current findings, oxidation and carboxymethylation emerge as the most frequently used functionalization techniques for hybridizing MONPs with nanocellulose. These processes introduce carboxylic acid and carboxymethyl groups, respectively, which act as capping agents that readily bond with MONPs. This results in high degrees of substitution (DS) and improved nanoparticle dispersion. Furthermore, hybridization enhances properties such as thermal stability, UV protection, antibacterial activity, adsorption capacity, and mechanical performance, underscoring its potential for diverse applications.

Keywords: nanocellulose; metal oxide; hybrid; nanocomposites; conjugation

7.1 Introduction

Over the past few decades, researchers have been increasingly dedicated to seeking sustainable and environmentally friendly end products derived from bioresources. This pursuit is driven by the unique properties and potential advantageous attributes of such products, including biodegradability, biocompatibility, low toxicity, high aspect ratio and affordability [1, 2]. Among the most captivating and consequential bioresources is nanocellulose. Thanks to its increased crystallinity, aspect ratio, and specific surface area, nanocellulose can form a percolation network that enhances energy transfer, making it valuable for a wide range of applications. Nanocellulose serves as a versatile platform for emerging biomedical applications and light-emitting devices. It can be classified into various forms, such as nanofibrillated cellulose (NFC), nanocrystalline cellulose (NCC), and bacterial nanocellulose (BNC), based on the method used to isolate cellulose from bioresources by removing lignin and hemicellulose [3]. Harnessing nanocellulose is critical for carbon capture and conservation, helping to combat global warming by reducing carbon dioxide (CO_2) emissions. It can replace plastic and metal substrates in numerous applications, reducing pollution. Furthermore, nanocellulose has garnered considerable interest among researchers due to its manifold advantages over synthetic cellulose [4]. For instance, it has superior tensile strength up to 50–100 GPa which stronger than E-glass (3,400 MPa) and X (3000 MPa) [5, 6], inexpensive isolation method, high crystallinity, low density, and easily modify properties due to abundance of–OH group on the outer layer of nanocellulose which help to widen their application by binding different functional group. Nanocellulose finds applications in a wide spectrum of fields, ranging from thermal insulation to flexible electronics, energy storage, biosensors, CO_2 absorption, materials for enzyme immobilization, wound healing, and biodegradable packaging [7].

Besides nanocellulose, advancements in nanotechnology have prompted research into synthesizing another nanoscale material with intriguing physiochemical properties in rapidly expanding interdisciplinary field within materials science and engineering which is known as metal oxide nanoparticles (MONPs). It has been reported that MONPs like zinc oxide (ZnO), titanium dioxide (TiO_2), and silicone dioxide (SiO_2) are the most produced in the world, with yearly averages of 28,000 tonnes, 40,000 tonnes, and 55,000 tonnes, respectively. This is because they are different from their bulk peers in many ways [8]. The surface energy, optical, thermal, and electrical properties of MONP vary from those of the bulk counterparts because of the factors including, their high surface to volume ratio, cellular penetration capabilities, and quantum effects [9]. The synthesis methodology and various parameters play an important part in defining the shape, surface area, size, crystallinity, and stability of MONPs. Their remarkable biocompatibility renders them non-toxic and devoid of systemic toxicity [9]. In addition, numerous researchers have recently hybridised MONPs with nanocellulose for a variety of applications. Despite the numerous benefits of MONPs, certain limitations are present. These encompass colloidal instability, leading to sedimentation, particularly when compared to their bulk equivalents. This propensity arises from the greater surface energy associated with smaller particle sizes, causing them to aggregate more readily than larger particles. This aggregation induced by a reduction in the system's free energy [10]. The terms "agglomerates" and "aggregates" are employed to describe weakly bound and tightly bound collections of MONPs respectively [11]. The latter, aggregates, prove challenging to disperse into primary particles through mechanical means. Biological risks also exist, as MONPs can lead to combustion explosions, and their synthesis presents difficulties due to the intricate task of maintaining particle size in solution.

One solution to the issue is functionalization, which can effectively produce hybrid plant-based nanocellulose/MONP composites with high stability, free from agglomeration or sedimentation. This is because functionalizing nanocellulose introduces specific chemical groups that attach to its surface, allowing it to serve a dual role as both a template and a capping agent for MONPs [12]. Notably, the–OH at the 6th carbon of the cellulose backbone is significantly more reactive than other–OH groups, making it ideal for modification through interactions with functional groups without altering the core structure of nanocellulose. This modification enables the nanocellulose to be tailored for specific applications [13, 14]. Once functionalized, the hybridization of nanocellulose with MONPs enhances their absorption capacity by altering their chemical and physical properties. This hybridization improves nanoparticle dispersion and promotes stronger interfacial interactions through chemical bonding, resulting in a nanocomposite with enhanced mechanical and thermal properties, UV-barrier capabilities, and several other advantageous features. Economically, hybridizing functionalized nanocellulose with MONPs is cost-effective. Nanocellulose is derived from abundant and renewable sources such as bamboo, kenaf, oil palm biomass, and rice straw, making it far more affordable than metal oxide nanoparticles. By hybridizing these two materials, the amount of MONPs used can be reduced and partially replaced by nanocellulose, which often exhibits comparable or superior qualities. Additionally, incorporating such high-performance nanomaterials into production can extend product lifespans by enhancing strength, stability, and biocompatibility, potentially resulting in higher-value products.

Recently, the application of the nanocellulose from various fields is increasing especially in the form of nanocomposite by incorporating different type of MONPs as shown in Figure 7.1. At where the recent advancements in nanotechnology have sparked renewed interest in utilizing metal oxide nanoparticles in combination with nanocellulose for various applications.

The primary aim of this paper is to highlights that–OH groups of nanocellulose can be precisely tailored for functionalization, which could making it an ideal capping agent for prominent MONPs such as silver oxide (Ag_2O), copper oxide (CuO), and magnesium oxide (MgO) and many more. Moreover, this review explores the available functionalization and the highlighted the most widely used functionalization. This review begins with an overview of plant-based nanocellulose and its isolation processes, discusses various types and benefits of nanocellulose functionalization, and examines the performance of hybrid functionalized nanocellulose/MONP nanocomposites. This review also underscores the importance of nanocellulose functionalization, elucidates underlying mechanisms, and highlights performance enhancements when functionalized nanocellulose is hybridized with MONPs. Ultimately, it emphasizes the vast potential of these hybrid materials for diverse sustainable applications.

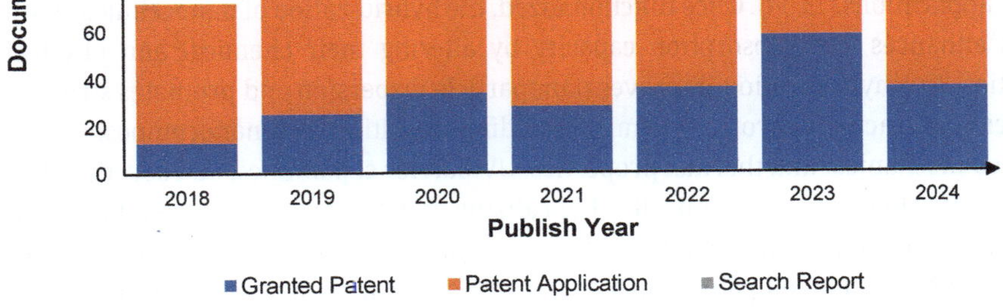

Figure 7.1: The result of patent document over time by LENS. Ong using the key word metal oxide nanoparticles with nanocellulose.

7.2 Overview on plant-based biomass

With a one trillion tonne yearly output, plant-based biomass is very prominent and environmentally friendly natural polymer which can be extracted from timber, agriculture wastes, bacterial cellulose and marine life like tunicates. A natural polymer that is enduring the most prominent industrial sources are timber or natural wood forestry leftovers such wood chips and branches. However, by concerns about sustainability and deforestation issue, the demand for non-wood cellulose materials such as agricultural crops or agricultural by-products becomes an increasing demand and an alternative source of cellulose [15]. Corn, wheat, rice, sorghum, barley, sugar cane, pineapple, bananas, oil palm and coconut crops are some other agricultural by-products that recently widely utilized in the form either NFC or NCC for variety purpose [16] which have different characteristic as shown in Table 7.1. NFC has flexible, longer, and wide nanofibers, whereas NCC has whiskers or needle like morphology. Besides that, NFC is produced through mechanical processing such as high-pressure homogenization, grinding or crushing to break down the larger cellulose fibres into nanocellulose [14]. The properties of the NFC are it has high surface area and aspect ratio, strong mechanical properties such as high tensile strength. However, NCC is produced mainly by hydrolysis with strong acids (sulfuric acid) to selectively remove the amorphous regions, leaving behind the crystalline cellulose structure only. It has strong, rigid, have high crystallinity index crystalline structure, high surface area and strong mechanical properties. Another significant properties of NCC and NFC is that since NFC is more flexible it is very suitable for the application which requiring high surface interaction such as in composite material, while NCC which has better mechanical strength and rigidity, it often used for reinforcing materials or optical application [17, 18]. In addition, NCC and NFC are produced from the plant-based cellulose, but bacterial nanocellulose (BNC) extracted from extracellular structures by bacteria belonging to the genera *Acetobacter, Agrobacterium, Alcaligenes, Pseudomonas, Rhizobium,* or *Sarcina* [19]. The most efficient producer of bacterial cellulose is the Gram-negative strain of acetic acid-producing bacteria *Acetobacter xylinum* (or *Gluconacetobacter xylinus*) [20].

In comparison to vegetable cellulose, bacterial cellulose does not contain any lignin, wax or hemicellulose. The restriction of bacterial cellulose is that its industrial manufacture is quite expensive and required to use hazardous chemical [23]. Consequently, plant-based nanocellulose synthesized from agricultural crops and waste materials offers a more sustainable approach to producing cellulose, owing to their remarkable properties and diverse range of applications. This has led to a substantial increase in research attention in recent decades. Nevertheless, a significant portion of agricultural residues is either incinerated, used as animal feed, or converted into biofuels. In contrast, agricultural by-products have the potential to serve as valuable sources of natural nanofibers due to their renewable nature. Table 7.2 provides an overview of recent applications for various types of agricultural waste.

Table 7.1: Classification of nanocellulose based on source, isolation method, and its physical properties. Reproduced from ref. [21, 22].

Type of nanocellulose	Source of cellulose	Isolation method	Average size
Nanofibrillated cellulose (NFC) 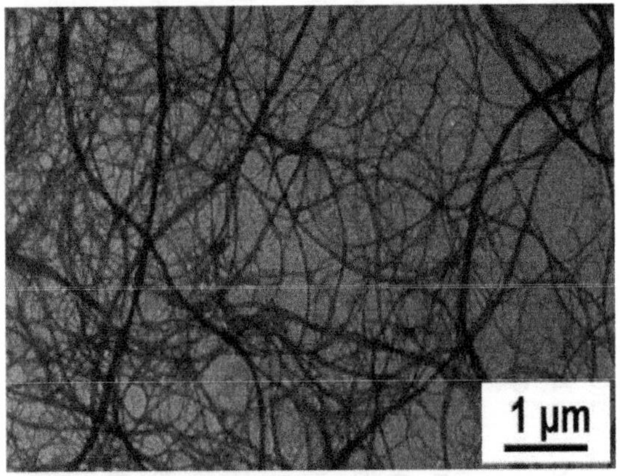 	Wood, sugar beet, potato tuber, hemp, flax	Mechanically or enzymatically	Diameter: 20–60 nm Length: Several micro lengths, longer Large aspect ratio
Nanocrystalline cellulose (NCC) 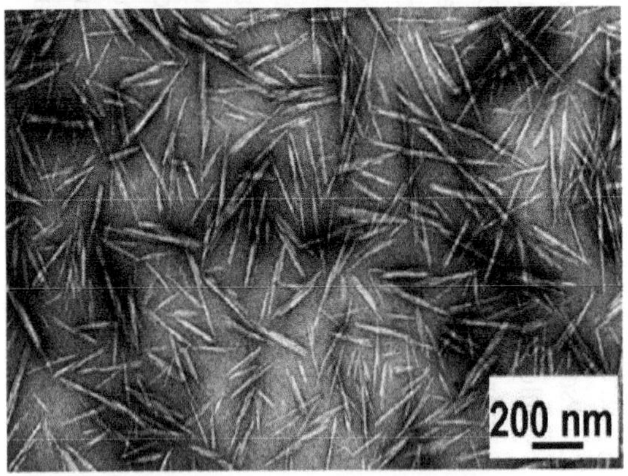 	Wood, cotton, hemp, flax, wheat straw, mulberry bark, ramie, avicel, tunicin, algae	Acid hydrolysis.	Diameter: 5–70 nm length: 100–250 nm Low aspect ratio:<100

Table 7.1: (continued)

Type of nanocellulose	Source of cellulose	Isolation method	Average size
Bacterial nanocellulose (BNC) 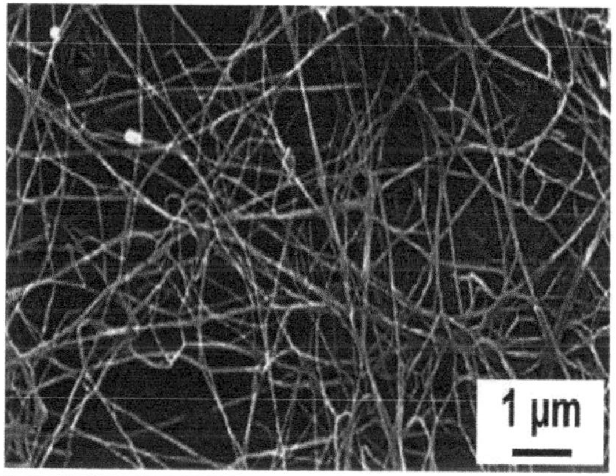 	Low-molecular weight sugars and alcohols, bacterial cellulose	Bacterial synthesis.	Diameter: 20–100 nm Length: 3D network

Table 7.2: The recent application of agriculture waste.

Type of agriculture waste	Application	Ref.
Rice straw	Paper making	[24]
	Fiber reinforcement composite	[25]
		[26]
		[27]
Coconut husk fibres	As reinforcing agents in nanocomposites,	[28]
	Improved compressive, fiber reinforced concreate,	[29]
	Cushioning material for transport packaging	[30]
	Road construction	[31]
Sugar palm	Bio-nanocomposite films	[32]
	Glass fibre reinforced thermoplastic polyurethane hybrid composites.	[33]
	High performance thermoplastic starch for packaging application	[34]
Sugarcane bagasse	Reinforced composite	[35]
	Cementitious composite	[36]
	Food packaging industry	[37]
	Absorbent for pollutant removal	[38]
Oil palm empty fruits bunch	Reinforced composite	[39]
	Papermaking fibre,	[40]
	Residual oil extraction,	
	Concrete and particleboard,	
	Fertilizer and planting media,	
	Alternative energy (charcoal and liquefied gas),	
Bamboo fibres	As a renewable textile fibre,	[41]
	Composite	[42]
Kenaf fibres	Composites,	[43]
	Cosmeceutical application	[44]
Corn starch	Bio-composite films,	[45]
	Thermoplastic starch	[46]

Based on the Table 7.2, it is obvious that nanocellulose mostly used as reinforced agent in composite and it is more preference to synthesis as NCC rather than NFC because it has high crystallinity index compared to NFC. NFC has web structure which induced to more complicated path, but NCC has more organized layered structure with fewer voids and high density which result in high intermolecular interaction between the hydrogen bonding between the nanocellulose chain and excellent mechanical properties thus it is suitable to be used as reinforce agent [47]. Furthermore, it has been reported that NCC films as a substrate exhibit greater efficiency compared to CNF, primarily because the NCC films are smoother, offer more uniform fibre distribution, and possess a lower degree of roughness [48].

Furthermore, plant-based biomass consists of cellulose, lignin, and a small percentage of extractives, as well as hemicellulose as shown in Figure 7.2. Lignin is linked to the hemicellulose polysaccharides inside the cell wall via carbon-carbon and ether linkages to form a 3-D network. It is the most difficult component in plant-based nanocellulose to breakdown, because it has complicated structure, large molecular weight, and it is insolubility. Moreover, cellulose is a type of carbohydrate which is inexpensive polymer contain D-glucopyranose units linked by β-1,4 glycoside bonds that form a linear structure with carbon (C), hydrogen (H), and oxygen (O). Each anhydroglucose unit of cellulose have three reactive–OH groups at carbon C2–OH, C3–OH, and C6–OH, making the nanocellulose reactive, high hydrophilicity, and negatively charged as illustrated in Figure 7.3. Moreover, dimers of glucose covalently bonded by glycosidic bond between the–OH group at C4 and C1 carbon atom of glucose unit [49].

The –OH group at C6–OH is approximately 10 times more reactive than the–OH group at C2 because it is attached to a carbon atom that is only bonded to one alkyl group, thereby reducing steric hindrance for each–OH group. In contrast, the–OH groups at C2 and C3 are relatively less reactive since they are attached to carbon atoms bonded to two alkyl groups [50]. These–OH groups on cellulose are capable of forming strong hydrogen

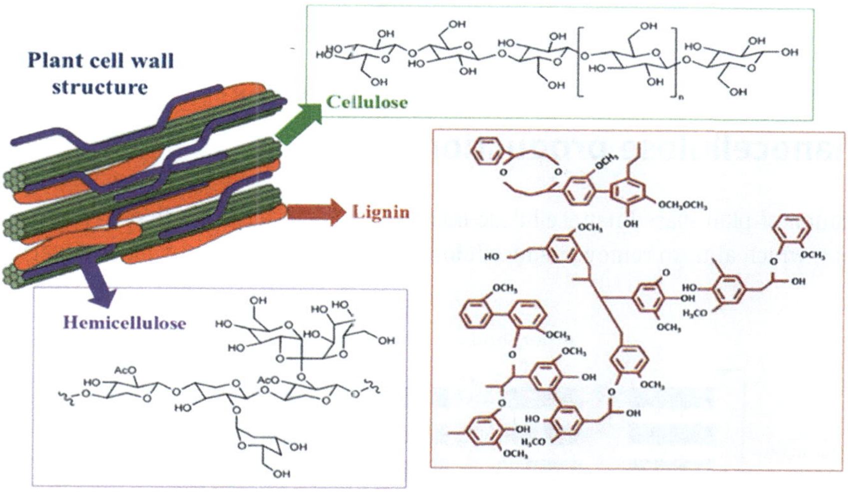

Figure 7.2: The components in plant-based biomass. Reproduced from ref. [7].

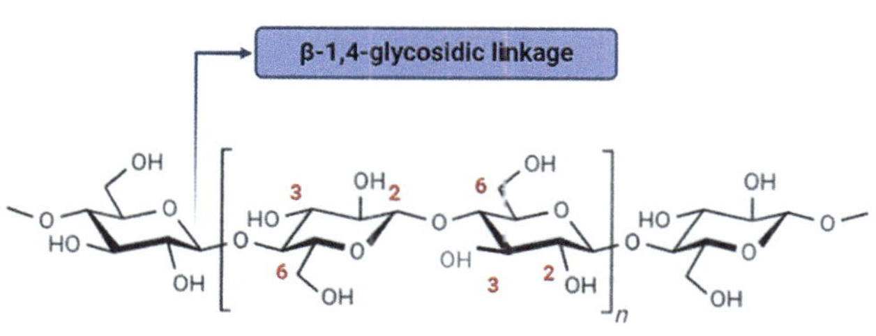

Figure 7.3: The 3 reactive–OH group in each anhydroglucose unit of nanocellulose.

bonds, rendering cellulose insoluble in both water and organic solvents [49]. Additionally, cellulose exhibits high hydrophilicity due to its abundant–OH groups, contributing to elevated thixotropic viscosities. This results in unique behaviours such as shear thinning and time-dependent increases in viscosity [51]. However, the nanoscale form of cellulose, known as nanocellulose, tends to self-agglomerate, which remains a challenge in its broader application [52]. To address this limitation, it is necessary to explore methods for rendering nanocellulose hydrophobic, thereby expanding its potential applications by modifying the–OH groups of nanocellulose.

The components of nanocellulose can further divided into two regions which is ordered and disorder region. The order region is known as crystalline region (cellulose) and the disorder structure known as amorphous region (lignin and hemicellulose) which has low density and less stability. As a result, during the pre-treatment by using acid amorphous region is easy degrade, while the order structure (cellulose) remains unchanged as demonstrated in Figure 7.4. Moreover, amorphous region is the cause to reduce the crystalline characteristic of the isolated cellulose. Thus, the removal of amorphous region is significant to get pure cellulose with superior mechanical properties [53].

7.3 Isolation of plant-based biomass for nanocellulose production

The isolation of plant-based nanocellulose involves two key steps. The first step is pretreatment, which aims to remove hemicellulose and lignin from the raw plant material

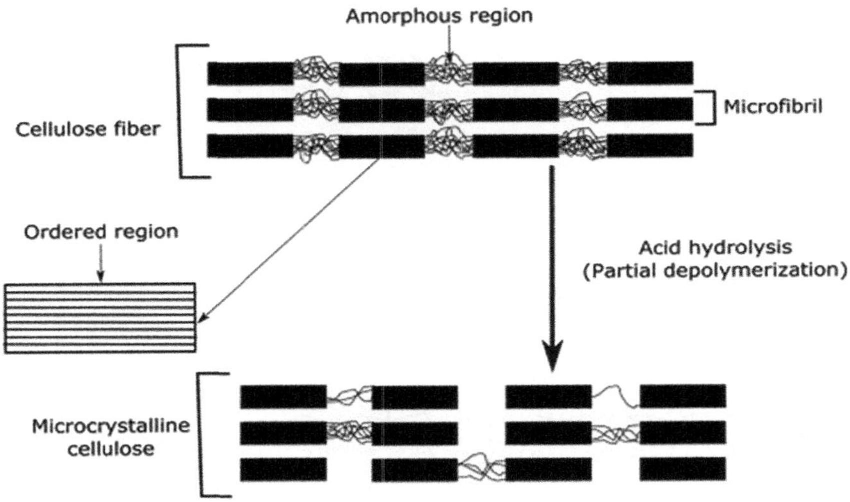

Figure 7.4: The impact of acid hydrolysis on the cellulose's amorphous region. Reproduced from ref. [54].

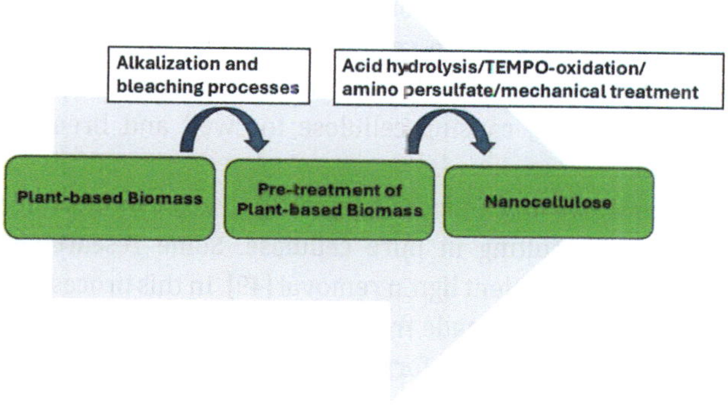

Figure 7.5: The flow chart of nanocellulose production.

while preserving the cellulose. The second step involves further decomposition to obtain nanostructured cellulose [55] as shown in Figure 7.5.

The efficiency of nanocellulose production relies heavily on the effective removal of non-cellulosic components such as hemicellulose and lignin during the pre-treatment process, as these components can significantly impact the properties of the final product. Prior to implementing the isolation method, it is crucial to assess the composition of wax, oil, lignin, or pectin in the source plant material. Different plant species exhibit variations in the proportions of cellulose, hemicellulose, and lignin. Additionally, the chain length and degree of polymerization play a pivotal role in influencing the properties of the resulting nanocellulose.

7.3.1 Pre-treatment of plant-based biomass

Pre-treatment techniques are crucial for breaking down bundles of fibres into individual fibres, aiming to enhance the characteristics of the final materials. This process involves disrupting the cell wall and modifying the interfibrillar cohesive forces [56]. The isolation of nanocellulose enhances accessibility by increasing surface area and porosity. Pre-treatment typically includes alkalization and bleaching processes to remove non-cellulosic components from the cellulose.

In alkaline pre-treatment, alkalis react with the ester bonds that link lignin to hemicellulose in lignocellulose, effectively removing lignin and a portion of hemicellulose that can negatively impact the thermal stability and crystallinity of nanocellulose [49]. This alkali-based delignification process, known as saponification, helps eliminate lignin, wax, and oil by using alkaline treatments, resulting in a rough surface topography. Alkalis like NaOH, potassium hydroxide (KOH), calcium hydroxide ($Ca(OH)_2$), ammonia (NH_3), and sodium carbonate (Na_2CO_3) are commonly used in alkaline pre-treatment. Ammonium hydroxide, such as NaOH, is often employed to remove lignin at elevated temperatures with mechanical stirring due to its non-corrosive, cost-effective, non-toxic, and easy-to-recover properties [57]. It is effective on various feedstocks and consumes less energy compared to other alkaline solutions [49]. The Na^+ cations from the NaOH diffuse through the intracrystalline spaces, causing cellulose to swell and breaking intermolecular hydrogen bonds.

Typically, pre-treatment combines alkalization and bleaching to achieve thorough removal of hemicellulose and lignin, resulting in pure cellulose. Some researchers employ sodium chlorite ($NaClO_2$) for more efficient lignin removal [49]. In this process, an acidic solution is prepared using a buffer, often made from NaOH and acetic acid. In the presence of the acetic buffer, $NaClO_2$ breaks down to form chlorine dioxide (ClO_2), which is released during bleaching. The use of bleaching reagents helps reduce halogenated residues and prevents the formation of chlorinated dioxins and furans [49]. The bleaching process is carried out multiple times to ensure complete removal of cementing components, significantly affecting the final properties of nanocellulose, resulting in improved thermal stability, morphology, and higher crystallinity. Successful bleaching is confirmed when the fibers appear white after washing for neutralization [49]. It is essential to note that the use of $NaClO_2$ in delignification should be conducted within a short timeframe due to potential health impacts and higher energy consumption.

7.3.2 Isolation of nanostructured cellulose

After alkalization followed by bleaching, the obtain cellulose will be further treated to obtain nanostructured cellulose. It can be conducted either by chemical or mechanical treatment separately or by the combination of above mention techniques to obtain the desired properties of nanocellulose. Acid hydrolysis, TEMPO-oxidation, amino persulfate and mechanical treatment or combination of these are the method than can be used to isolate the cellulose to nanocellulose [1]. Nanocellulose isolation via acid hydrolysis involve dilute and concentrated acids. The widely used acids by researchers are sulfuric acid (H_2SO_4), hypochloric acid (HClO), hydrochloric acid (HCl), phosphoric acid (H_3PO_4), formic acid (CH_2O_2), and maleic acid ($C_4H_4O_4$) [58]. Concentration of acid, cellulose to acid ratio, temperature, and reaction time are factors that determining morphology and reinforcing capacity of nanocellulose. However, the most widely used acid for acid hydrolysis is H_2SO_4 with the aid of hydrolysis of glycosidic bond by esterification of–OH group to form sulphate half-ester side group on the surface of the cellulose [1, 59]. Moreover, it is commonly used method for nanocellulose synthesis due to the stable suspension of nanocellulose which mainly caused by the electrostatic repulsion between the anionic charges of sulphate group generated on the surface of the nanocellulose [60]. Besides H_2SO_4, other acid such as HCI also can be used to hydrolyse the cellulose to nanocellulose but without the esterifying the surface of nanocellulose which does not generate anionic charges on its surface, thus it has a great potential to aggregate [61]. However, the drawback of nanocellulose synthesis using acid hydrolysis via H_2SO_4 is it can result in low thermal stability nanocellulose [62]. This drawback can overcome by neutralize the cellulose by using distilled water or by using NaOH [63]. Another author claim that, present of sulphate group is highly preferable for dispersing nanocellulose in polar aprotic solvent [64].

Next, the oxidation of cellulose is achieved through TEMPO (2,2,6,6-tetramethylpiperidinyl-1-oxyl)-mediated oxidation in the presence of NaBr and NaOCl. This method is considered a surface modification technique as it converts the hydroxymethyl group ($–CH_2–OH$ group) on cellulose into their carboxylic form [65]. The resulting anionic charges on the nanocellulose surface create strong electrostatic repulsion forces when dispersed in water. This oxidation reaction is environmentally friendly and straightforward to implement. There are two different conditions for TEMPO-mediated oxidation: in highly alkaline water, TEMPO/NaBr/NaClO is used, where TEMPO and NaBr dissolve first, and the oxidation commences with the addition of NaClO. Under neutral or weak acid conditions, TEMPO/NaClO/$NaClO_2$ is utilized to oxidize the nanocellulose [59]. Furthermore, ammonium persulfate (APS) has gained attention for nanocellulose extraction due to its favourable attributes, such as cost-effectiveness, high water solubility, and minimal long-term toxicity, compared to its potassium and sodium counterparts. APS isolation serves as a green and robust alternative to acid hydrolysis for producing nanocellulose with improved homogeneity. The APS method is versatile and can process various cellulosic fibres without requiring prior hemicellulose and lignin hydrolysis. The use of APS yields highly carboxylated nanocellulose, in contrast to sulfuric acid-based methods, which produce sulphated nanocellulose [66].

Additionally, mechanical methods exist for nanocellulose isolation, breaking down cellulose fibres into nano-sized units with diameters less than 100 nm or producing cellulosic nanofibers without the use of toxic organic solvents. Mechanical techniques encompass high-pressure homogenization, grinding, ball milling, extrusion, sonication, and the cryocrushing method. High-pressure homogenization allows for efficient nanocellulose isolation without the need for hazardous organic solvents [67]. The cryocrushing process involves treating fibres with liquid nitrogen to generate ice crystals that exert higher pressure within the cell wall, resulting in cell wall breakdown and the formation of microfibrillated cellulose (MFC). Mechanically obtained nanocellulose exhibits a gel-like characteristic in water due to a strongly entangled network with crystalline and amorphous domains [68].

Following nanocellulose isolation, the final stage involves the drying process. In larger-scale applications, drying methods include freeze-drying and spray-drying [51]. These techniques are widely used to manage particle size distribution, moisture content, and binding ability of nanocellulose. Effective implementation of freeze-drying, spray-drying, and supercritical drying helps prevent re-aggregation and aids in the redispersion of resulting nanocellulose in aqueous solutions. However, it's worth noting that while freeze drying is a quick and cost-effective approach, it often leads to highly aggregated masses of material that are challenging to disperse in aqueous fluids or polymer matrices. Supercritical drying can be costly to install and maintain, making it less common in practice [54].

7.4 Chemical approaches for surface functionalization of nanocellulose

The extracted nanocellulose contains a significant number of–OH group, rendering it highly hydrophilic and limiting its widespread application. This hydrophilic nature poses challenges in dispersing nanocellulose in hydrophobic matrices, such as low-polarity and non-polar chemical solvents, most polymer matrices, and MONPs in nanocomposite preparation [69]. Poor dispersion of nanocellulose in non-polar matrices hinders interfacial interactions, leading to aggregation and phase separation. Therefore, it is imperative to employ further surface modification of nanocellulose using physical and/or chemical techniques to expand its applications in various industrial sectors.

Numerous approaches can be adopted to alter the interfacial chemistry of nanocellulose, achieving a desirable hydrophilic-hydrophobic balance since achieving a homogeneous nanocellulose dispersion is essential for producing products with exceptional mechanical properties. Researchers have identified several strategies to enhance nanocellulose dispersion and promote interfacial contact between nanocellulose and MONPs. These approaches include the use of surfactants, physical methods, enzymatic processes, surface coatings, masterbatches, and mechanical treatments, among others, to address the aforementioned limitations [70]. Chemical functionalization is one of the most widely used methods for generating stable electrostatic charges, either positive or negative, to prevent nanocellulose aggregation. This approach is particularly effective because chemically modified nanocellulose exhibits enhanced stability, with chemical bonds that are significantly stronger than van der Waals forces or hydrogen bonds. For instance, carboxylated nanocellulose, which introduces carboxyl groups to the surface, can substantially affect material properties even in small amounts, and these adsorptions are often irreversible due to the strong interactions involved. In contrast, physical modifications tend to offer lower shear resistance, and substances absorbed through these methods are more likely to disperse when exposed to external forces [71].

Recent research has predominantly emphasized chemical functionalization as a means to modify the surface and structure of nanocellulose while preserving its bulk structure. The degree of substitution (DS) is a key parameter that gauges the effectiveness of such modifications, and it signifies the extent to which the–OH groups on anhydroglucose units have reacted. A maximum DS value of 3 is achieved when all–OH groups have undergone modification. It's worth noting that nanocellulose with a DS value close to 1 tends to disperse more effectively in polar solvents, while a DS value close to 3 results in improved dispersibility in hydrophobic, non-polar solvents [72]. Chemical functionalization not only allows for the transformation of nanocellulose properties from hydrophilic to hydrophobic but also facilitates the enhancement of dispersion and interfacial interactions through chemical bonding between nanocrystals and matrices [73]. Prior to incorporating nanocellulose into metal oxide nanoparticles (MONPs), surface modification is considered an effective method for controlling surface chemistry while preserving the core crystalline structure. The primary C6–OH group can be readily modified by interacting with functional groups such as sulfate groups ($-OSO_3^-$), carboxyl groups (-COO), aldehyde groups (-CHO), and amino groups ($-NH_2$) to tailor the material for specific applications. These incorporated compounds can enhance the absorption capacity of ionic species or MONPs by altering chemical and physical properties.

Functionalization serves two primary objectives: the introduction of anionic charges on the nanocellulose surface and the creation of a hydrophobic surface. These goals can be achieved through techniques such as phosphorylation, carboxymethylation, oxidation, sulfonation for anionic charges, and acetylation, silylation, urethanization and amidation for rendering the surface hydrophobic [74]. The subsequent section provides an overview of functionalization, delving into the underlying mechanisms and the distinctive attributes of each functionalization approach.

7.4.1 Phosphorylation

The addition of phosphate ester groups (PO_4^{-3}) on the surface of the nanocellulose by phosphorylation is an effective method for producing superior materials for improvements in a variety of industries, including biomedical sciences, fuel cells, wastewater treatment, biochemical segregations, and ion exchange efficiency [58]. Some of the compounds used to functionalize nanocellulose include $POCl_3$ (phosphorus oxychloride), P_2O_5 (phosphorus pentoxide), $(NH_4)_2HPO_4$ (diammonium hydrogen phosphate), H_3PO_4 (phosphoric acid) and organophosphates with either DMF, pyridine or urea, which is to make the cellulose fibre to swell [75]. But the main phosphate precursor is H_3PO_4. The mechanism of the phosphorylation is started with addition 85 % of H_3PO_4 in a system with nanocellulose at 373 K for 30 min as shown in Figure 7.6. The reaction can be performed in two manners which is by heterogeneously and homogeneously. Regardless of which technique is used, the reaction will result in different phosphate products [75]. There are two possible tautomeric product of phosphates cellulose structure as shown in Figure 7.6. The reaction can be generally expressed by Eq. (7.1) [75]:

$$\text{Cellulose}-OH + H_3PO_4 \rightarrow \text{Cel}-P(O)(OH)_2 + H_2O \tag{7.1}$$

It has been demonstrated that the oxygen content of nanocellulose increases after it has been functionalized. This suggests that the original–OH group in the cellulose has been replaced by a phosphate group, whose incorporation is directly correlated with pH value [75], without any effect on the crystallinity or degree of polymerization (DP) of the nanocellulose [77]. Nanocellulose materials are more flammable and have lower thermal stability, yet after modification to produce C=C stable bonds that promote flame

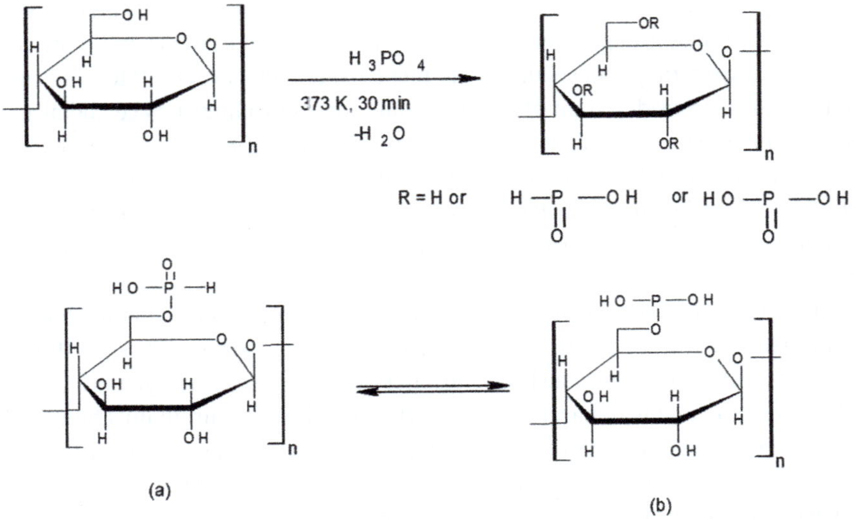

Figure 7.6: The two tautomeric product of phosphorylation (a) phosphate group and (b) phosphite. Reproduced from ref. [76].

retardation, sulphuric acid and phosphate esters catalyse the dehydration of nanocellulose as shown in Figure 7.7. Moreover, the factor that can affect the DS of phosphate ester group is depend on the type of phosphorylating agent, reaction time and also temperature.

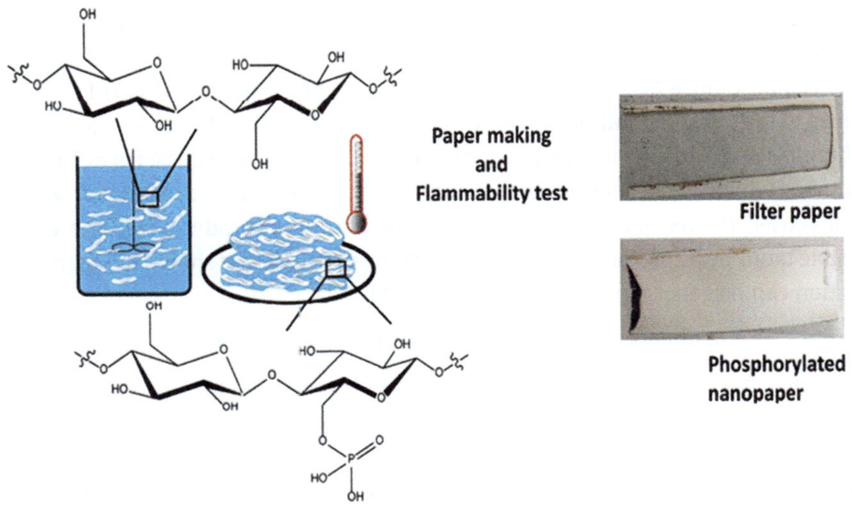

Figure 7.7: The application of the phosphorylated nanocellulose polymer for flame retardant filter paper. Reproduced from ref. [78].

The resulting functionalized phosphorylated nanocellulose polymer can be employed for large-scale applications and might be added to other polymeric chains to improve their properties. This phosphate nanocellulose is utilised in the textile industry as a flame retardant for cellulose-based materials and as a biomaterial for the treatment of diseases involving the transfer of calcium ions [76]. In a report stated that bamboo based phosphorylated cellulose with mechanical grinding show excellent mechanical properties and at the same time combination of phosphorus containing group with lignin endows show exceptional intrinsic flame retardant performance due to the formation of protective layer composed of PxOy compound and carbon layer provide effective heat and oxygen barrier effect [79]. Phosphorylation of cotton cellulose with baker's yeast (hexokinase) and phosphoryl donor adenosine-5′-triphosphate (ATP) result in improvement in flame resistance [80]. Moreover, phosphate esterification process begins after the water start evaporation [81] and filtration step should avoided in order to synthesis high charge content [77]. The modify nanocellulose with phosphate ester group is very capable to bind with metal ions such as Ag^+, $Fe_3{}^+$ and $Cu_2{}^+$, which have been used to remove metal ions for industrial water treatment. It proven that the modify nanocellulose significantly improved the sorption velocity and sorption capacity [82].

7.4.2 Carboxymethylation

Following that, surface functionalization of nanocellulose occurs by replacing the–OH group on surface of nanocellulose to carboxymethyl groups (–C H_2–COOH) via the carboxymethylation process by using chloroacetic acid, or its sodium salt reacts with alkaline cellulose which increases the negative charge on the surface of nanocellulose improve the dispersibility due to electrostatic repulsion. The mechanism of carboxymethyl involves alkalinization followed by etherification [76]. In the alkalinization stage, cellulose is mixed with ethanol and sodium hydroxide solution whereas in second stage, sodium chloroacetate added and mix at 353 K, as shown in Figure 7.8.

The benefit of carboxymethyl cellulose is carboxymethyl groups, which promote the electrostatic repulsion prevents aggregation and electrostatic attraction promote strong interfacial bonding [71], improve material decomposition into nanosized particles and undergo further reactivity for surface modification to tailor the properties of the nanomaterial for downstream applications [76]. Numerous industrial sectors

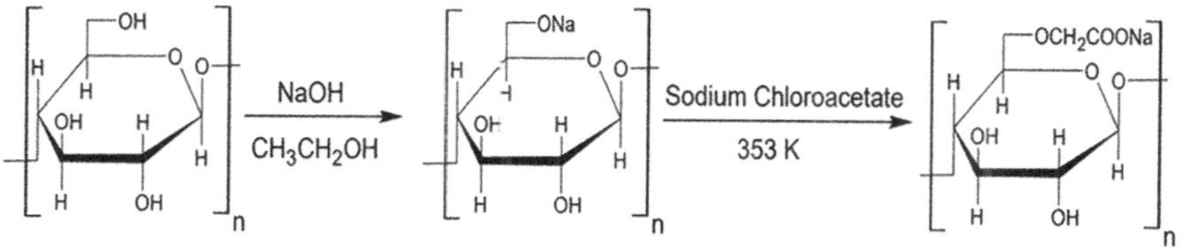

Figure 7.8: The mechanism of carboxymethylation process. Reproduced from ref. [76].

employ this carboxymethyl cellulose, including in food industry, as a transparent film and thickeners [83] or viscosity modifier for stabilising emulsions, and in petroleum industry as a component of drilling mud [76]. Additionally, it is an ingredient in many goods, including detergents, toothpaste, water-based paints, textile sizing, and several paper products [76, 84]. The electric charge on the surface of the cellulose is completely altered by this carboxymethylation, which transforms it into an ion exchanger. The addition of more carboxymethyl groups causes the zeta potential value to rise. The carboxymethylated nanocellulose shows a valuable role in removing harmful substances from aqueous phase, and this process is extremely affected by the interaction between the adsorbate and adsorbent and pH value. Moreover, the degree of substitution (DS) of carboxymethylation process can achieved more than 1 because the three–OH group on the nanocellulose can react with chloroacetic acid. As the temperature increased (60–80 °C), the DS rate also increased between 1.5 and 2 within 2 h [85]. In addition, another factor that can influence the DS of carboxymethylation process of nanocellulose is amount of NaOH and ethanol (H_2O). The DS reduced as the amount of NaOH increased and began to increase as the alcohol concentration increased [86]. Drawbacks include the use of toxic halocarbons and sometimes increasing the hydrophilicity of the resulting nanocellulose [74]. Resulting cation on the surface of nanocellulose increase resistance to fungus attack and thermal stability. Etherification has also been used in order to insert cationic charges on the nanocellulose surface. Moreover, due to no toxicity property, carboxymethyl nanocellulose widely used in food packaging as good film forming ability. It also has water solubility and reported that it can improve thermal properties of the film [71, 87].

7.4.3 Oxidation

Furthermore, there are several methods for producing carboxylated nanocellulose, including various oxidative hydrolysis processes or a method consisting of two stages of acidic hydrolysis followed by oxidation of cellulosic material as shown in Figure 7.9. However, the most popular way to synthesis carboxylated nanocellulose is 2,2,6,6-tetramethyl-1-piperidinyloxy (TEMPO) mediated oxidation reaction through oxidation of–OH group at C6 nanocellulose which been adopted as a post-production strategy produced from the acid hydrolysis of cellulosic sources [1].

The mechanism of the TEMPO oxidation is started with active sites of primary hydroxyl group at (C6–OH) on surface of nanocellulose to aldehyde group by the TEMPO-hydrolysed free radicals, which are further oxidized carboxylic acid –COOH group, while the C2 and C3 –OH alcoholic–OH group maintain by (TEMPO)-mediated oxidation reaction [88] without changing the original crystallinity [89]. In order to convert the (–C H_2–OH) group on nanocellulose to –COOH group and also to increase the rate of reaction, a simple oxidation pathway with two reagents which is sodium bromide (NaBr) and sodium hypochlorite (NaOCl) known as nitroxyl radicals to specifically

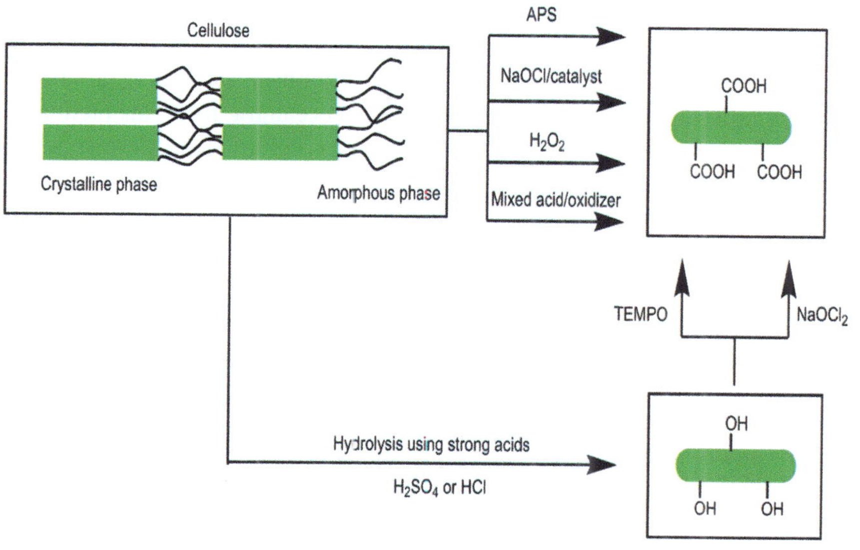

Figure 7.9: The general ways to synthesis carboxylated nanocellulose from cellulose. Reproduced from ref. [1].

oxidize the main ($-CH_2-OH$) group while maintaining the integrity of the secondary (-OH) groups as revealed in Figure 7.10. Under aqueous and mild conditions, the utilisation of TEMPO radicals as an oxidative pre-treatment prior to mechanical treatment is a

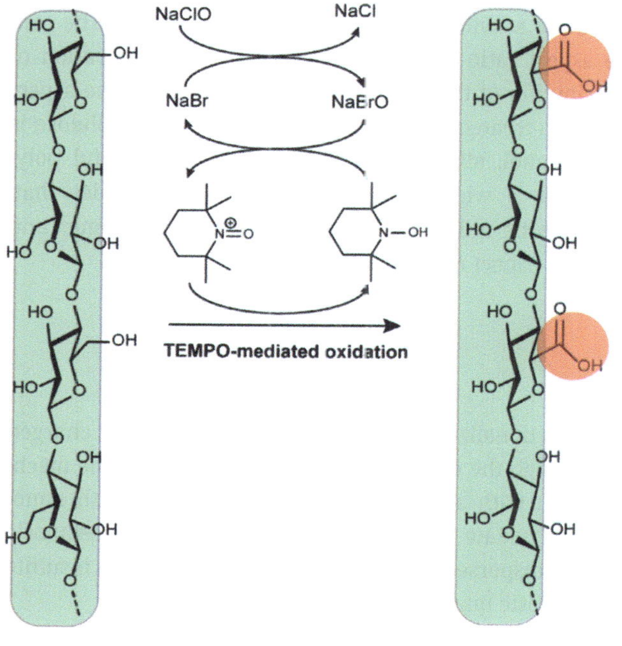

Figure 7.10: The TEMPO-oxidation of cellulose. Reproduced from ref. [72].

promising surface modification approach for introducing carboxylate and aldehyde functional groups into solid native celluloses. TEMPO-mediated oxidation is employed in mechanical processing to transform cellulose into nanofibers, facilitating a one-pot manufacturing procedure for the reengineering of nanocellulose [88]. This method is considered both safe and efficient, as it accelerates the defibrillation process while conserving energy through reduced processing cycles.

Introduction of carboxylate groups led to better dispersion in aqueous phase, and it help to produce multifunctional hybrid nanomaterials coupled with MONPs [55]. The nanocellulose created by the TEMPO-oxidation process has a negative charge density that can reach up to 1,500 μmol/g. thus have ability to disperse in water [68, 90]. Moreover, nanocellulose prepared by TEMPO-sonomechanical treatments synthesis more uniform size distribution and have higher anionic charge compared to nanocellulose prepared by acid hydrolysis [1]. However, TEMPO-oxidation method has a limitation which is inability to dissolve the amorphous domains of cellulose directly from nanocellulose, thus as an alternative to TEMPO-oxidation, ammonium persulfate (APS) method was created, which involved a one-step oxidation process to destroy the amorphous part and give the anionic carboxylates group by replacing primary–OH group of cellulose. This procedure is less hazardous and less expensive. Moreover, APS approach produced more crystallinity and enhanced thermal stability than the TEMPO-oxidation procedure [91]. But APS method's disadvantage is the substantial APS and long reaction time are required [92]. Next, carboxylated nanocellulose can be synthesis via full consumption of hydrogen peroxide (H_2O_2) which is the simplest and greenest oxidizing agent [1]. Besides that, transition metal-catalysed oxidation process which involved three steps using sodium hypochlorite (NaOCl)/catalyst and mixed acid solution with oxidizers are also available method to synthesis carboxylated nanocellulose at where one-pot procedure is created. The concentration of NaOCl influence aspect ratio and distribution. As the concentration of NaClO escalated, the width distribution of the nanoparticles narrowed and the aspect ratio increased [49]. The NaOH/urea-treated, TEMPO/NaBr/NaClO oxidized cellulose is water soluble but insoluble in ethanol, allowing for the production of useful polyelectrolytes and intermediates with a wide range of functions [93]. Besides that, carboxylated nanocellulose also can be utilized as reinforcing agent in polymer and nanoparticles anchor due to its high aspect ratio [94].

7.4.4 Sulfonation

Sulfonation approach is used to functionalize nanocellulose imparting anionic charged onto the nanocellulose surface through the esterification of the–OH groups. The mechanism of cellulose functionalization with sulfur incorporation is started with nanocellulose oxidation with sodium periodate ($NaIO_4$) result in formation of cellulosic material 2,3-dialdehyde. Then, it will dispersed in water and treated with sodium bisulfite ($NaHSO_3$), creating sulfonated cellulosic materials as shown in Figure 7.11.

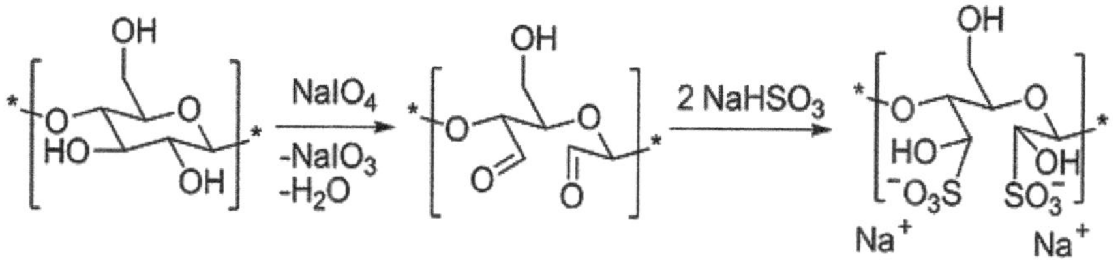

Figure 7.11: The mechanism of sulfonation of cellulosic materials. Reproduced from ref. [96].

These sulphate esters generate highly stable colloidal suspensions at the surface of the nanocellulose which after carefully monitored the drying process and improve its capabilities to well disperse in water [74]. Moreover, the resulting anionic chargers on the nanocellulose show better thermal stability compared to hydrolysed cellulose with sulfuric acid. It is green method as it not required halogenated waste [96]. The authors studied the effects of varying the sulfuric acid concentration (55–62 wt%), treatment temperature (35–65 °C), and reaction time (20–60 min) and discovered that using 55 wt% sulfuric acid at 45 °C for 40 min could result in a high yield of sulfonation (91.6 %) with a sulphur content of 3.8 wt%. The nanocellulose created under these condition had an 81.3 % crystallinity and showed superior thermal characteristics compared to the conventional techniques [72]. This type of functionalization of nanocellulose help to improve the hydrophilicity of cellulose surface and most commonly employed in the removal contaminants from aqueous medium [76]. Moreover, it is suggested that oxidation step is important after sulfonation to produce nanocellulose with high dispersibility in aqueous medium and also the presence of the sulfonic group proven that can provide an increase value of water retention in the cellulosic material structure. It was discovered that using this oxidation/sulfonation process, cellulosic material's water absorption may be increased by 8.0–199.0 % [97].

7.4.5 Acetylation

Acetylation of nanocellulose process is simple and non-toxic, at where acetyl group (-COC H_3) will be introduced on the surface of the nanocellulose by series of condensation reactions between (-OH) groups on nanocellulose and ester functional groups (-COO). Typically, it involves the addition of $C_4H_6O_3$ (acetic anhydride) as an acyl donor which is a common reagent to functionalize cellulose mainly for surface hydrophobization such as dispersive/re-dispersive behaviour improvement and properties enhancement [52]. Besides acetic anhydride, dry CH_3COOH (acetic acid) with either H_2SO_4 or $HClO_4$ (perchloric acid) also used as reaction catalyst. Organic solvent is widely used to enhance the DS during the acetylation of nanocellulose [88]. "One pot reaction" technique used to simplify the challenging and complicated processes of surface functionalization by acetylation as shown in Figure 7.12 [98]. It is successfully introducing acetyl functional group using a combination of hydrochloric acid (HCl) and acetic acid with Fischer esterification technique and produce cellulose nano-whiskers that can disperse well in chemical solvent as opposed to water in this manner. Recently, it is discovered that, Fischer esterification in conjunction with acid hydrolysis has been reported as an effective way for nanocellulose isolation for plant-based cellulose and for the surface functionalization which in one route process [52].

The mechanism of acetylation can be done via two different ways, including the heterogeneous process and the homogeneous process with the presence of acetic acid and sulfuric acid as catalyst. The heterogeneous route is used to produce cellulose esters but, homogeneous acetylation use to create more advantage in term of reproducibility, excellent end product recovery and better control over DS [52]. The ratio of acetic anhydride and acetic acid is very crucial because it may affect the DS value. Moreover, when the reaction time increase the acetylation process also increased, however, it is reported that beyond the 8 h the degree of acetylation decrease drastically [99]. Results revealed that acetylation of sulfuric-acid-hydrolysed nanocellulose might produce higher efficiency and DS values (2.1) than hydrochloric acid hydrolysis (1.8). This is because the hydrolysis treatment with the sulfuric acid result in presence of many sulphate ester groups on the surface of sulfuric-acid-hydrolysed nanocellulose which imparted electrostatic repulsion between neighbouring of nanocellulose and this process increased the number of (-OH) groups available for the acetylation reaction [72]. Another

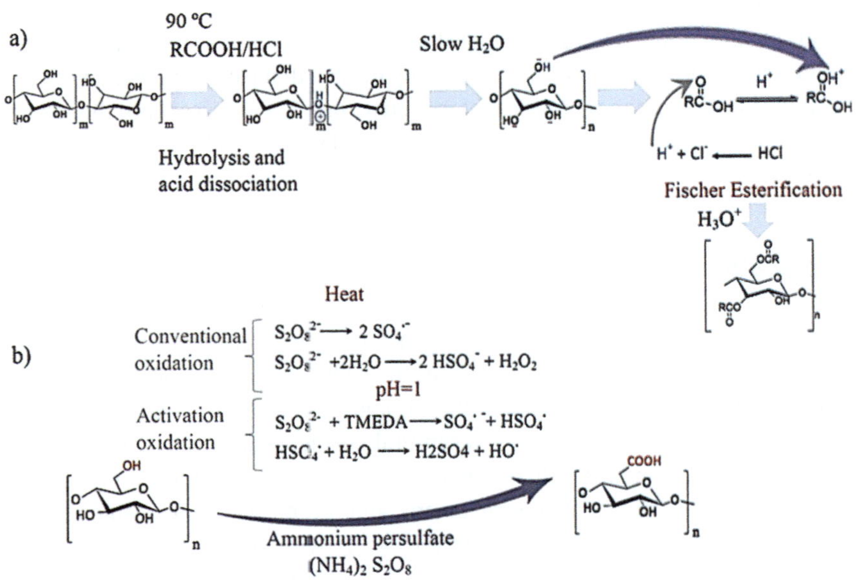

Figure 7.12: The mechanism of a) cellulose hydrolysis and Fischer esterification of hydroxyl group using mixture of acetic and hydrochloric acid b) carboxylated cellulose with ammonium persulfate (APS) hydrolysis. Reproduced from ref. [98].

factor that can affect the acetylation reaction is the ultrasonic treatment. A simple mixture with acetic and sulfuric acid without the aid of ultrasonic had DS value of 0.22, however with the aid of the ultrasonic treatment the DS value was 0.46. Acetylated cellulose with greater DS was shown to be more hydrophobic and thermally resistant. In addition, acetylated nanocellulose with greater DS was discovered to have good filler-matrix interactions with hydrophobic matrices, resulting in more mechanical reinforcement in composites manufactured with it [52]. Acetylated nanocellulose able to disperse in organic solvent such as ethyl acetate and toluene and more compatible with hydrophobic polymer matrices when act as reinforcing nanofiller [72]. The shorter the reaction time, the higher the DS value of esterification of nanocellulose [100]. Acetylated nanocellulose configuration has been demonstrated to be more porous and capable of retaining its nanoscale dimension [52].

7.4.6 Amidation

Amidation of nanocellulose is most common and straightforward method. This method targets the carboxylic groups (-COOH) of preoxidized nanocellulose to produce a stable amide bond. EDAC ($CH_{17}N_3$) or N-ethyl-N-(3-(dimethylamino)-propyl)carbodiimide hydrochloride is commonly used carbodiimide for nanocellulose amidation reaction. The bifunctional amines of small alkyl chain length by EDC/N-hydroxysuccimide (NHS)-mediated coupling furnished the desired aminated nanocellulose via amide linkage [101]. First of all, the carboxylate groups will be first introduced onto the surface of nanocellulose via the previous discuss functionalization which is TEMPO-mediated oxidation. Then the reaction of surface-carboxylated nanocellulose will mixed with 2-(N-morpholino)-ethanesulfonic acid buffer (20 mL, pH = 5) and the resulting suspension was sonicated for 20 min EDC-HCl and NHS added to the suspension and the mixture was stirred at room temperature for 15 min. After this, there are two options which either add 1,4-diaminobutane to produce Amine-1 or 1,2-diaminoethane to produce amine-2 stirred for 24 h at room temperature as shown in Figure 7.13. The mixture obtain will be first dialyzed against saturated sodium chloride solution for 1 day and then against distilled water until no more free diamines were detected in the dialysate [102]. Purified aminated nanocellulose samples were obtained by freeze-drying. Both simple steps involved (oxidation and amidation steps) is very simple with increasing zeta potential value but there is a little agglomeration detected upon amine decorated nanocellulose [102].

7.4.7 Silylation

Silylation is a process by which silyl groups, R_3 Si, are inserted onto the surface of nanocellulose using a substitution technique to improve compatibility with on polar matrices [52]. A study on effective of variable aminoalkyl chain, including 3-aminopropyl

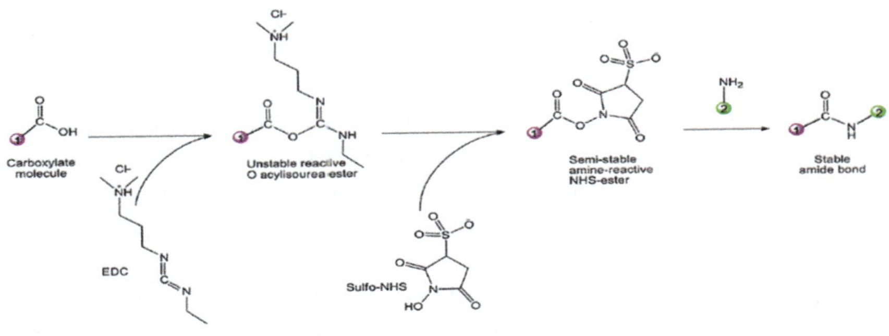

Figure 7.13: The surface functionalization of nanocellulose to synthesis aminated nanocellulose. Reproduced from ref. [104].

trimethoxysilane (APMS), 2-aminoethyl 3-aminopropyl trimethoxysilane (DAMS) and 3- 2-(2-aminoethylamino) ethylamino propyl-trimethoxysilane (TAMS) on chemical grafting on cellulose nanofibre to observe their antimicrobial activity as shown in Figure 7.14. It is proven that immobilized of TAMS (3 amines)-treated nanofibre was the best candidate which show highest stability and available for grafting in the form of silanol group for longer period of time and to be utilized as a filler material in food packaging product because it has excellent resistance toward Gram positive bacteria of *Bacillus Subtilis* and *Staphylococcus aureus* [104]. Silanes have a high sensitivity for (-OH) groups even at low temperatures, which makes them identified as efficient coupling agents. Example of coupling agent are alkoxy silane and trimethylchlorosilane (TMCS) which are most effective coupling agent in transforming interface of nature fibre-matrix [105]. Only after the curing process, permanent surface grafting through condensation of silanol (Si–OH) with C–OH will occur. A good leaving group leads the mechanisms of silyation reaction, which means a leaving group with low basicity, the capacity to stabilise negative charges in the transitional state, and minimal or no back bonding between the leaving group and silicon atom. The mechanism involves the substitution of a trimethylsilyl group for the active hydrogens in–OH, –COOH, –NH, –NH_2, and–SH groups. When there is a good leaving group, it will improve the silylation process via nucleophilic attack SN2. In the intermediate phase of the reaction mechanism, this leads in the formation of a bimolecular transition state.

Surface functionalization of the nanocellulose material using the silylation approach is an easy way to increase the hydrophobicity of the hydrophilic cellulosic surface. The silylation process produces hydrophilic fibres and less harmful to the environment than esterification since it may be carried out using a solid gas reaction without the use of solvent alkyl silyl ethers. Not only that, recently researchers are also worked on elucidating the interfacial adhesion of silane-modified CNF film coating onto glass and aluminium [52]. Moreover, researchers have proven that silane modified nanocellulose is capable to be used as adhesive with improved qualities [106]. It is stated that, many

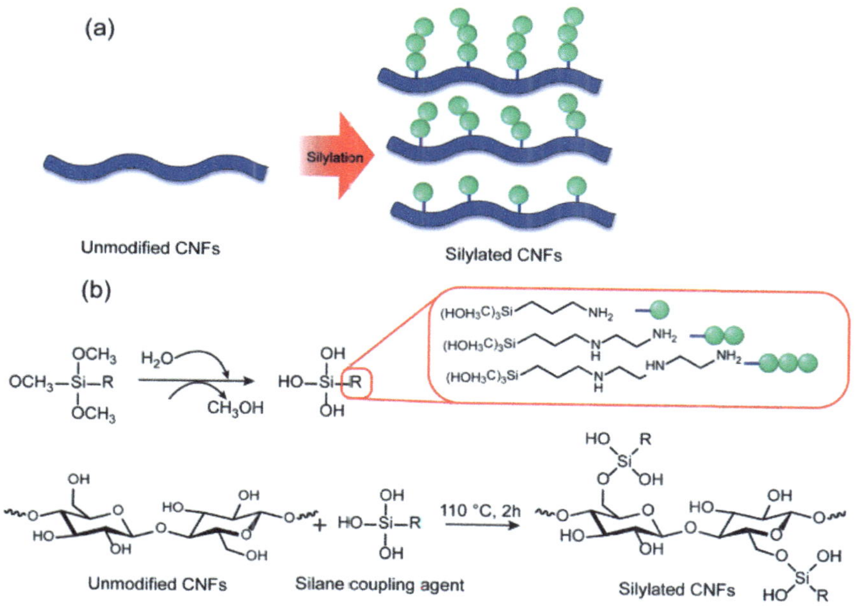

Figure 7.14: The a) chemical synthetic route b) acetylation of nanocellulose with three different different aminosilanes. Reproduced from ref. [72].

hydroxyl and amino compounds regarded as non-volatile or unstable at 200–300 °C. However, is proven that after the silylation process, silylated derivatives are more volatile, more stable and thus yielding narrow and symmetrical peaks according to gas chromatography analysis (GC) [107]. Moreover, it also improve interfacial stability [13] improves physicochemical properties of composite materials and provide additional reinforcement properties.

7.4.8 Carbamation/Urethanization

Urethanization method involves the reaction of isocyanate (R–N=C=O) and oxygen from–OH groups of nanocellulose, which results in the creation of urethane linkage as shown in Figure 7.15. The isocyanate functional group (-NCO) reacts strongly with other functional groups such as–N H_2, –OH, and–COOH. Urethanization can be carried out either with absence of a catalyst with an excess of n-octadecyl isocyanate or with present of catalyst such as n-butyltindilaurate [88]. The surface modified nanocellulose via urethanization result in high hydrophobicity [88].

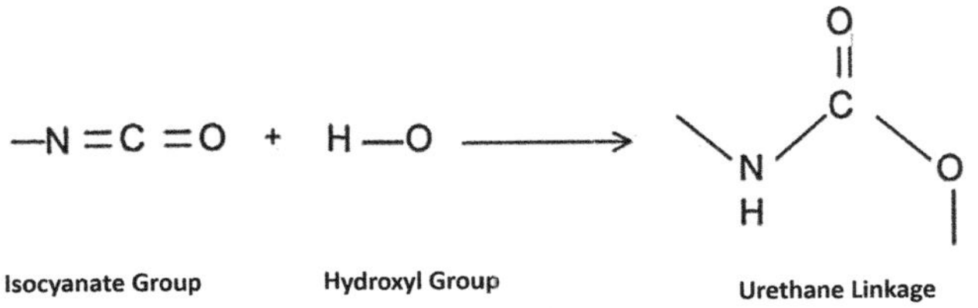

Figure 7.15: The reaction isocyanate and hydroxyl groups of nanocellulose to form urethane linkage. Reproduced from ref. [108].

7.5 Hybridization of nanocellulose with metal oxide nanoparticles

It is crucial that MONPs disperse efficiently within nanocellulose without forming aggregates that might impact the final properties. Additionally, they should exhibit a highly narrow size distribution to enable the synthesis of nanocomposites with exceptional properties. Finding effective methods for large-scale production while maintaining MONP dispersion in the nanocellulose matrix is of paramount importance. MONPs can be synthesized separately and then added to nanocellulose (*ex situ* processes *ex situ*), or they can be synthesized using nanocellulose as a template (*in-situ* processes *in-situ*). There are several approaches to hybridize nanocellulose with MONPs. Figure 7.16 illustrates four methods for producing hybrid nanocellulose/MONPs nanocomposites.

The initial approach is referred to as simple blending, which is notably straightforward, involving the immersion or mixing of pre-prepared MONPs with cellulose materials. This method is known as the *ex situ* synthesis of nanoparticles, followed by their integration with nanocellulose. While simple blending is uncomplicated, its drawback lies in the potential for non-uniform distribution of MONPs within the cellulose matrix during the subsequent drying process, potentially resulting in aggregate formation. For example, to synthesize ZnO particles in the size range of 20–40 nm, various methods such as electro-spinning, sol-gel, ultrasonication, precipitation, and spray pyrolysis are employed. In contrast, composite preparations, such as hydrothermal and microwave methods, are conducted to provide larger ZnO-NPs. When creating composite materials of nanocellulose and MONPs, chemical approaches are typically favoured over physical methods. This preference arises from the utilization of uncomplicated equipment, more gentle reaction conditions, and easier scalability [109]. The synthesis of MONPs is carried out using cellulose materials and heat technology, without the addition of external reducing or capping agents, relying mostly on the abundance of–OH groups on the cellulose surface. The–OH groups serve as templates for metal precursors through ion-dipole interactions, and the formed MONPs are stabilized to achieve well-dispersed MONPs on the nanocellulose surface. A variety of techniques, including the hydrothermal process, UV radiation, and microwave radiation, activate the surface–OH groups to

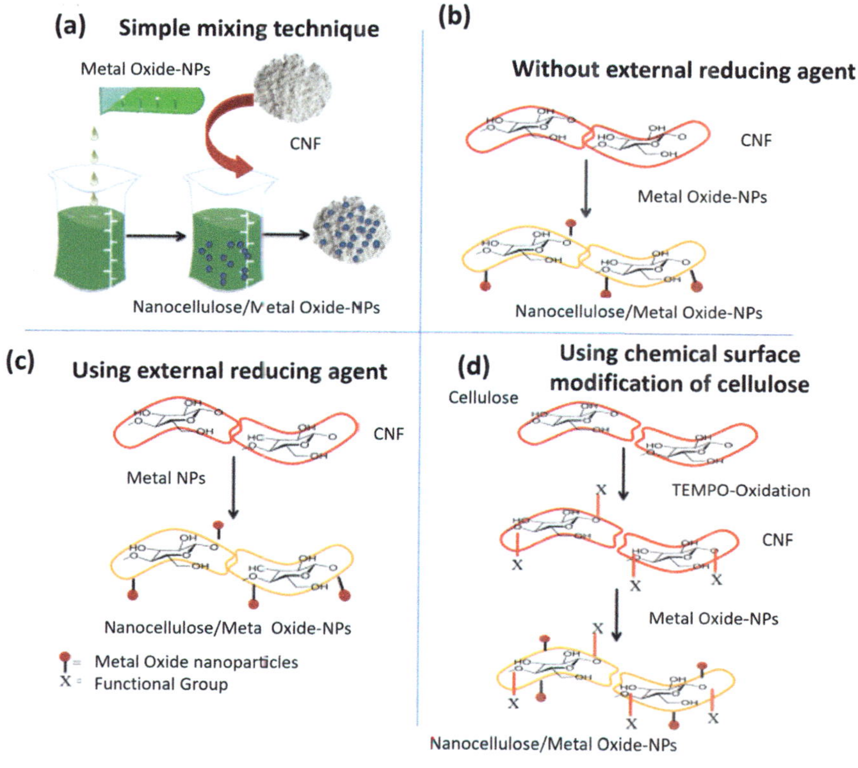

Figure 7.16: The 4 step to hybridize the pre-treated nanocellulose with MONPs (a) by simple blending of cellulose and MONPs, (b) synthesis MONPs without any external reducing or capping agents (c) by using external reducing agent (d) via surface modification of nanocellulose. Reproduced from ref. [109].

significantly reduce and bind metal precursors (metal ions) to the cellulose material's surface. Recently, dissolved cellulose with an ionic liquid (NaOH-urea) has been used as a green technique to create cellulose/inorganic nanohybrids, such as cellulose/ZnO-NPs and cellulose/CuO-NPs. This approach has been employed to produce functional textiles with antibacterial properties.

The next method for hybridizing nanocellulose with MONPs involves the addition of an external reductant, such as sodium borohydride, ascorbic acid, sodium citrate, triethanolamine, or plant extracts, to the nanocomposite. This external reductant is used to reduce metal ions to metallic nanoparticles. In this case, the cellulose material acts as an additional stabilizer to prevent agglomeration of inorganic nanoparticles (MONPs). Cellulose fibers, with their distinct structure and the presence of ether and (-OH) groups, serve as efficient nano-reactors for the *in-situ* synthesis of nanocellulose. The ether and hydroxyl functionalities stabilize the as-prepared nanocellulose through surface interactions following reduction and tightly anchor the MONPs onto the fibers through

Table 7.3: The type of surface modification with specific purpose.

Purpose of the functionalization	Type of process	Functional group	Reagent
To create anionic charges	Carboxymethylation	Carboxymethyl groups ($-CH_2-COOH$)	Chloroacetic acid, or its sodium salt reacts with alkaline cellulose.
	TEMPO-oxidation	Carboxylic acid group ($-COOH^{-}$)	2,2,6,6-Tetramethylpiperidine-1-oxyl (TEMPO) mediated oxidation
	Sulfonation	Sulphate groups ($-SO_4^{-2}$)	Concentrated sulfuric acid (H_2SO_4)
	Phosphorylation	Phosphate ester group (PO_4^{-3})	Phosphorus salt/Phosphoric acid (H_3PO_4),
To give hydrophobicity	Esterification/ Acetylation	Acetyl group ($-COCH_3$)	Addition of acetic anhydride and dry acetic acid with either sulfuric or perchloric acid catalyst
	Silylation	Silyl groups (R_3Si)	Alkyldimethylchlorosilanes alkoxy silane, n-octadecyl
	Urethanization	Urethane linkage	Isocyanate ($C_{19}H_{37}NO$), isocynaic acids
	Amidation	Amide bond	Carbodiimid

ion-dipole interactions. The macromolecular chains within the host material play a significant role as templates in the creation of nanocellulose, improving their dispersion within the cellulose matrix and preventing aggregation formation. Moreover, the polymer chains contribute significantly to providing the MONPs in nanocellulose with a narrow size distribution and well-defined shape. Finally, the last method for modifying the surface of cellulose material involves attaching functional groups capable of reducing and stabilizing the metal precursor to MONPs. This surface modification, also known as the functionalization of nanocellulose, as discussed in a previous section, can be performed during the preparation step or after nanocellulose production. Surface functionalization slightly alters the surface of the nanocellulose material during pretreatment with sulfuric acid (H_2SO_4) and phosphoric acid (H_3PO_4), providing charges for the aqueous dispersion and isolation processes. Table 7.3 shows the list of functionalization and its purpose.

7.5.1 Influences of metal oxide nanoparticle size on their properties

The uniform dispersion of MONPs is a critical feature to consider for certain applications such as electrical, optical, and magnetic characteristics. Commercial usage of MONPs as sensors, ceramics, absorbents, and catalysts is mostly related to their physicochemical properties which influenced by the size of the MONPs [110]. MONPs is purely synthesis from the metal's precursors and performs a fascinating function in a wide variety of sectors including physics, chemistry, and materials sciences due to their nano size features. Chemical methods such as chemical precipitation, hydrothermal synthesis, solvothermal synthesis, sol-gel, thermal decomposition, microwave radiation and including recent popular method which is green synthesis gained prominence for the synthesis of MONPs because of their mono dispersive and non-agglomerative properties, which could be achieved by stabilising MONPs. The important of MONPs astonishing when researchers found that MONPs majority of their physiochemical properties mainly influence by the size [111, 112]. MONPs which are nano size will have high surface area to volume ratio. As a result, MONPs will interact with the environment more efficiently than bulk particles, resulting in an increase in surface energy and consequently, result in poor stability [8]. The interaction of a surface with its surroundings is defined by surface energy. MONPs which has high surface energy, thus it exhibits high reactivity which also lead to poor stability and can easily oxidized with environment exposure [113].

Furthermore, the quantum effects which only influence by the size of the MONPs at nanoscale are responsible for distinct features of MONPs. The amount of surface atoms in bulk materials is negligible, and their properties are completely decided by the interior atoms. However, in the case of MONPs, surface atom contribution increases with decreasing particle size, implying that surface atoms have a strong influence on MONP properties. The smaller the size of the nanostructure, the greater the difference between energy levels, hence resulting in a spectrum of discrete energies. Next, the effect of nano size MONPs can relate to electrical properties. The electrical properties are usually associated with charge carrier mobility. When a material's dimensions reduce to the nanoscale range, the quantum-size effect and quantum confinement effect are unavoidable [113]. Metal such as coper (Cu), aluminium (Al) or gold (Ag) are good conductor because it has 'loose' valence electrons bound to the atom therefore they can be available to conduct electric, whereas materials such as glass, rubber or ceramic are insulator because they prevent electrical conduction due to their high resistance [114].

In this case, oxide is insulator and then MONPs have the properties of semiconductor because it possesses electrical charge transport capabilities that distinguish them from ordinary covalent semiconductors and the non-stoichiometric oxides having a cation metallic element such as Zn or Cu that is tightly bonded by an ionic connection with negative (anionic) oxygen ions. This class of materials has a high level of conductivity sensitivity in conjunction with better thermal and chemical stability [114]. MONPs' electrical conductivity is lower than that of bulk materials, because the band gap energy increases as the nanomaterials' particle sizes decrease as illustrated in Figure 7.17. MONPs have different band gap such as ZnO (3.2eV), CuO (1.7eV), TiO_2 (3.2 eV), and Fe_2O_3 (2.14 eV) which will influence their optical and electronic properties for high temperature and electronic applications [115]. The two types of semiconductors include intrinsic and extrinsic. Whereas the intrinsic semiconductor is a pure semiconductor, the extrinsic semiconductor is a semiconductor that has been doped with impurities to modify its electrical properties. Extrinsic semiconductors are further divided into two type which is n-type semiconductor and p-type semiconductor.

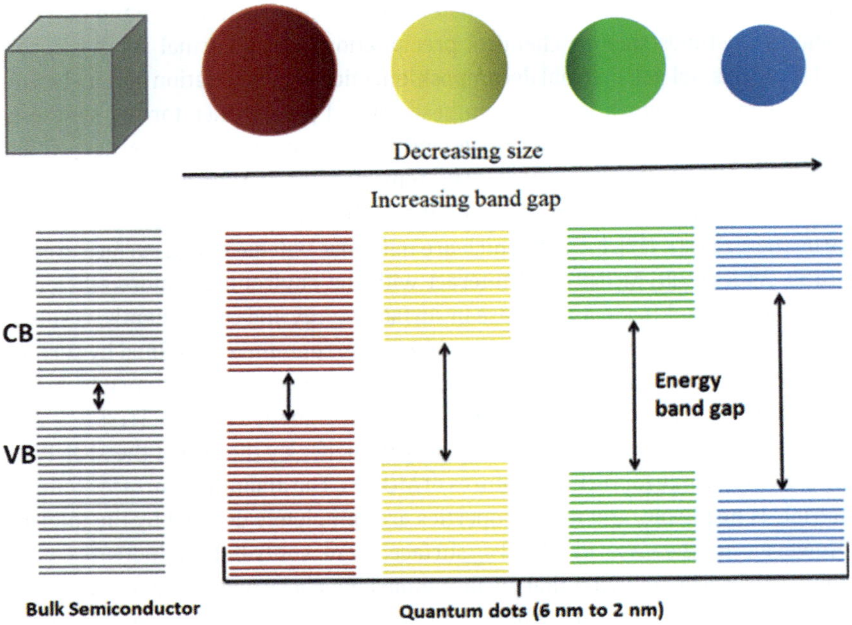

Figure 7.17: The effect of size on the energy band gap. Reproduce from ref. [117].

In addition, n-type semiconductor which doped with pentavalent element provides extra electron (doner atom) such as phosphorus (P), arsenic (As) or antimony (Sb) used to increase the electron for conduction. Example of n-type semiconductor are silicon (Si) and germanium (Ge) which is belong to 4th group in the periodic table, and the doping is from the 5th group of period table ensuring a close size between the atoms [114]. The '*n*' word in the n-type semiconductor stands for the negative charge [114]. For instant, when Si atoms doped with pentavalent atom, the 4 electrons of pentavalent atom will bond with 4 Si atom while the 5th electron will remain loosely and denoted as negative charge on the loose electron due to doping. This loose electron required very small ionization energy to be free and move at room temperature and it consider as majority current carriers. In this n-type semiconductor the majority charge carrier are electron and minority charge carriers are holes [117]. The electron density is higher than hole density. ZnO, indium gallium zinc oxide (IGZO), and tin dioxide (SnO_2) are of most popular n-type oxide semiconductors [114].

On the contrary, p-type semiconductor doped with trivalent element which provide extra holes (acceptor atom) such as boron or indium [118]. It denoted as positive charge due to addition of impurity atoms with trivalent to silicon. In the p-type semiconductor the majority charge carrier are holes whereas minority charge carriers are electron. The hole density is greater than electron density. For example, tin monoxide (SnO) and CuO are examples of common p-type oxide semiconductors. Due to their distinctive characteristics and many applications, semiconductors, in particular n-type metal oxide semiconductors, are considerable interest. Their significance can be observed in nearly every aspect of life, including the development of electrical devices, environmental applications, solar and hydrogen cells, as well as sensor technology. MONPs could be used as doping materials to improve the sensitivity of sensors, as well as the photocatalytic activity of semiconductors in fuel cells, solar cells, and environmental applications.

Furthermore, MONPs such as ZnO-NPs is a inorganic n-type semiconductor compound, have remarkable photocatalytic activity in eliminating water contaminants due to their great chemical stability [119]. Its low-cost production, abundance, non-toxic nature, UV-blocking properties, and potent antibacterial effects are attributed to a substantial micro-quantum effect [120]. It also been recognised by the Food and Drug Administration (FDA) as a novel and effective antitumor treatment due to its intrinsic anticancer potential [9]. Apart from that, CuO-NPs is a p-type semiconductor compound offer excellent chemical and physical properties such as s semiconductor nature [121], solar energy efficiency, stability, low cost, and antibacterial activity [122]. It also can be used as catalyst in batteries due to its high electrochemical capabilities. Another widely used MONPs is AgO which has excellent antibacterials activity. Furthermore, AgO-NPs usually used in appliances such as refrigerators and washing machines, as well as sensor applications [8]. Furthermore, TiO_2-NPs high photocatalytic activity, and chemical stability [119] due to its potential to stimulate cell migration, adhesion, osseointegration, and wound healing, it is also widely employed in biomedical applications for bone and tissue engineering [9]. Another non-toxic and widely employed in biomedical applications as a drug delivery mechanism for anticancer therapy is MgO-NPs.

7.6 Performance of hybrid nanocellulose/metal oxide nanocomposites

The hybrid nanocomposite emphasizes significant difference in numerous aspects. The researchers in material science are very attached to the development of hybrid nanocomposites by incorporating the organic and inorganic which have different properties. An organic component, such as nanocellulose gives high flexibility, dielectric, and piezoelectric capacity to hybrid nanocomposites, while an inorganic component such as Fe_3O_4-NPs contribute magnetic and electric properties [123]. Notably, nanocellulose has shown great promise as a sustainable and eco-friendly nanomaterial. It serves not only as a reinforcing agent but also plays a pivotal role in advanced electrochemical energy conversion and conservation devices for green and renewable electronics The functionalized nanocellulose/MONPs nanocomposite hybridization yields valuable properties suitable for a wide range of applications. Numerous studies have underscored the significant enhancement of end product qualities. This is achieved by avoiding the mutual restrictions often encountered with various fillers from different origins and by promoting compatibility at the interface between the matrix and MONPs, even at low loading levels [81].

7.6.1 Advancements in energy conservation devices

In today's world, addressing the increasing demand for energy is of paramount importance, and nanotechnology has significantly impacted energy storage and conversion. Recent years have witnessed remarkable progress in electrochemical energy generation and storage technologies, encompassing batteries and supercapacitors, with promising applications spanning small-scale devices to automobiles. Supercapacitors, in particular, offer a high power density, making them capable of storing more energy than regular capacitors but less than traditional batteries. However, the widespread commercial adoption of energy storage technologies, despite their substantial potential, remains unsatisfactory. In recent years, carbon-based materials like graphene, CNTs, and fullerenes have gained widespread use in energy storage and conversion applications. In addition to these carbon materials, metallic nanoparticles have emerged as an alternative due to their higher intrinsic conductivity compared to carbon-based conductive materials. Nevertheless, their poor dispersion onto substrates presents a challenge [124]. Researchers have turned to environmentally friendly and sustainable methods, such as coating or hybridizing metal nanoparticles with nanocellulose substrates. Nanocellulose stands out as an excellent candidate for electrodes because of its high surface area. A larger surface area facilitates increased ion absorption, enhancing energy storage capacity.

Cellulose fibres, known for their porous structures and electrolyte absorption capabilities, have shown great promise as substrates for energy materials in storage devices. For instance, cellulose-derived carbon fibre/SnO_2 electrodes have exhibited impressive reversible capacity and cycling retention, outperforming standard SnO_2 nanoparticles. Transition metal oxides, with their high electrochemical capacities, are also being explored as negative electrode materials. NiO, in particular, has demonstrated its suitability as an electrode option due to its low cost, high theoretical capacity, and non-toxicity. However, its low electrical conductivity poses challenges for rate performance and capacity retention [125]. To address these challenges, researchers are exploring the creation of porous morphologies and hybridizing NiO materials with a carbonaceous matrix. High porous structures have proven to be an excellent substrate for MONPs. Natural cellulose-rich cotton fibre, for instance, has been employed as a carbon source and template, resulting in high reversible specific capacity and stable cycling performance, making it a promising lithium-ion anode material [126]. Furthermore, incorporating carbon nanotubes CNTs into cellulose fibre-based electrodes, along with MnO_2, has shown excellent supercapacitor performance with impressive cycling durability. This indicates that cellulose fibres can play a pivotal role as substrates for supercapacitors [127].

7.6.2 Good adsorption capacity

In recent developments, functionalized nanocellulose, incorporating functional groups such as carboxylated groups (COO-), sulfate esters (SO^{-}_{3}), and amine groups ($-NH_2$), has gained significant attention for applications in water treatment. These applications extend to the removal of heavy metals like Zn^{+}, Cu^{2+}, and Pb^{2+}, dye removal, air purification, as well as the decontamination of microorganisms and viruses in water. Additionally, the treatment of oil-contaminated wastewater has emerged as a critical environmental concern in recent years, with far-reaching impacts on the environment, aquatic plants, and wildlife. Biopolymeric aerogels have emerged as a green and sustainable solution to address this issue, demonstrating superior selective oil absorption capacity compared to previous methods. In a study, the hybridization of iron oxide nanoparticles within cellulose-based aerogels was reported to significantly enhance oil adsorption capacity, as depicted in Figure 7.18. To ensure the compatibility of these two materials and create uniformly magnetic aerogels, the introduction of a SiO_2 shell was found to be essential for modifying Fe_3O_4 NPs [128]. This modification not only increased the hydrophobicity and oleophilicity of the nanocellulose aerogel but also enhanced its mechanical stability and surface roughness.

7.6.3 Improvement in thermal stability

Concentration and type of MONPs used to hybrid nanocellulose will influence the thermal stability of nanocomposite [129]. In recent research, extracted cellulose from

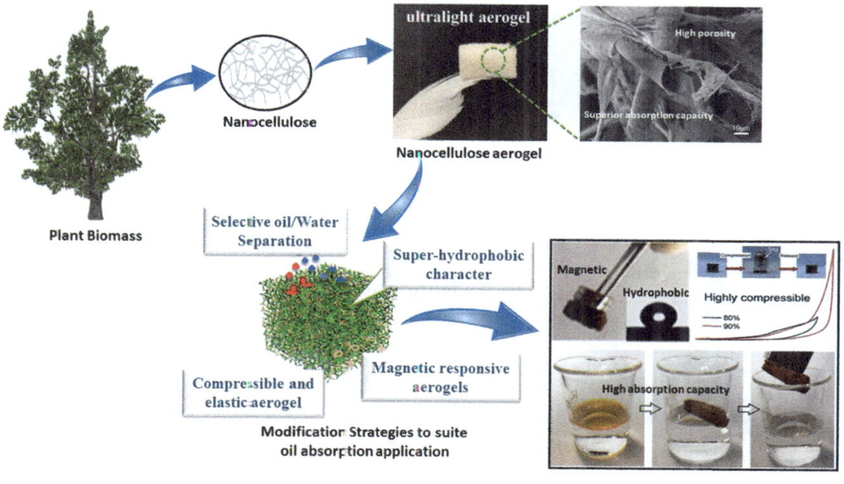

Figure 7.18: The nanocellulose based aerogel with iron oxide nanoparticles for oil absorption. Reproduced from ref. [128].

Eucalyptus tree wood is functionalized chemically using 3 % aqueous solution of sulfuric acid and then with urea before being combined with MONPs such as ZnO and CuO which are gaining popularity due to their outstanding physical and chemical properties. The hybrid cellulose/MONPs was synthesized by *in situ* method and the successful grafting of amine group on the cellulose was detected at spectra band around 3530 cm^{-1} and presence of wide band around 3,330 cm^{-1} which related to the symmetrical valence vibration of the N–H bond during the urea treatment confirm by the Fourier transform infrared spectroscopy (FTIR) test. Based on the scanning electron microscope (SEM) the hybrid nanocellulose/MONPs was observed which is not smooth and contain microactivities due to the addition of MONPs. The MONPs is randomly attracted to the surface of the cellulose formed a nonwoven mat as shown in Figure 7.19. The thermal degradation of modified cellulose, hybrid cellulose/ZnO-NPs and hybrid cellulose/CuO-NPs nanocomposite is 319.75 °C, 324.16 °C and 253.54 °C, respectively. It is clear that the addition of different MONPs shows different thermostability of the hybrid materials. Hybrid cellulose/ZnO-NPs nanocomposite have higher thermal stability compared to the hybrid cellulose/CuO nanocomposite due to the stronger interaction of Zn^{2+} ion associated with modified cellulose matrix [130].

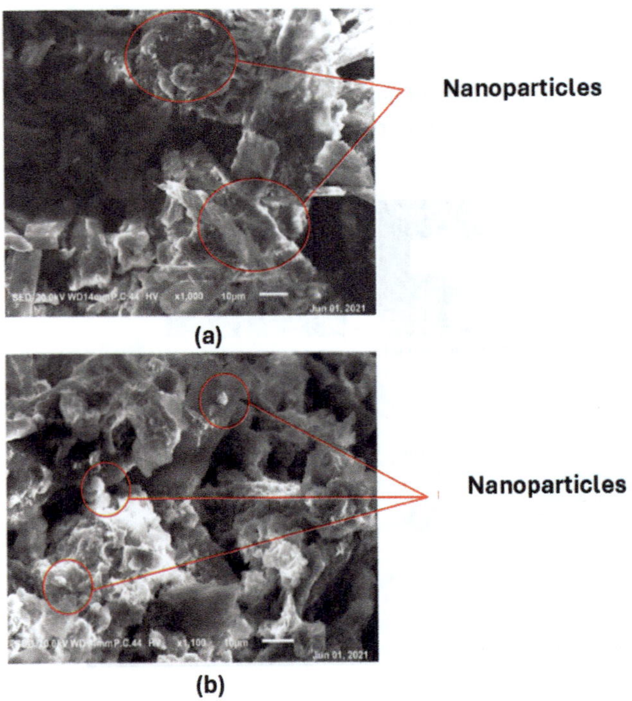

Figure 7.19: The SEM of (a) cellulose/CuO and (b) cellulose/ZnO. Reproduced from ref. [130].

Besides that, different concentration of MONPs also can influence the thermal stability of the hybrid nanocomposite. This is proven by Yu et al., at where the author investigates effect of various Zn^{2+} ions concentration to the thermal stability of the hybrid nanocomposite [131]. The final result indicates that hybrid nanocellulose/Zn-0.5 composite has the best thermal stability compared nanocellulose/Zn –1.0 and nanocellulose/Zn-1.5. This is due to the reaction between the modify nanocellulose via one-pot which rich in carboxyl group, act as scaffold and carrier for stabilization of Zn^{2+} ions through electrostatic interaction thus provide narrow size distribution with 15–85 nm compared to other hybrid nanocomposite nanocellulose/Zn-1.0 (50–210 nm) and nanocellulose/Zn-1.5 (55–265 nm) [131]. Narrow size distribution shows well dispersion. When there is homogeneous dispersion of MONPs with oxygen atom in modify nanocellulose, it will provide strong interaction between nanocellulose and MONPs, thus thermal stability also will increase. In addition, the thermal stability of different type of MONPs such as ZnO, SiO_2, TiO_2, Al_2O_3 and Fe_2O_3 incorporation with nanocellulose was investigated by Goikuria et al. [132]. It is proven that the thermal stability of nanocomposite overall increased after the addition of 10 wt% Fe_2O_3 and ZnO by a simple low-cost environmentally friendly approach to overcome the relatively poor thermal stability of nanocellulose when extracted by sulfuric acid assisted hydrolysis of cellulose as shown in Figure 7.20. The novel method is fabrication of nanocellulose by simple co-assembly method using water as dispersing agent.

However, following the incorporation of MONPs, there is an overall increase in the degradation temperature, particularly pronounced in the cases of nanocellulose/Fe_2O_3-NPs and nanocellulose/ZnO-NP, where the temperature elevates by 72 and 75 °C, respectively. On the other hand, the addition of SiO_2, TiO_2, and Al_2O-NPs to nanocellulose results in an improvement in thermal stability by 5, 14, and 15 °C, respectively. The

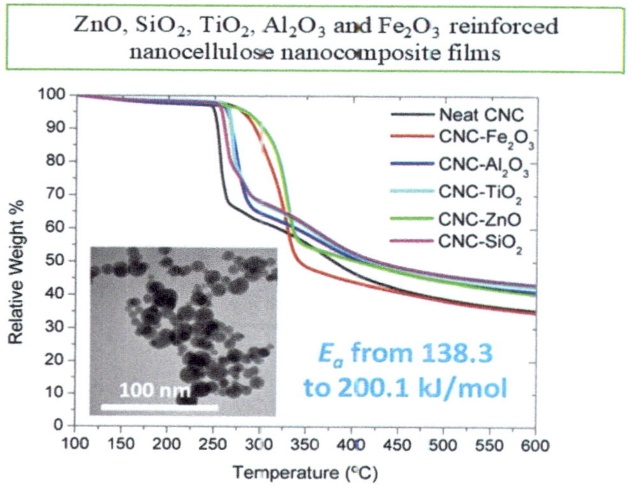

Figure 7.20: Thermal stability of nanocomposite film. Reproduced from ref. [132].

enhanced thermal stability observed with the inclusion of Fe_2O_3, ZnO, and Al_2O_3 compared to TiO_2 and SiO_2 can be attributed to their greater chemical similarity and non-homogeneous particle distribution within the composite matrix, which may reduce the effectiveness of MONPs in enhancing thermal stability. The negatively charged surfaces of nanocellulose amplify the impact of certain MONPs on thermal stability, particularly at higher concentrations, resulting in a dual effect. This effect promotes the disposition of MONPs at the polymer surface, encouraging a more even distribution of particles throughout the polymer matrix. As a result, MONPs exhibit increased effectiveness in acting as a barrier and delaying the degradation process [132].

One of the tests that aids in assessing the long-term thermal performance for heat transfer applications is the colloidal stability test. Given that MONPs are susceptible to aggregation, which can lead to sedimentation and clogging, ultimately resulting in a reduction of thermal conductivity, it becomes crucial to examine their colloidal behaviour. Surface charge plays a pivotal role in determining colloidal behaviour and is commonly quantified in terms of zeta potential. The magnitude of the zeta potential reflects the degree of electrostatic repulsion between adjacent particles carrying similar charges in a dispersion. Furthermore, the value of zeta potential is influenced by factors such as particle size, the composition of the medium, and the pH value. In zeta potential analysis, as a general guideline, particle zeta potentials exceeding |30| are considered colloidally stable, or in other words, zeta potential values should exceed 30 mV to maintain stability and prevent nanoparticle agglomeration [110]. An outcome greater than 30 mV indicates a stable dispersion, while a result of 60 mV or more suggests excellent stability owing to interparticle electrostatic repulsion. Conversely, if the zeta potential is less than 30 mV, it implies that the suspension tends to agglomerate, as detailed in Table 7.4 [133].

7.6.4 Improvement in mechanical properties

Dental resin composite is a widely used restorative material for treating dental caries, a prevalent oral disease caused by the continuous demineralization from acid produced by bacteria. However, dental resin composites have a drawback in that they tend to

Table 7.4: The zeta potential result and the colloidal stability [134].

Zeta Potential (mV)	Colloidal Stability
0 to ± 5	Rapid agglomeration
± 10 to ± 30	Incipient stability
± 30 to ± 40	Moderate stability
± 40 to ± 60	Good stability
> ± 60	Excellent stability

accumulate biofilm and plaque, increasing the risk of secondary caries and potentially leading to restoration failure. To address this issue, nanocellulose is incorporated into ZnO to harness its antibacterial properties. The process involves the removal of the amorphous regions (lignin and hemicellulose) from nanocellulose using HCI and citric acid. Subsequently, surface modification takes place through Fischer esterification, facilitating the interaction between the–COOH from citric acid and the–OH from cellulose. Any unreacted–COOH from citric acid remaining on the nanocellulose surface can adsorb Zn^{2+} through electrostatic attraction, as depicted in Figure 7.21 [134]. The resulting hybrid nanocellulose/ZnO-NP nanocomposite is employed to reinforce dental resin composites due to the excellent mechanical properties it imparts to the composite even with the addition of a small quantity of the hybrid. SEM imaging reveals the absence of noticeable agglomeration among the ZnO-NPs, indicating the successful dispersion of nanocellulose. The nanocellulose, featuring surface functional groups like–COOH, effectively absorbs Zn^{2+} to form the hybrid nanocomposite. Furthermore, the appearance of new peaks at 1736 cm^{-1} and 458 cm^{-1} in the FTIR spectrum can be attributed to the stretching vibrations of C=O and Zn, respectively, signifying the incorporation of MONPs into the nanocellulose. Moving on to the mechanical properties, the dental resin composite containing 2 wt% of the hybrid nanocellulose/ZnO-NPs nanocomposite exhibits a high compressive strength of 371.0 MPa and a flexural modulus of 8.3 GPa, surpassing the reference sample [134]. In addition to enhancing mechanical performance, the inclusion

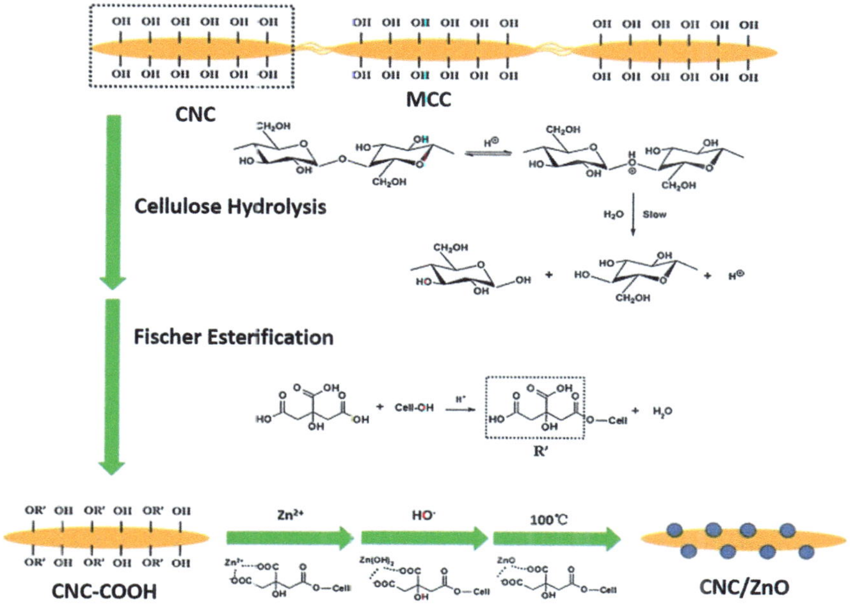

Figure 7.21: The process to synthesis nanocellulose/ZnO-NP hybrid nanocomposite. Reproduced from ref. [134].

of 2 % of the nanocellulose/ZnO-NPs hybrid nanocomposite significantly elevates the antimicrobial properties, resulting in a remarkable 78 % reduction in bacterial counts.

7.6.5 UV sensor and antimicrobial activity

Besides improvement in thermal stability and mechanical properties, hybrid nanocellulose/MONPs shows great antimicrobial activity and stable UV barrier properties [109]. Among all inorganic photo-catalytic materials, ZnO-NPs is reported to have high photo-catalytic effectiveness and to be far more biocompatible than TiO_2 and have excellent semiconductor MONPs with 3.47 eV band gap [109]. The strong UV respond of ZnO films is depend on the morphology and nature of the surface, however in the case of ZnO powder form it depend on surface area and crystalline size. Recent studies have shown that incorporation of nanocellulose with ZnO-NPs and nanocomposite can enhance the UV sensing capabilities of ZnO-NPs [109]. A study showed that cellulose was essential for the growth of the ZnO-NPs surface's porosity, and the higher porosity resulted in a better ability to absorb UV rays. The hybrid nanocellulose ZnO-NPs nanocomposite were therefore eagerly expected for use in extremely sensitive UV sensors. It proven that the hybrid nanocellulose/Zn-NPs nanocomposite film has stable response over on/off cycle of UV light exposure and shows responsivity and sensitivity of nanocomposite film with 60 % ZnO is 36.5 mA/W and 247 respectively. Even in curved device the response and sensitivity are still high which can be used as flexible UV sensors [135]. UV sensitivity of carboxymethylated nanocellulose with CeO_2 and SiO_2-NPs was tested by Abitbol et al. At 365 nm illumination, the extent of UV-screening is diminished, with the film that contains 25 % CeO_2 NPs (100/0 Ce/Si) absorbing the most UV-light [136]. UV screening in the UVA spectral region can be improved by increasing film thickness and/or CeO_2 NP content. The hybrid coatings and films being used in barrier applications, photovoltaics, cosmetic formulations such as sunscreens, and for the maintenance and preservation of wood and glass surfaces, as well as other surfaces that require a smooth, hard, and transparent finish as well as UV protection [136].

One of the drawbacks of MONP is it tend to aggregate due to high surface energy and large surface area. As a result, a capping agent such as nanocellulose should be incorporated to improve the dispersion. Recently, ZnO-NPs has been hybridize with nanocellulose materials as templates because of their plentiful OH groups [68]. Moreover, nanocellulose hosted ZnO-NPs nanostructures have emerged as a significant field in material science, engineering, biomedical, catalysis, sensor, and electronics [12]. Hybrid nanocellulose/ZnO-NPs nanocomposite showed better antibacterial and photocatalytic activity than pure ZnO–NPs [137]. In recent research, nanocellulose was synthesis by utilizing oil palm fruits empty bunches which was used as host polymer for the synthesis of ZnO-NPs through *in-situ* solution casting method. The resultant nanocellulose/ZnO nanocomposite shows more photocatalytic activity than pure ZnO nanostructure upon degradation of methylene blue. Not only that this resultant hybrid nanocellulose/ZnO nanocomposite proven that it improves the antibacterial activity against Gram-positive *S. aureus* and Gram-negative *Escherichia coli* as compared to Zn-NPs only [138]. Besides that, hybridization of nanofibres with CuO shows strong antimicrobial effect when tested against some bacteria such as *E. coli*, *S. aureus* and *Candida albicans* [139].

Referring to Table 7.5, it becomes evident that, in the hybridization of nanocellulose/MONPs nanocomposites, the most frequently employed functionalization methods

Table 7.5: The advantage of hybridization of nanocellulose/MONPs.

Type of nanocellulose	Type of MONPs	Type of functionalization	Advantage	Ref.
Eucalyptus fibres (sulphonated cellulose)	ZnO and CuO	Sulfonation	High thermal stability	[130]
Cellulose nanocrystal	ZnO	Esterification	High antibacterial, photocatalytic activity and improvement in thermal stability	[131]
Cellulose nanocrystals	ZnO, SiO_2, TiO_2, Al_2O_3 and Fe_2O_3	Sulfonation	High thermal stability	[132]
Carboxymethylated nanocellulose	CeO_2 and SiO_2	Carboxymethylation	UV-light absorption	[136]
Oil palm empty fruit bunches	ZnO	Oxidation	Photocatalytic and antibacterial activity	[138]
Cellulose nanocrystal	ZnO	Sulfonation	Good antimicrobial activity and thermal properties	[140]
TEMPO-oxidized bamboo pulp	Carbon/ZnO	Oxidation	Photocatalytic activity	[141]
Carboxymethylated cellulose	ZnO	Carboxymethylation	Excellent antibacterial activity	[142]
Sulphonated nanocellulose	Graphene doped TiO_2	Sulfonation	Photocatalytic and antibacterial activity	[143]
Cellulose	CuO	Carboxymethylation	Strong antibacterial activity.	[121]
Cellulose nanocrystal	CNC/ZnO nanohybrids	Esterification	Strong antibacterial	[134]
Wood	Fe_2O_3	Carboxymethylation	Excellent conservation of magnetic characteristics on storage	[144]
Carboxylated cellulose	Cu_2O	Oxidation	Antimicrobial properties and capacity in dry removeable	[145]
Carboxylated cellulose	ZnO	Oxidation	Inhibit microbial growth	[108]
Cellulose	Fe_2O_3	Carboxymethylation	Superparamagnetic behaviour at room temperature	[146]
Carboxymethyl cellulose	CeO_2	Carboxymethylation	Higher photocatalytic activity, generate reactive oxygen species which can degrade organic pollutants	[147]
Cellulose nanocrystal	ZnO/CuO	Sulfonated	Better larvicidal activity against malaria vector	[128]

include sulfonation, carboxymethylation, and oxidation [126]. These techniques impart charges to nanocellulose. In addition, esterification/acetylation methods are commonly used to confer a hydrophobic nature by reducing surface energy. What these approaches have in common is the incorporation of specific functional groups onto nanocellulose, namely carboxymethyl groups (–CH_2–COOH), carboxylic acid groups (-COOH), and esterification (-$COCH_3$), which contain carbonyl groups (C=O), also known as oxo groups. These functional groups exhibit strong adsorption capacity for various cations such as Al^{3+}, Cu^{2+}, MO^{2+}, Zn^{2+}, and more. Consequently, these ions readily anchor onto the nanocellulose surface, resulting in high DS values. A high DS value enhances the dispersion of MONPs in nanocellulose, leading to lower viscosity due to the prevalence of electrostatic repulsion forces, as desired. In contrast, nanocomposites with low DS may experience aggregation because of the predominance of hydrogen bonding interactions between–OH groups and ammonium groups. Moreover, hybrid nanocellulose/MONPs nanocomposites exhibit robust thermal stability because MONPs possess greater heat resistance compared to cellulose. These nanocomposites also showcase outstanding UV barrier properties, antibacterial characteristics, and exceptional mechanical performance.

7.7 Conclusions

In conclusion, it is evident that addressing the challenges related to the dispersion of MONPs in nanocellulose requires functionalization or chemical modification of nanocellulose to produce high-quality nanocomposites. However, before functionalization, the raw biomass-derived materials must undergo pretreatment to remove lignin and hemicellulose prior to nanocellulose production. This pretreatment can involve chemical, mechanical, or combined methods to effectively produce nanocellulose. The integration of functionalized nanocellulose with MONPs in novel nanohybrids offers a promising basis for sustainable, long-term advancements. There are two main categories of functionalization methods, each serving distinct purposes: (1) Functionalization methods such as phosphorylation, carboxymethylation, oxidation, and sulfonation introduce ionic charges to the nanocellulose surface, while (2) methods like acetylation, silylation, urethanization, and amidation modify the primary–OH groups of nanocellulose to create a hydrophobic surface. These modifications are crucial, as nanocellulose's inherent hydrophilicity, due to its–OH groups, can result in poor interfacial adhesion in hydrophobic matrices. Thus, pre-incorporation surface modification of nanocellulose via chemical processes is essential.

This review demonstrates that incorporating functionalized nanocellulose with MONPs such as ZnO, CuO, and TiO_2 improves MONP compatibility and dispersion, preventing agglomeration. Consequently, the resulting nanocomposites exhibit enhanced properties, including superior thermal stability, improved mechanical characteristics, increased adsorption capacity, effective UV barrier properties, photocatalytic activity, antibacterial efficacy, and strong energy storage capabilities. Among the various functionalization methods, oxidation and carboxymethylation are particularly noteworthy, as they introduce carboxylic acid and carboxymethyl groups to the nanocellulose surface. These functional groups readily bond with MONPs, resulting in a high DS, which enhances dispersion. Additionally, the properties of the final nanocomposite are significantly influenced by factors like the type and size of MONP, nanocellulose loading, and DS values. A higher DS in functionalized nanocellulose enhances hybridization with MONPs, providing a clear economic advantage over the use of MONPs alone in production. Expanding the application of these nanocomposites to areas such as energy storage, solar technology, water purification, and other suspension-based applications holds significant promise for further development.

Acknowledgements: The authors would like to thank the editors S.M. Sapuan, Mohd Roshdi Hassan, Eris Elianddy Supeni and Azizan As'arry for their guidance and review of this article before its publication

References

1. Lam E, Hemraz UD. Preparation and surface functionalization of carboxylated cellulose nanocrystals. Nanomaterials 2021;11. https://doi.org/10.3390/nano11071641.
2. Wu Q, Meng Y, Wang S, Li Y, Fu S, Ma L, et al. Rheological behavior of cellulose nanocrystal suspension: influence of concentration and aspect ratio. J Appl Polym Sci 2014;131:1–8.
3. Islam MT, Alam MM, Patrucco A, Montarsolo A, Zoccola M. Preparation of nanocellulose: a review. AATCC J Res 2014;1:17–23.
4. Sun B, Zhang M, Shen J, He Z, Fatehi P, Ni Y. Applications of cellulose-based materials in sustained drug delivery systems. Curr Med Chem 2017;24. https://doi.org/10.2174/0929867324666170705143308
5. Hashemzehi M, Mesic B, Sjöstrand B, Naqvi M. A comprehensive review of nanocellulose modification and applications in papermaking and packaging: challenges, technical solutions, and perspectives. BioResources 2022;17. https://doi.org/10.15376/biores.17.2.hashemzehi.
6. Jasmania L. Preparation of nanocellulose and its potential application. Int J Nanomater Nanotechnol Nanomed 2018;4:014–21.
7. Mateo S, Peinado S, Morillas-Gutiérrez F, La Rubia MD, Moya AJ. Nanocellulose from agricultural wastes: products and applications–a review Processes 2021;9. https://doi.org/10.3390/pr9091594.
8. Alhalili Z. Metal oxides nanoparticles: general structural description, chemical, physical, and biological synthesis methods, role in pesticides and heavy metal removal through wastewater treatment. Molecules 2023;28. https://doi.org/10.3390/molecules28073086.
9. Negrescu AM, Killian MS, Raghu SNV, Schmuki P, Mazare A, Cimpean A. Metal oxide nanoparticles: review of synthesis, characterization and biological effects. J Funct Biomater 2022;13. https://doi.org/10.3390/jfb13040274.
10. Zhang W. Nanoparticle aggregation: principles and modeling. Adv Exp Med Biol 2014;811:20–43.
11. Shrestha S, Wang B, Dutta P. Nanoparticle processing: understanding and controlling aggregation. Adv Colloid Interface Sci 2020;279:102162.
12. Prusty K, Sahu D, Swain SK. Nanocellulose as a template for the production of advanced nanostructured material. Sawston, United Kingdom: Woodhead Publishing; 2017.

13. Morant-Miñana MC. Surface modification for improving the stability of nanoparticles in liquid media. Encycl Biomed Polym Polym Biomater 2016;27:5569–84.
14. Omran AA, Mohammed AA, Sapuan SM, Ilyas RA, Asyraf MR, Rahimian Koloor SS, et al. Micro-and nanocellulose in polymer composite materials: a review. Polymers. 2021;13:231.
15. Worku LA, Bachheti A, Bachheti RK, Rodrigues Reis CE, Chandel AK. Agricultural residues as raw materials for pulp and paper production: overview and applications on membrane fabrication. Membr (Basel) 2023; 13:1–17.
16. Panthapulakkal S, Sain M. The use of wheat straw fibres as reinforcements in composites. In: Biofiber reinforcements in composite materials. Sawston, UK: Woodhead Publishing; 2015:423–53 pp.
17. Santos RF, Ribeiro JCL, Franco de Carvalho JM, Magalhães WLE, Pedroti LG, Nalon GH, et al. Nanofibrillated cellulose and its applications in cement-based composites: a review. Constr Build Mater 2021;288:123122.
18. Oyohwose UA, Omoko VI. Synthesis , characterization and application of nanocrystalline cellulose : a review; 2023:0–9 p.
19. Lahiri D, Nag M, Dutta B, Dey A, Sarkar T, Pati S, et al. Bacterial cellulose: production, characterization and application as antimicrobial agent. Int J Mol Sci 2021;22:1–18.
20. Aswini K, Gopal NO, Uthandi S. Optimized culture conditions for bacterial cellulose production by Acetobacter senegalensis MA1. BMC Biotechnol 2020;20:1–16.
21. Fornari A, Rossi M, Rocco D, Mattiello L. A review of applications of nanocellulose to preserve and protect cultural heritage wood, paintings, and historical papers. Appl Sci 2022;12. https://doi.org/10.3390/app122412846.
22. Solhi L, Guccini V, Heise K, Solala I, Niinivaara E, Xu W, et al. Understanding nanocellulose-water interactions: turning a detriment into an asset. Chem. Rev. 2023;123:1925–2015.
23. Musino D, Rivard C, Landrot G, Novales B, Rabilloud T, Capron I. Hydroxyl groups on cellulose nanocrystal surfaces form nucleation points for silver nanoparticles of varying shapes and sizes. J Colloid Interface Sci 2021;584:360–71.
24. Kaur D, Bhardwaj NK, Lohchab RK. Prospects of rice straw as a raw material for paper making. Waste Manag 2017;60:127–39.
25. Yao F, Wu Q, Lei Y, Xu Y. Rice straw fiber-reinforced high-density polyethylene composite: effect of fiber type and loading. Ind Crops Prod 2008;28:63–72.
26. Bassyouni M, Waheed Ul Hasan S. The use of rice straw and husk fibers as reinforcements in composites. In: Biofiber reinforcements in composite materials. Amsterdam, Netherlands: Elsevier; 2015;385–422 pp.
27. Xu H, Dun M, Zhang Z, Zhang L, Shan W, Wang W. A new process of preparing rice straw-reinforced LLDPE composite. Polym (Basel) 2022;14. https://doi.org/10.3390/polym14112243.
28. Rosa MF, Medeiros E, Malmonge J, Gregorski K, Wood D, Mattoso L, et al. Cellulose nanowhiskers from coconut husk fibers: effect of preparation conditions on their thermal and morphological behavior. Carbohydr Polym 2010;81:83–92.
29. Ede A, Agbede J. Use of coconut husk fiber for improved compressive and flexural strength of con-crete. Int J Sci Eng Res 2015;6:968–74.
30. Pinho da Costa Castro CD, de Assis Fonseca Faria J, Bassani Hellmeister Dantas T. Testing the use of coconut fiber as a cushioning material for transport packaging. Mater Sci Appl 2012;03:151–6.
31. Lay Ting T, Putra Jaya R, Abdul Hassan N, Yaacob H, Sri Jayanti D. A review of utilization of coconut shell and coconut fiber in road construction. J. Teknol. 2015;76:121–5.
32. Ilyas RA, Sapuan S, Ibrahim R, Abral H, Ishak M, Zainudin E, et al. Effect of sugar palm nanofibrillated cellulose concentrations on morphological, mechanical and physical properties of biodegradable films based on agro-waste sugar palm (Arenga pinnata (Wurmb.) Merr) starch. J Mater Res Technol 2019;8: 4819–30.
33. Atiqah A, Jawaid M, Sapuan SM, Ishak MR, Ansari MNM, Ilyas RA. Physical and thermal properties of treated sugar palm/glass fibre reinforced thermoplastic polyurethane hybrid composites. J Mater Res Technol 2019;8:3726–32.

34. Ilyas RA. Properties of sugar palm nanocellulose fibre-reinforced biopolymer composite [Doctoral thesis]. Malaysia: Universiti Putra Malaysia; 2019.
35. Mahmud MA, Anannya FR. Sugarcane bagasse – a source of cellulosic fiber for diverse applications. Heliyon 2021;7:e07771.
36. Amin MN, Ashraf M, Kumar R, Khan K, Saqib D, Ali SS, et al. Role of sugarcane bagasse ash in developing sustainable engineered cementitious composites. Front. Mater. 2020;7:1–12.
37. Pereira L, Mafalda R, Marconcini JM, Mantovani GL. The use of sugarcane bagasse-based green materials for sustainable packaging design. Smart Innov Syst Technol 2015;35:113–23.
38. Wang D, Tian J, Guan J, Ding Y, Wang ML, Tonnis B, et al. Valorization of sugarcane bagasse for sugar extraction and residue as an adsorbent for pollutant removal. Front Bioeng Biotechnol 2022;10:1–12.
39. Faizi MK, Shahriman A, Abdul Majid M, Shamsul B, Ng Y, Basah S, et al. An overview of the Oil Palm Empty Fruit Bunch (OPEFB) potential as reinforcing fibre in polymer composite for energy absorption applications. MATEC Web Conf 2016;90. https://doi.org/10.1051/matecconf/20179001064.
40. Windiastuti E, Suprihatin Y, Bindar, Hasanudin U. Identification of potential application of oil palm empty fruit bunches (EFB): a review. IOP Conf Ser Earth Environ Sci 2022;1063. https://doi.org/10.1088/1755-1315/1063/1/012024.
41. Nayak L, Mishra SP. Prospect of bamboo as a renewable textile fiber, historical overview, labeling controversies and regulation. Fash Text 2016;3. https://doi.org/10.1186/s40691-015-0054-5.
42. Kaur N, Saxena S, Gaur H, Goyal P. A review on bamboo fiber composites and its applications. 2017 Int Conf Infocom Technol Unmanned Syst Trends Futur Dir ICTUS 2017 2018:843–9. https://doi.org/10.1109/ICTUS.2017.8286123.
43. Ismail NF, Mohd Radzuan NA, Sulong AB, Muhamad N, Che Haron CH. The effect of alkali treatment on physical, mechanical and thermal properties of kenaf fiber and polymer epoxy composites. Polym (Basel) 2021;13. https://doi.org/10.3390/polym13122005.
44. Chu CC, Chew SC, Nyam KL. Recent advances in encapsulation technologies of kenaf (Hibiscus cannabinus) leaves and seeds for cosmeceutical application. Food Bioprod Process 2021;127:99–113.
45. Csiszár E, Kun D, Fekete E. The role of structure and interactions in thermoplastic starch–nanocellulose composites. Polym (Basel) 2021;13:1–14.
46. Othman SH, Nordin N, Azman NAA, Tawakkal ISMA, Basha RK. Effects of nanocellulose fiber and thymol on mechanical, thermal, and barrier properties of corn starch films. Int J Biol Macromol 2021;183:1352–61.
47. Iwamoto S, Lee SH, Endo T. Relationship between aspect ratio and suspension viscosity of wood cellulose nanofibers. Polym J 2014;46:73–6.
48. Li A, Xu D, Luo L, Zhou Y, Yan W, Leng X, et al. Overview of nanocellulose as additives in paper processing and paper products. Nanotechnol Rev 2021;10:264–81.
49. Dhali K, Ghasemlou M, Daver F, Cass P, Adhikari B. A review of nanocellulose as a new material towards environmental sustainability. Sci Total Environ 2021;775:145871.
50. Heise K, Koso T, King AW, Nypelö T, Penttilä P, Tardy BL, et al. Spatioselective surface chemistry for the production of functional and chemically anisotropic nanocellulose colloids. J Mater Chem A 2022;10. https://doi.org/10.1039/d2ta05277f.
51. Sinquefield S, Ciesielski PN, Li K, Gardner DJ, Ozcan S. Nanocellulose dewatering and drying: current state and future perspectives. ACS Sustain Chem Eng 2020;8:9601–15.
52. Chin KM, Sung Ting S, Ong HL, Omar M. Surface functionalized nanocellulose as a veritable inclusionary material in contemporary bioinspired applications: a review. J Appl Polym Sci 2018;135. https://doi.org/10.1002/app.46065.
53. Horvat A, A study of the uncertainty associated with tar measurement and an investigation of tar evolution and composition during the air-blown fluidised bed gasification of torrefied and non- torrefied grassy biomass A thesis submitted for the Degree of Doctorate; 2016:1–193 pp.

54. Trache D, Hussin MH, Hui Chuin CT, Sabar S, Fazita MN, Taiwo OF, et al. Microcrystalline cellulose: isolation, characterization and bio-composites application—a review. Int J Biol Macromol 2016;93: 789–804.
55. Wang X, Lei Q, Luo J, Wang P, Xiao P, Ye Y, et al. Application of nanocellulose in oilfield chemistry. ACS Omega 2021;6:20833–45.
56. Viel M, Collet F, Lanos C. Development and characterization of thermal insulation materials from renewable resources. Constr Build Mater 2019;214:685–97.
57. Mihalca V, Kerezsi AD, Weber A. Gruber-Traub C, Schmucker J, Vodnar DC, et al. Protein-based films and coatings for food industry appl cations. Polymers 2021;13. https://doi.org/10.3390/polym13050769.
58. Thakur V, Guleria A, Kumar S, Sharma S, Singh K. Recent advances in nanocellulose processing, functionalization and applications: a review. Mater. Adv. 2021;2:1872–95.
59. Phanthong P, Reubroycharoen P, Hao X, Xu G, Abudula A, Guan G. Nanocellulose: extraction and application. Carbon Resour Convers 2018;1:32–43.
60. Michelin M, Gomes DG, Romaní A, Polizeli MLTM, Teixeira JA. Nanocellulose production: exploring the enzymatic route and residues of pulp and paper industry. Molecules 2020;25:1–36.
61. Yu H, Qin Z, Liang B, Liu N, Zhou Z, Chen L. Facile extraction of thermally stable cellulose nanocrystals with a high yield of 93% through hydrochloric acid hydrolysis under hydrothermal conditions. J Mater Chem A 2013;1:3938–44.
62. Vanderfleet OM, Reid MS, Bras J, Heux L, Godoy-Vargas J, Panga MKR, et al. Insight into thermal stability of cellulose nanocrystals from new hydrolysis methods with acid blends. Cellulose 2019;26:507–28.
63. Ilyas RA, Sapuan S, Ibrahim R, Atikah M, Atiqah A, Ansari M, et al. Production, processes and modification of nanocrystalline cellulose from agro-waste: a review. Nanocrystalline Mater 2020;3–32. https://doi.org/10.5772/intechopen.87001.
64. Voronova MI, Surov OV, Rubleva NV, Kochkina NE, Zakharov AG. Dispersibility of nanocrystalline cellulose in organic solvents. Khimiya Rastit Syr'ya 2019:39–50. https://doi.org/10.14258/jcprm.2019014240.
65. Islam MT, Alam MM, Zoccola M. Review on modification of nanocellulose for application in composites. Int. J. Innov. Res. Sci. Eng. Technol. 2013;2:5444–51.
66. Dawwam GE, Al-Shemy MT, El-Demerdash AS. Green synthesis of cellulose nanocrystal/ZnO bio-nanocomposites exerting antibacterial activity and downregulating virulence toxigenic genes of food-poisoning bacteria. Sci Rep 2022;12:1–18.
67. Patil TV, Patel DK, Dutta SD, Ganguly K, Santra TS, Lim KT. Nanocellulose, a versatile platform: from the delivery of active molecules to tissue engineering applications. Bioact Mater 2022;9:566–89.
68. Oun AA, Shankar S, Rhim JW. Multifunctional nanocellulose/metal and metal oxide nanoparticle hybrid nanomaterials. Crit Rev Food Sci Nutr 2020;60:435–60.
69. Kalia S, Dufresne A, Cherian BM, Kaith BS, Avérous L, Njuguna J, et al. Cellulose-based bio- and nanocomposites: a review. Int J. Polym. Sci. 2011;2011. https://doi.org/10.1155/2011/837875.
70. Ferreira FV, Pinheiro IF, Gouveia RF, Thim GP, Lona LMF. Functionalized cellulose nanocrystals as reinforcement in biodegradable polymer nanocomposites. Polym Compos 2018;39:E9–29.
71. MC Li, C Mei, X Xu, S Lee, Q Wu. Cationic surface modification of cellulose nanocrystals: toward tailoring dispersion and interface in carboxymethyl cellulose films 2016;107. https://doi.org/10.1016/j.polymer.2016.11.022.
72. Ghasemlou M, Daver F, Ivanova EP, Habibi Y, Adhikari B. Surface modifications of nanocellulose: from synthesis to high-performance nanocomposites. Prog Polym Sci 2021;119:101418.
73. Kamiya H, Iijima M. Surface modification and characterization for dispersion stability of inorganic nanometer-scaled particles in liquid media. Sci Technol Adv Mater 2010;11. https://doi.org/10.1088/1468-6996/11/4/044304.
74. Thomas B, Raj MC, B AK, H RM, Joy J, Moores A, et al. Nanocellulose, a versatile green platform: from biosources to materials and their applications. Chem. Rev. 2018;118:11575–625.

75. Bezerra RDS, Silva M, Morais A, Osajima J, Santos M, Airoldi C, et al. Phosphated cellulose as an efficient biomaterial for aqueous drug ranitidine removal. Mater (Basel) 2014;7:7907–24.
76. Bezerra RDS, Teixeira PRS, Teixeira ASNM, Eiras C, Osajima JA, Filho ECS. Chemical functionalization of cellulosic materials – main reactions and applications in the contaminants removal of aqueous medium. Cellul – Fundam Asp Curr Trends 2015. https://doi.org/10.5772/61431.
77. Rol F, Sillard C, Bardet M, Yarava JR, Emsley L, Gablin C, et al. Cellulose phosphorylation comparison and analysis of phosphorate position on cellulose fibers. Carbohydr Polym 2020;229. https://doi.org/10.1016/j.carbpol.2019.115294.
78. Ghanadpour M, Carosio F, Larsson PT, Wågberg L. Phosphorylated cellulose nanofibrils: a renewable nanomaterial for the preparation of intrinsically flame-retardant materials. Biomacromolecules 2015;16:3399–410.
79. Zhang S, Li SN, Wu Q, Li Q, Huang J, Li W, et al. Phosphorus containing group and lignin toward intrinsically flame retardant cellulose nanofibril-based film with enhanced mechanical properties. Compos. Part B Eng. 2021;212:108699.
80. Tzanov T, Stamenova M, Cavaco-Paulo A. Phosphorylation of cotton cellulose with baker's yeast hexokinase. Macromol Rapid Commun 2002;23:962–4.
81. Noguchi Y, Homma I, Matsubara Y. Complete nanofibrillation of cellulose prepared by phosphorylation. Cellulose 2017;24:1295–305.
82. Liu P, Borrell PF, Božič M, Kokol V, Oksman K, Mathew AP. Nanocelluloses and their phosphorylated derivatives for selective adsorption of Ag+, Cu2+ and Fe3+ from industrial effluents. J Hazard Mater 2015;294:177–85.
83. Kono H, Tsukamoto E, Tajima K. Facile post-carboxymethylation of cellulose nanofiber surfaces for enhanced water dispersibility. ACS Omega 2021;6:34107–14.
84. Rahman MS, Hasan MS, Nitai AS, Nam S, Karmakar AK, Ahsan MS, et al. Recent developments of carboxymethyl cellulose. Polym (Basel) 2021;13. https://doi.org/10.3390/polym13081345.
85. Salmi T, Valtakari D, Paatero E, Holmbom B, Sjöholm R. Kinetic study of the carboxymethylation of cellulose. Ind Eng Chem Res 1994;33:1454–9.
86. Bhandari PN, Jones DD, Hanna MA. Carboxymethylation of cellulose using reactive extrusion. Carbohydr Polym 2012;87:2246–54.
87. Wang J, Han X, Zhang C, Liu K, Duan G. Source of nanocellulose and its application in nanocomposite packaging material: a review. Nanomaterials 2022;12. https://doi.org/10.3390/nano12183158.
88. Mbakop S, Nthunya LN, Onyango MS. Recent advances in the synthesis of nanocellulose functionalized-hybrid membranes and application in water quality improvement. Processes 2021;9. https://doi.org/10.3390/pr9040611.
89. Isogai A, Saito T, Fukuzumi H. TEMPO-oxidized cellulose nanofibers. Nanoscale 2011;3:71–85.
90. Bhat AH, Dasan YK, Khan I, Soleimani H, Usmani A. Application of nanocrystalline cellulose: processing and biomedical applications. Cellul Nanofibre Compos Prod Prop Appl 2017:215–40. https://doi.org/10.1016/B978-0-08-100957-4.00009-7.
91. Oun AA, Rhim JW. Characterization of carboxymethyl cellulose-based nanocomposite films reinforced with oxidized nanocellulose isolated using ammonium persulfate method. Carbohydr Polym 2017;174:484–92.
92. Leung ACW, Hrapovic S, Lam E, Liu Y, Male KB, Mahmoud KA, et al. Characteristics and properties of carboxylated cellulose nanocrystals prepared from a novel one-step procedure. Small 2011;7:302–5.
93. Tang Z, Li W, Lin X, Xiao H, Miao Q, Huang L, et al. TEMPO-Oxidized cellulose with high degree of oxidation. Polym (Basel) 2017;9:3–4.
94. Hastuti N, Kanomata K, Kitaoka T. Characteristics of TEMPO-oxidized cellulose nanofibers from oil palm empty fruit bunches produced by different amounts of oxidant. IOP Conf Ser Earth Environ Sci 2019;359. https://doi.org/10.1088/1755-1315/359/1/012008.

95. Liimatainen H, Visanko M, Sirviö J, Hormi O, Niinimäki J. Sulfonated cellulose nanofibrils obtained from wood pulp through regioselective oxidative bisulfite pre-treatment. Cellulose 2013;20:741–9.
96. Norizan MN, Shazleen SS, Alias AH, Sabaruddin FA, Asyraf MRM, Zainudin ES, et al. Nanocellulose-based nanocomposites for sustainable applications: a review. Nanomaterials 2022;12:1–51.
97. Zhang J, Jiang N, Dang Z, Elder TJ, Ragauskas AJ. Oxidation and sulfonation of cellulosics. Cellulose 2008;15: 489–96.
98. Braun B, Dorgan JR. Single-step method for the isolation and surface functionalization of cellulosic nanowhiskers. Biomacromolecules 2009;10:334–41.
99. Fahma F, Takemura A, Saito Y. Acetylation and stepwise solvent-exchange to modify hydrophilic cellulose whiskers to polychloroprene-compatible nanofiller. Cellulose 2014;21:2519–27.
100. Wang X, Wang N, Xu B, Wang Y, Lang J, Lu J, et al. Comparative study on different modified preparation methods of cellulose nanocrystalline. Polym (Basel) 2021;13. https://doi.org/10.3390/polym13193417.
101. Eyley S, Thielemans W. Surface modification of cellulose nanocrystals. Nanoscale 2014;6:7764–79.
102. Hemraz UD, Boluk Y, Sunasee R. Amine-decorated nanocrystalline cellulose surfaces: synthesis, characterization, and surface properties. Can J Chem 2013;91:974–81.
103. Bart J, Tiggelaar R, Yang M, Schlautmann S, Zuilhof H, Gardeniers H. Room-temperature intermediate layer bonding for microfluidic devices. Lab Chip 2009;9:3481–8.
104. Saini S, Belgacem MN, Bras J. Effect of variable aminoalkyl chains on chemical grafting of cellulose nanofiber and their antimicrobial activity. Mater Sci Eng C 2017;75:760–8.
105. Sharma A, Thakur M, Bhattacharya M, Mandal T, Goswami S. Commercial application of cellulose nano-composites – a review. Biotechnol. Reports 2019;21:e00316.
106. Gadhave RV, Dhawale PV, Sorate CS. Surface modification of cellulose with silanes for adhesive application: review. Open J Polym Chem 2021;11:11–30.
107. Orata F. Derivatization reactions and reagents for gas chromatography analysis. Adv. Gas Chromatogr – Prog Agric Biomed Ind Appl 2012;91:83–108.
108. Hassan A. Evaluation of a nanostructured MWCNT based RAM coating applied to a glass fiber composite. Int. J. Technol. Eng. Stud. 2021;7:1–9.
109. Farooq A, Patoary MK, Zhang M, Mussana H, Li M, Naeem MA, et al. Cellulose from sources to nanocellulose and an overview of synthesis and properties of nanocellulose/zinc oxide nanocomposite materials. Int J Biol Macromol 2020;154:1050–73.
110. Corr SA. Metal oxide nanoparticles. SPR Nanosci. 2016;3:31–56.
111. Khan I, Saeed K, Khan I. Nanoparticles: properties, applications and toxicities. Arab J Chem 2019;12: 908–31.
112. Abu-dief AM. Development of metal oxide nanoparticles as semiconductors. J Nanotechnol Nanomater 2020:4–10. https://doi.org/10.33696/nanotechnol.1.002.
113. Asha AB, Narain R. Nanomaterials properties. In: Polymer science and nanotechnology. Amsterdam, Netherlands: Elsevier Inc.; 2020:343–59 pp.
114. Albarakati AJ, Matter IA. N-type metal oxide semiconductor: materials and their environmental applications. Biointerface Res Appl Chem 2023;13:1–23.
115. Oyewo OA, Elemike EE, Onwudiwe DC, Onyango MS. Metal oxide-cellulose nanocomposites for the removal of toxic metals and dyes from wastewater. Int J Biol Macromol 2020;164:2477–96.
116. Ramalingam G, Kathirgamanathan P, Ravi G, Elangovan T, Arjun kumar B, Manivannan N, et al. Quantum confinement effect of 2D nanomaterials. Quan Dots – Fundam Appl 2020. https://doi.org/10.5772/intechopen.90140.
117. Bhari B, Ali A. Overview of metal contacts technlogy on semiconductor. In: Optical fiber laser technology: series 2. Malaysia: Penerbit UTHM; 2019:77–88 pp.
118. Authors NCERT. Chapter Fourteen: semiconductor electronics: materials, devices and simple circuits. Physicstextb. Cl 2014;12:468–512.

119. Naseem T, Durrani T. The role of some important metal oxide nanoparticles for wastewater and antibacterial applications: a review. Environ Chem Ecotoxicol 2021;3:59–75.
120. Bi J, Mo C, Huang M, Lin Y, Yuan P, et al. Immunotoxicity of metal and metal oxide nanoparticles: from toxic mechanisms to metabolism and outcomes. Biomater Sci 2023;11:4151–83.
121. Shankar S, Wang LF, Rhim JW. Preparation and properties of carbohydrate-based composite films incorporated with CuO nanoparticles. Carbohydr Polym 2017;169:264–71.
122. Chaudhary RG, Bhusari GS, Tiple AD, Rai AR, Somkuvar SR, Potbhare AK, et al. Metal/metal oxide nanoparticles: toxicity, applications, and future prospects. Curr Pharm Des 2019;25:4013–29.
123. Janićijević A, Pavlović VP, Kovačević D, Perić M, Vlahović B, Pavlović VB, et al. Structural characterization of nanocellulose/Fe3O4 hybrid nanomaterials 2019;55:1164–76.
124. Hsu HH, Zhong W. Nanocellulose-based conductive membranes for free-standing supercapacitors: a review. Membr (Basel) 2019;9. https://doi.org/10.3390/membranes9060074.
125. Oh SI, Kim JC, Kim DW. Cellulose-derived tin-oxide-nanoparticle-embedded carbon fibers as binder-free flexible Li-ion battery anodes. Cellulose 2019;26:2557–71.
126. Tan Y, Li Q, Lu Z, Yang C, Qian W, Yu F. Porous nanocomposites by cotton-derived carbon/NiO with high performance for lithium-ion storage. J Alloys Compd 2021;874:159788.
127. Gui Z, Zhu H, Gillette E, Han X, Rubloff GW, Hu L, et al. Natural cellulose fiber as substrate for supercapacitor. ACS Nano 2013;7:6037–46.
128. Elfeky AS, Salem SS, Elzaref AS, Owda ME, Eladawy HA, Saeed AM, et al. Multifunctional cellulose nanocrystal/metal oxide hybrid, photo-degradation, antibacterial and larvicidal activities. Carbohydr Polym 2020;230:115711.
129. Ul-Islam M, Khattak WA, Ullah MW, Khan S, Park JK. Synthesis of regenerated bacterial cellulose-zinc oxide nanocomposite films for biomedical applications. Cellulose 2014;21:433–47.
130. Bakkardouch FE, Atmani H, El Khalloufi M, Jouaiti A, Laallam L. Modified cellulose-based hybrid materials: effect of ZnO and CuO nanoparticles on the thermal insulation property. Mater Chem Phys 2021;271: 124881.
131. Yu HY, Chen GY, Wang YB, Yao JM. A facile one-pot route for preparing cellulose nanocrystal/zinc oxide nanohybrids with high antibacterial and photocatalytic activity. Cellulose 2015;22:261–73.
132. Goikuria U, Larrañaga A, Vilas JL, Lizundia E. Thermal stability increase in metallic nanoparticles-loaded cellulose nanocrystal nanocomposites. Carbohydr Polym 2017;171:193–201.
133. Pate K, Safier P. Chemical metrology methods for CMP quality. Adv Chem Mech Planarizat 2016:1–325. https://doi.org/10.1016/B978-0-08-100165-3.00012-7.
134. Wang Y, Hua H, Li W, Wang R, Jiang X, Zhu M. Strong antibacterial dental resin composites containing cellulose nanocrystal/zinc oxide nanohybrids. J Dent 2019;80:23–9.
135. Komatsu H, Kawamoto Y, Ikuno T. Freestanding translucent ZnO–cellulose nanocomposite films for ultraviolet sensor applications. Nanomaterials 2022;12. https://doi.org/10.3390/nano12060940.
136. Abitbol T, Ahniyaz A, Álvarez-Asencio R, Fall A, Swerin A. Nanocellulose-based hybrid materials for UV blocking and mechanically robust barriers. ACS Appl Bio Mater 2020;3:2245–54.
137. Azizi S, Mohamad R, Rahim RA, Moghaddam AB, Moniri M, Ariff A, et al. ZnO-Ag core shell nanocomposite formed by green method using essential oil of wild ginger and their bactericidal and cytotoxic effects. Appl Surf Sci 2016;384:517–24.
138. Lefatshe K, Muiva CM, Kebaabetswe LP. Extraction of nanocellulose and in-situ casting of ZnO/cellulose nanocomposite with enhanced photocatalytic and antibacterial activity. Carbohydr Polym 2017;164:301–8.
139. Barua S, Das G, Aidew L, Buragohain AK, Karak N. Copper-copper oxide coated nanofibrillar cellulose: a promising biomaterial. RSC Adv 2013;3:14997–5004.
140. S Azizi, M Ahmad, M Mahdavi, S Abdolmohammadi. Preparation, characterization, and antimicrobial activities of ZnO nanoparticles/cellulose nanocrystal nanocomposites. 2013;8:1841–51.
141. Xiao H, Zhang W, Wei Y, Chen L. Carbon/ZnO nanorods composites templated by TEMPO- oxidized cellulose and photocatalytic activity for dye degradation. Cellulose 2018;25:1809–19.

142. Yadollahi M, Gholamali I, Namazi H, Aghazadeh M. Synthesis and characterization of antibacterial carboxymethyl cellulose/ZnO nanocomposite hydrogels. Int J Biol Macromol 2014;74:136–41.
143. Ikram M, Rasheed F, Haider A, Naz S, Ul-Hamid A, Shahzadi A, et al. Photocatalytic and antibacterial activity of graphene oxide/cellulose-doped TiO2 quantum dots: in silico molecular docking studies. Nanoscale Adv 2022;4:3764–76.
144. Breijaert TC, Daniel G, Hedlund D, Svedlindh P, Kessler V, Granberg H, et al. Self-assembly of ferria – nanocellulose composite fibres. Carbohydr Polym 2022;291:119560.
145. Valencia L, Kumar S, Nomena EM, Salazar-Alvarez G, Mathew AP. In-situ growth of metal oxide nanoparticles on cellulose nanofibrils for dye removal and antimicrobial applications. ACS Appl Nano Mater 2020;3:7172–81.
146. Luna-Martínez JF, Reyes-Melo E, González-González V, Guerrero-Salazar C, Torres-Castro A, Sepúlveda-Guzmán S. Synthesis and characterization of a magnetic hybrid material consisting of iron oxide in a carboxymethyl cellulose matrix. J Appl Polym Sci 2013;127:2325–31.
147. Ko JW, Il Lee B, Chung YJ, Park CB. Carboxymethyl cellulose-templated synthesis of hierarchically structured metal oxides. Green Chem 2015;17:4167–72.

Mohammed Hassouna Alaaeddin, Salit Mohd Sapuan*,
Mohd Zuhri Mohamed Yusoff, Edi Syams Zainudin, Faris M. AL-Oqla and
Abir Khan

8 Potential development of thermally stable polycrystalline photovoltaic modules utilizing biocomposite materials

Abstract: Solar energy has become a vital source for generating and harvesting electricity. Enhancing the efficiency of solar modules is an essential factor for successful transition to a sustainable and green energy production. Since natural fiber composites (NFCs) possess various outstanding properties, they can be used as an adequate alternative to synthetic fibers and environmentally harmful materials that are used in photovoltaic (PV) applications and module components. Solar modules are believed to undergo a decline in efficiency when exposed to excessive sun radiation, which eventually caused by a gradual escalation in temperature. Therefore, utilizing appropriate NFCs in the rear components and backsheets of solar modules could significantly contribute to the overall efficiency of these modules and enhance their thermal stability.

Keywords: natural fiber; natural fiber composites; synthetic fibers; photovoltaics applications; outstanding properties; thermal stability

---

***Corresponding author: Salit Mohd Sapuan**, Advanced Engineering Materials and Composites Research Centre (AEMC), Department of Mechanical and Manufacturing Engineering, Universiti Putra Malaysia, 43400 UPM Serdang, Selangor, Malaysia; and Laboratory of Biocomposite Technology, Institute of Tropical Forestry and Forest Products (INTROP), Universiti Putra Malaysia, 43400 UPM Serdang, Selangor, Malaysia, E-mail: sapuan@upm.edu.my. https://orcid.org/0000-0002-4598-6358

Mohammed Hassouna Alaaeddin, Mohd Zuhri Mohamed Yusoff and Edi Syams Zainudin, Advanced Engineering Materials and Composites Research Centre (AEMC), Department of Mechanical and Manufacturing Engineering, Universiti Putra Malaysia, 43400 UPM Serdang, Selangor, Malaysia; and Laboratory of Biocomposite Technology, Institute of Tropical Forestry and Forest Products (INTROP), Universiti Putra Malaysia, 43400 UPM Serdang, Selangor, Malaysia. https://orcid.org/0000-0002-1069-7345 (M.Z. Mohamed Yusoff)

Faris M. AL-Oqla, Department of Mechanical Engineering, The Hashemite University, 13133 Zarqa, Jordan

Abir Khan, Advanced Engineering Materials and Composites Research Centre (AEMC), Department of Mechanical and Manufacturing Engineering, Universiti Putra Malaysia, 43400 UPM Serdang, Selangor, Malaysia; and National Institute of Textile Engineering and Research (NITER), Nayarhat, Savar, Dhaka, 1350, Bangladesh

As per De Gruyter's policy this article has previously been published in the journal Physical Sciences Reviews. Please cite as: M. H. Alaaeddin, S. M. Sapuan, M. Z. Mohamed Yusoff, E. S. Zainudin, F. M. AL-Oqla and A. Khan "Potential development of thermally stable polycrystalline photovoltaic modules utilizing biocomposite materials" *Physical Sciences Reviews* [Online] 2024. DOI: 10.1515/psr-2024-0026 | https://doi.org/10.1515/9783111234762-008

Abbreviations

APP	ammonium polyphosphate
DSSCs	dye-sensitized solar cells
NF	natural fiber
NFCs	natural fiber composites
OPV	organic photovoltaic
PCE	power conversion efficiency
PLA	polylactic acid
PV	photovoltaic
PVDF	polyvinylidene fluoride
SSPF	short sugar palm fiber
TGA	thermogravimetric Analysis

8.1 Introduction

Elevated temperatures in solar modules cause a remarkable drop in their electrical performance and a decline in efficiency [1, 2]. Additionally, solar modules and solar cells encounter further drawbacks that contribute to efficiency losses and yield reductions, such as using full aluminium rear-side metallization, which might result in cell bowing, cell breakage, increased series resistance in larger cells, breakage of fragile and thin cells, differences in thermal expansion of components, and losses in the interconnection material [3]. Moreover, the durability of solar module backsheets is a key factor in the long-term performance of these modules [4]. Despite their commercial use, backsheets still face various issues, such as low degradation resistance, especially in materials like polyethylene terephthalate (PET) under prolonged ultra violet (UV) exposure; yellowing due to photo-oxidation in PET; cracking caused by thermal cycling and UV exposure in brittle polymers; and delamination in adhesive layers, often due to moisture ingress and poor adhesive quality [5, 6].

In recent years, natural fiber composites (NFCs) have emerged as an alternative to synthetic or glass fibers. Fortunately, the rise in NFCs and their applications has opened new possibilities for more sensitive and advanced applications [7]. Since solar modules' backsheets are characterized by their thermal, optical, mechanical, and technical properties, certain types of NFCs possess qualities that make them suitable for improving the performance of these backsheets. This improvement ultimately contributes to the overall efficiency and thermal stability of the modules. Thus, introducing novel lightweight, thermally stable, and durable backsheets to solar modules is crucial to address the existing challenges [8]. A more effective utilization of these green and ecologically friendly materials is of paramount importance to enhance the performance of various applications and to replace synthetic fibers [9].

Recent studies have primarily discussed technological advancements, materials, and coatings in photovoltaic and solar cell technologies as listed in Table 8.1. These studies

Table 8.1: Recent studies in photovoltaic technologies.

Ref	Main focus	Materials discussed	Applications
[10]	Solar cell layers technology and performance improvements.	Copper indium gallium selenide (CIGS), hydrophobic coatings, silicon wafer.	Solar cell layers, especially in dusty regions.
[11]	Comparative analysis of four generations of photovoltaic cells.	Silicon, gallium arsenide, organic-based nanomaterials (graphene, carbon nanotubes), metal nanoparticles.	Solar cells across various generations, with emphasis on novel designs and materials.
[12]	Self-cleaning coatings for photovoltaic modules.	Super-hydrophobic (organosilicon, fluorinated polymers) and super-hydrophilic (TiO_2) coatings.	Photovoltaic modules, particularly in desert and plateau regions where dust is a concern.

focused on aspects such as enhancing efficiency through anti-dust coatings [10], comparing generations of photovoltaic cells [11], and the use of self-cleaning coatings in photovoltaic modules [12]. In contrast, this review emphasizes the use of eco-friendly NFCs to replace synthetic materials in solar module components, which is not a primary concern in the other studies. Therefore, this review aims to focus on the utilization of NFCs for the improvement of efficiency and sustainability of solar modules, particularly the thermal stability.

8.2 Utilizations of natural fiber composites (NFCs)

The excessive utilization of industrial and finite materials has led to environmental damage and the depletion of natural resources. However, there is a growing interest in long-term sustainability and increased awareness of environmental concerns. This interest emphasizes the proper utilization of natural resources as alternatives to synthetic materials. As a result, the advanced selection of materials has become pivotal in engineering to sustain production and achieve both successful and sustainable designs for various applications [9]. The need for natural fiber-composites (NFCs) has never been more prevalent than it is today. Natural fibers offer significant advantages, including reduced density and cost savings when compared to synthetic fibers with similar properties [13]. The use of natural fibers in composites has become one of the most widely studied topics for several reasons. These reasons include pollution control, waste elimination, lower production energy consumption, and the ability to competently replace environmentally harmful synthetic fibers. Natural fibers are eco-friendly and bio renewable, providing several advantages over synthetic fibers, particularly in reinforcing composites. As a result, they can be effectively used in a wide range of applications [14]. Table 8.2 shows the most of common advantages and disadvantages of NFCs. In recent years, NFCs have been utilized in various applications with different uses and

Table 8.2: Common advantages and disadvantages of NFCs [15].

Advantages	Disadvantages
High specific strength, stiffness and low density	Higher variability in their properties
Less hazard in manufacturing processes	Limited matrix options due to lower processing temperature
Renewable, environmentally friendly, require less energy consumption. Whilst returning oxygen to the environment; involves CO_2 absorption	Lower impact strength compared to synthetic fiber composites
Less abrasive damage to processing equipment	Higher moisture absorption
Produced at lower cost compared to synthetic fibers	Lower durability but can be improved considerably with treatment
Lower emission of toxic fumes during incineration at end of life or when subjected to heat	

functionalities. These include instrument panels, oil and air filters, packaging, paneling, construction materials, and notably, in the automotive and aircraft industries [15].

Natural fibers, derived from different types of plants, come in various forms and possess distinct properties. Examples include kenaf, hemp, flax, sisal, henequen, abaca, jute, sugar palm, oil palm, ramie, pineapple leaf, banana pseudo-stem, coir, bamboo, bagasse, rice husk, wood, cotton, and silk. These fibers have been reported for use in the preparation of NFCs for various applications [16]. In the photovoltaic (PV) and renewable energy industries, there is emerging potential for using biocomposite materials in the fabrication of solar modules and their components. For instance, lightweight and durable biocomposites are being introduced for solar module backsheets. It is anticipated that using short sugar palm fiber (SSPF) reinforced polyvinylidene fluoride (PVDF) in solar module backsheets could enhance their thermal stability. This is due to the integration of natural fibers, which possess excellent thermal properties, into the polymer matrix, thus improving the overall performance of the modules [8]. PVDF, a fluoropolymer, is one of the important pyroelectric polymers that can contribute to the phenomenon of heat-generated electricity. As a fluoropolymer, PVDF contains fluorine atoms in its structural formula, which allows for a wide range of implementations [17, 18]. Furthermore, organic materials have become a significant trend in the third generation of photovoltaic technology, including dye-sensitized solar cells (DSSCs). This trend opens up prospects for non-toxic emissions during the manufacturing process, thermal stability, design flexibility, and rapid efficiency enhancement. Dyes play a major role in enhancing the power conversion efficiency (PCE) in solar cells by absorbing photon flux. Tremendous types of dye molecules are being developed and used in DSSCs [19]. Organic photovoltaic (OPV) devices are currently poised to fill a niche for low-power and low-cost options in the solar cell market [20]. Scientists are increasingly encouraged to develop biodegradable polymer composite materials from natural and environmentally friendly resources. Biocomposite materials have gained researchers' attention due to their favorable

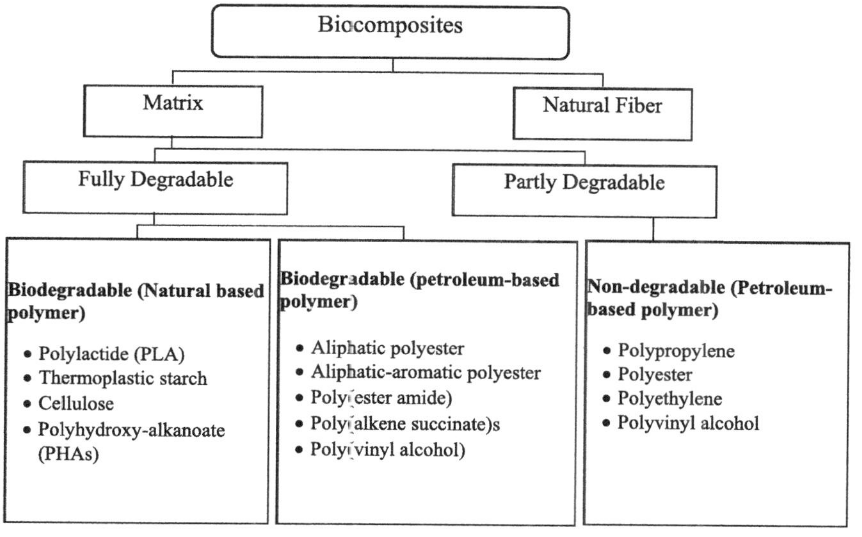

Figure 8.1: Classifications of biocomposites [21].

properties and environmental attributes [21]. Figure 8.1 depicts classifications of biocomposites.

8.3 Potential contributions to thermally stable photovoltaic modules

Interest in fiber-based photovoltaic (PV) development has significantly increased in recent years. This growth has led to notable advancements in solar energy. Specifically, the use of natural fibers (NFs) in PV modules shows promise for more reliable and efficient energy harnessing. NFs offer an eco-friendly alternative to synthetic materials commonly used in solar modules. NFs have become crucial as reinforcement materials in polymer matrices and are widely used as biocomposites in filling, coating, and laminating forms. One key advantage of short NFs is their ability to adjust the dielectric response of selected polymer matrices. This property makes them suitable for sensitive applications where treatment and modifications can influence the electrical, mechanical, and biological properties of the natural fiber composites (NFCs) [22]. In photovoltaic applications, for instance, short sugar palm fiber (SSPF) has been used to reinforce polyvinylidene fluoride (PVDF), creating a biocomposite material recommended as backsheets for solar modules. This biocomposite enhances the thermal stability, durability, and mechanical properties of the modules [8, 23]. The PVDF-SSPF biocomposites demonstrate high thermal stability, a homogeneous structure, and strong interfacial bonding [8]. The dynamic mechanical analysis (DMA) of PVDF-SSPF composites shows that the storage modulus increases with fiber content, indicating improved stiffness. For example, at higher fiber loading, the maximum endothermic peak occurred at 168.10 °C, as observed in the differential scanning calorimetry (DSC) analysis. The thermal stability of the composites was confirmed by the thermogravimetric analysis (TGA), with mass loss occurring gradually at elevated temperatures. These properties contribute to their high compatibility and strength, as well as favorable mechanical properties. Additionally, the biocomposites show excellent resistance to water absorption, reduced moisture content, and minimal thickness swelling, indicating their potential for green production [23]. Alaaeddin et al. [24] have evaluated the use of PVDF-SSPF in a wide range of engineering applications. The authors employed the analytic hierarchy process (AHP), technique for order of preference by similarity to the ideal solution (TOPSIS), and elimination and choice expressing reality (ELECTRE). In the context of the research, the AHP process was used to prioritize polymers by breaking down the decision criteria into a structured hierarchy, allowing for pairwise comparisons among the materials based on their properties. TOPSIS was then applied to rank the polymers by evaluating how close each material was to the ideal solution, considering the best and worst alternatives. Finally, ELECTRE helped refine the rankings by using pairwise comparisons to eliminate less favorable options. These methods combined to ensure a comprehensive evaluation of polymers, considering various criteria such as thermal stability, mechanical properties, and cost-effectiveness, ultimately leading to the selection of PVDF as the top material for short sugar palm fiber composites. A recent study by Alaaeddin et al. [25] introduced novel developed polycrystalline-solar module with novel backsheet-based biocomposite materials, the solar module consisted of various layers, the PVDF-SSPF backsheet is embedded within module's main components and can be clearly seen as in Figure 8.2.

In the developed solar module, the major novelty can be spotted at the backsheet of the module where the biocomposite materials are involved. The developed solar module consists of eight main parts, the backsheet included the PVDF-SSPF composite within its

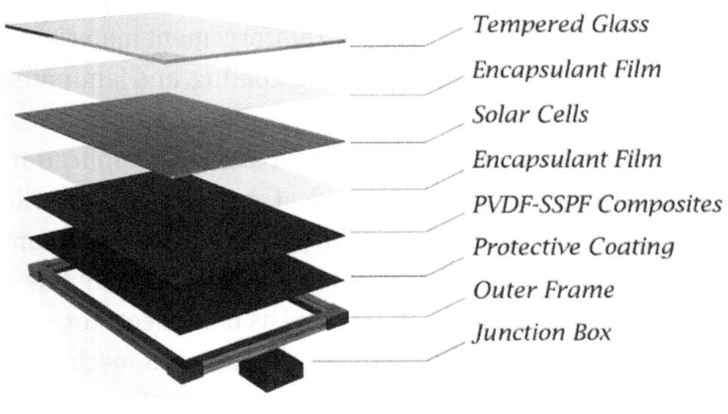

Figure 8.2: Schematic diagram of the developed solar module [25].

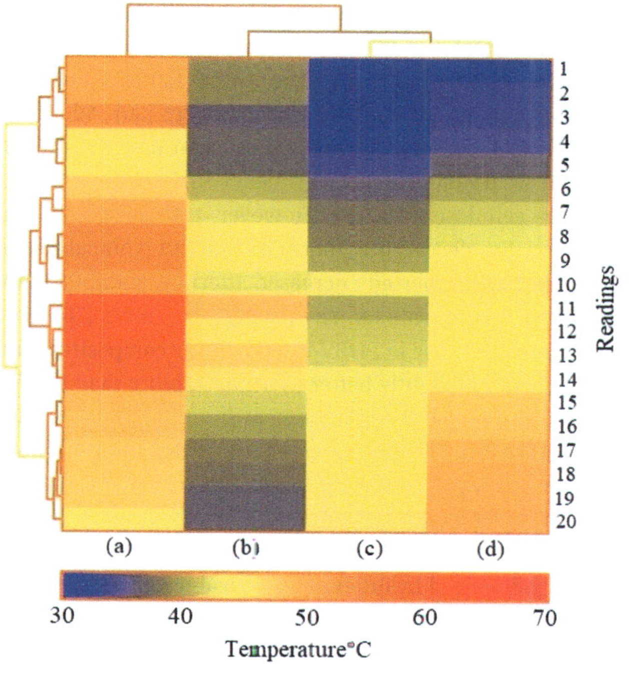

Figure 8.3: Heatmap illustration depicting the thermal performance of (a) surface temperature of the developed module, (b) backsheet temperature of the developed module. (c, d) Simultaneous comparison between the developed backsheet temperature (c), and the conventional backsheet temperature (d) [25].

structure. The module has been thermally compared to commercial ones, observation and thermal analysis proved thermal enhancement on the developed solar module, which can be attributed to the addition of PVDF-SSPF composites within its backsheet. The thermal enhancement led to a better thermal stability in the solar module and therefore higher electrical outcome, Figure 8.3 provides a heatmap analysis demonstrating the thermal performance of the developed module against commercial ones [25].

In the heatmap analysis, the PVDF-SSPF backsheet performed well in tests (a & b). Notably, it showed outstanding results in tests (c & d) when compared to conventional backsheets. Column (c) demonstrated excellent thermal performance, with enhanced stability and reduced heat absorption, outperforming the conventional backsheet in column (d). The study attributed these improvements to the specific thermal properties of SSPF and the inherent stability of PVDF, a polymer recommended for complex, nano, and advanced applications.

NFCs or biocomposites present a viable alternative to existing glass fiber-reinforced composites, especially in industries such as construction and automotive. Polylactic acid (PLA) is an effective choice for replacing petroleum-based synthetic polymers, offering benefits like low production energy, biodegradability, and reduced greenhouse gas emissions. Incorporating natural fillers into the PLA matrix enhances its overall properties. To further improve the thermal stability of NFCs, the addition of ammonium polyphosphate (APP) has been recommended. APP effectively enhances flame retardancy while preserving the environmentally friendly nature of biocomposites [26]. Additionally, chemical treatments, such as NaOH, are commonly used to clean fibers' surfaces and enhance their properties. NaOH treatment provides a rougher surface topography, reveals fibrils, and alters the native cellulose structure. However, this treatment can weaken hydrogen bonding, potentially decreasing thermal stability compared to untreated fibers [27]. Other studies have reported increased thermal stability with treated fibers, though higher alkaline concentrations may damage the fibers [28]. The thermal behavior of natural fibers, such as those in PHBV-bamboo fiber composites, can vary. For example, these composites exhibit slightly better thermal stability than PHBV-wood fiber composites, possibly due to the lower lignin content in bamboo fibers [29]. Understanding the thermal properties of natural fibers is essential for determining their suitability for specific applications. Thermogravimetric analysis (TGA) helps assess thermal stability, glass transition temperature, and other crucial factors, guiding the use of fibers in appropriate applications [30]. TGA and thermal stability in fiber and composite structure can be significantly influenced by different chemical treatments. TGA is essential to determine the ignition temperature, drying temperature, composite's structure and purity. However, to determine the effectiveness of chemical treatment for the purpose of synergizing the chemical bonding of NFCs or between natural fiber and their respective polymer; it is essential to understand compounds stability temperature, temperature dependence and weight [31]. Chemical treatments, including alkali and mixed (alkali + silane) treatments, have been shown to increase the thermal stability of NFCs. These treatments modify the morphology of natural fibers, with mixed treatments resulting in a thin coating layer and alkali treatments creating a rougher surface [32]. In conclusion, considering multiple natural fibers in hybrid composites, rather than relying on a single fiber, may lead to optimal properties, including improved thermal, physical, and mechanical characteristics [33].

8.4 Conclusions

The utilization of thermally enhanced natural fiber composites (NFCs) in the manufacturing of solar modules, particularly in their backsheets, holds significant promise. Solar modules often experience a decline in efficiency when exposed to excessive sunlight, primarily due to the rise in temperature. By improving the features and properties of a module's components, it is possible to positively influence the thermal behavior of solar modules, thereby enhancing their efficiency and overall performance.

The growing interest in natural and organic biocomposite materials across various engineering and renewable energy sectors underscores their potential as replacements for environmentally harmful compounds, pollutants, and synthetic materials. This shift in focus is driven by the increasing recognition of these materials' capability to contribute to sustainable and efficient energy solutions.

References

1. Häberlin H. Photovoltaics: system design and practice. Chichester: John Wiley & Sons; 2012.
2. Lynn PA. Electricity from sunlight: an introduction to photovoltaics. Chichester: John Wiley & Sons; 2011.
3. Spath M, De Jong PD, Bennett I, Visser T, Bakker J. A novel module assembly line using back contact solar cells. In: 2008 33rd IEEE photovoltaic specialists conference. IEEE; 2008:1–6 pp.
4. Gambogi W, Heta Y, Hashimoto K, Kopchick J, Felder T, MacMaster S, et al. Weathering and durability of PV backsheets and impact on PV module performance. In: Reliability of photovoltaic cells, modules, components, and systems VI. International Society for Optics and Photonics; 2013, vol 8825:88250B p.
5. Gu X, Krommenhoek PJ, Lin C-C, Yu L-C, Nguyen T, Watson SS. Depth profiling of mechanical degradation of PV backsheets after UV exposure. In: Reliability of photovoltaic cells, modules, components, and systems VIII. International Society for Optics and Photonics; 2015, vol 9563:956305 p.
6. Lin C-C, Krommenhoek PJ, Watson SS, Gu X. Depth profiling of degradation of multilayer photovoltaic backsheets after accelerated laboratory weathering: cross-sectional Raman imaging. Sol Energy Mater Sol Cell 2016;144:289–99.
7. Abdulrahman KO, Abed AM, Bayode A, Bhowmick S, Dey S, Hien TD, et al. Hierarchical composite materials: materials, manufacturing, engineering. Berlin: Walter de Gruyter GmbH & Co KG; 2018.
8. Alaaeddin M, Sapuan S, Zuhri M, Zainudin E, AL-Oqla FM. Lightweight and durable PVDF–SSPF composites for photovoltaics backsheet applications: thermal, optical and technical properties. Materials 2019;12:2104.
9. Al-Oqla FM, Sapuan S. Natural fiber composites. In: Kenaf fibers and composites. Boca Raton: CRC Press; 2018.
10. Alarifi IM. Advanced selection materials in solar cell efficiency and their properties – a comprehensive review. Mater Today Proc 2023;81:403–14.
11. Luceño-Sánchez JA, Díez-Pascual AM, Peña Capilla R. Materials for photovoltaics: state of art and recent developments. Int J Mol Sci 2019;20. https://doi.org/10.3390/ijms20040976.
12. Qin J, Lu H. A review of self-cleaning coatings for solar photovoltaic systems: theory, materials, preparation, and applications. Environ Sci Pollut Control Ser 2023;30:91591–616.
13. Westman MP, Fifield LS, Simmons KL, Laddha S, Kafentzis TA. Natural fiber composites: a review. Richland, WA, United States: Pacific Northwest National Lab.(PNNL); 2010.
14. Sanjay M, Madhu P, Jawaid M, Senthamaraikannan P, Senthil S, Pradeep S. Characterization and properties of natural fiber polymer composites: a comprehensive review. J Clean Prod 2018;172:566–81.
15. Pickering KL, Efendy MA, Le TM. A review of recent developments in natural fibre composites and their mechanical performance. Compos Appl Sci Manuf 2016;83:98–112.
16. Bachtiar D, Sapuan S, Hamdan M. Flexural properties of alkaline treated sugar palm fibre reinforced epoxy composites. Int J Automot Mech Eng 2010;1:79–90.
17. Alaaeddin M, Sapuan S, Zuhri M, Zainudin E, AL-Oqla FM. Properties and common industrial applications of Polyvinyl fluoride (PVF) and polyvinylidene fluoride (PVDF). IOP Conf Ser Mater Sci Eng 2018;409:012021.
18. Alaaeddin M, Sapuan S, Zuhri M, Zainudin E, AL-Oqla FM. Polyvinyl fluoride (PVF); its properties, applications, and manufacturing prospects. IOP Conf Ser Mater Sci Eng 2019;538:012010.
19. Boro B, Gogoi B, Rajbongshi B, Ramchiary A. Nano-structured TiO2/ZnO nanocomposite for dye-sensitized solar cells application: a review. Renew Sustain Energy Rev 2018;81:2264–70.
20. Gledhill SE, Scott B, Gregg BA. Organic and nano-structured composite photovoltaics: an overview. J Mater Res 2005;20:3167–79.

21. Sahari J, Sapuan S. Natural fibre reinforced biodegradable polymer composites. Rev Adv Mater Sci 2011;30: 166–74.
22. Alaaeddin M, Sapuan S, Zuhri M, Zainudin E, Al-Oqla FM. Photovoltaic applications: status and manufacturing prospects. Renew Sustain Energy Rev 2019;102:318–32.
23. Alaaeddin M, Sapuan S, Zuhri M, Zainudin E, Al-Oqla FM. Physical and mechanical properties of polyvinylidene fluoride-short sugar palm fiber nanocomposites. J Clean Prod 2019;235:473–82.
24. Alaaeddin M, Sapuan S, Zuhri M, Zainudin E, Al-Oqla FM. Polymer matrix materials selection for short sugar palm composites using integrated multi criteria evaluation method. Compos B Eng 2019;176:107342.
25. Alaaeddin M, Sapuan S, Zuhri M, Zainudin E, AL-Oqla FM. Development of photovoltaic module with fabricated and evaluated novel backsheet-based biocomposite materials. Materials 2019;12:3007.
26. Shukor F, Hassan A, Islam MS, Mokhtar M, Hasan M. Effect of ammonium polyphosphate on flame retardancy, thermal stability and mechanical properties of alkali treated kenaf fiber filled PLA biocomposites. Mater Des 2014;54:425–9.
27. El-Shekeil Y, Sapuan S, Khalina A, Zainudin E, Al-Shuja'a O. Effect of alkali treatment on mechanical and thermal properties of Kenaf fiber-reinforced thermoplastic polyurethane composite. J Therm Anal Calorim 2012;109:1435–43.
28. Mohammed L, Ansari MN, Pua G, Jawaid M, Islam MS. A review on natural fiber reinforced polymer composite and its applications. Int J Polym Sci 2015;2015. https://doi.org/10.1155/2015/243947.
29. Singh S, Mohanty AK, Sugie T, Takai Y, Hamada H. Renewable resource based biocomposites from natural fiber and polyhydroxybutyrate-co-valerate (PHBV) bioplastic. Compos Appl Sci Manuf 2008;39:875–86.
30. Asim M, Paridah MT, Chandrasekar M, Shahroze RM, Jawaid M, Nasir M, et al. Thermal stability of natural fibers and their polymer composites. Iran Polym J 2020;29:625–48.
31. Nurazzi NM, Asyraf MRM, Rayung M, Norrrahim MNF, Shazleen SS, Rani MSA, et al. Thermogravimetric analysis properties of cellulosic natural fiber polymer composites: a review on influence of chemical treatments. Polymers 2021;13:2710.
32. Neto JSS, Lima RAA, Cavalcanti DKK, Souza JPB, Aguiar RAA, Banea MD. Effect of chemical treatment on the thermal properties of hybrid natural fiber-reinforced composites. J Appl Polym Sci 2019;136:47154.
33. Krishnasamy S, Thiagamani SMK, Kumar CM, Nagarajan R, Shahroze RM, Siengchin S, et al. Recent advances in thermal properties of hybrid cellulosic fiber reinforced polymer composites. Int J Biol Macromol 2019;141:1–13.

Mohammad Zaid Hasan, Mohd Sapuan Salit*, Nor Salwa Hamdan, Faieza Abdul Aziz and Abir Khan

9 Environmental life cycle analysis of natural fiber composites in energy sector

Abstract: Environmental impact evaluation across a product's complete journey is achieved through Life Cycle Assessment (LCA), a systematic approach that examines each stage from beginning to end. This methodology considers everything from obtaining initial materials to final disposal, including manufacturing processes, distribution networks, consumer usage, and end-of-life handling through recycling or waste management. The goal is to comprehend the overall environmental footprint. Several challenges hinder the adoption of bio-fiber composites in the energy sector, despite their environmental advantages. This chapter identifies key challenges, such as inefficiencies in raw material extraction, high energy consumption during production, and waste management issues during disposal. The objective is to outline the principles of LCA, review recent research, and propose technological and policy solutions to address these challenges. Case studies from various energy sectors are also examined. In conclusion, the chapter emphasizes the significance of ongoing research and development to enhance the lifecycle sustainability of natural fiber composites. The analysis reveals key information that helps drive the shift to environmentally friendly and renewable energy solutions.

Keywords: Life Cycle Assessment (LCA); natural fiber composites (NFCs); sustainable energy applications

9.1 Introduction

The energy sector is constantly evolving, driven by the dual imperatives of meeting the growing demand and mitigating environmental impacts. A key development in

***Corresponding author: Mohd Sapuan Salit**, Department of Mechanical and Manufacturing Engineering, Advanced Engineering Materials and Composites (AEMC) Research Centre, Universiti Putra Malaysia, 43400 UPM Serdang, Selangor, Malaysia, E-mail: sapuan@upm.edu.my. https://orcid.org/0000-0002-4598-6358
Mohammad Zaid Hasan, Department of Mechanical and Manufacturing Engineering, Advanced Engineering Materials and Composites (AEMC) Research Centre, Universiti Putra Malaysia, 43400 UPM Serdang, Selangor, Malaysia, E-mail: ham123zaid@gmail.com
Nor Salwa Hamdan, Department of Skills Development, Ministry of Human Resources Malaysia, Level 7-8, Block D4, Complex D, 62530 Putrajaya, Malaysia, E-mail: salwa.hamdan@mohr.gov.my
Faieza Abdul Aziz, Department of Mechanical and Manufacturing Engineering, Universiti Putra Malaysia, 43400 UPM Serdang, Selangor, Malaysia
Abir Khan, National Institute of Textile Engineering and Research (NITER), Nayarhat, Savar, Dhaka 1350, Bangladesh, E-mail: akhan@niter.edu.bd

As per De Gruyter's policy this article has previously been published in the journal Physical Sciences Reviews. Please cite as: M. Z. Hasan, M. S. Salit, N. S. Hamdan, F. A. Aziz and A. Khan "Environmental life cycle analysis of natural fiber composites in energy sector" *Physical Sciences Reviews* [Online] 2024. DOI: 10.1515/psr-2024-0027 | https://doi.org/10.1515/9783111234762-009

sustainable materials is the growing use of composites made from natural fibers [1]. By utilizing renewable plant-based materials, these innovative composites present an eco-conscious option to replace traditional materials currently used in energy generation and efficiency applications. However, to fully realize their potential, it is crucial to understand their life-cycle sustainability [2, 3]. These innovative composites combine naturally occurring fibers with a binding matrix that may be derived from either biological or artificial sources. The fiber components, sourced from plant or animal materials, enhance the material's structural integrity and rigidity. Meanwhile, the matrix component serves dual purposes: it acts as a binding agent to hold the fibers in place while also shielding them from environmental damage [4, 5]. Figure 9.1 illustrates the key phases in the life cycle of natural fiber-reinforced composites, from material acquisition to end-of-life. Several factors drive the use of these materials in the energy sector. The advantages of these materials are threefold: their renewable source materials decrease reliance on petroleum-based resources; their eco-friendly nature stems from being both decomposable and producing fewer carbon emissions; and their distinctive characteristics – particularly their lightweight yet robust structure – make them ideal for diverse energy applications [6, 7]. One notable application is in wind energy systems, where these natural fiber-based materials are being explored as substitutes for traditional glass and carbon fiber components. Specifically, they show promise in constructing turbine blades

Figure 9.1: Life cycle of bio-fiber based composites [11].

and nacelle covers, offering an environmentally responsible alternative to conventional non-decomposable materials [8]. Similarly, hybrid jute-glass fiber composites have shown promise in optimizing solar parabolic troughs by reducing wind load effects and improving structural efficiency under various configurations [9]. Additionally, cellulose-based composites with carbon nanofibers are being developed for battery applications, demonstrating high specific capacity and durability as anode materials in lithium-ion batteries [10]. These advancements highlight the versatility and potential of natural fiber composites across multiple energy technologies.

Despite the significance of life-cycle sustainability of natural fiber composites, this concept is not well understood. A comprehensive environmental analysis called Life Cycle Assessment (LCA) examines how products and systems affect the environment throughout their complete journey – starting from gathering raw materials, moving through manufacturing and usage phases, and ending with final disposal [12, 13]. Using Life Cycle Assessment to evaluate natural fiber composites reveals their environmental performance and areas where they could be enhanced. This chapter explores an in-depth Life Cycle Assessment examining how these composites function within energy applications. We begin by introducing these innovative materials and exploring their potential uses in energy generation and efficiency. The discussion then moves to examining the key principles and methods of Life Cycle Assessment, specifically focusing on how this tool measures material sustainability. Real-world examples demonstrate how this assessment approach has been implemented across different energy technologies. The analysis then shifts to examining future developments, including both technological advances and regulatory frameworks that could enhance the sustainability of these bio-based composites in energy applications. This examination emphasizes ongoing research needs to address current limitations and maximize these materials' benefits. Through detailed analysis of the complete lifecycle sustainability aspects of these composites, this exploration emphasizes their crucial role in developing more environmentally conscious energy systems. It demonstrates how implementing these materials can minimize environmental impact in energy technology, helping achieve broader sustainability targets.

The knowledge and practical examples provided here serve as a resource for the scientific community, engineering professionals, and decision-makers. This information aims to encourage innovative thinking and joint efforts to fully realize how natural fiber composites can shape an environmentally sustainable energy landscape.

9.2 Natural fiber composites

Bio-fiber composites are materials made by combining naturally occuring fibers such as jute, hemp, flax, and kenaf with a polymer matrix to produce a strong, lightweight, and eco-friendly material [14]. Natural fiber composites are gaining widespread attention for their environmental benefits – they break down naturally, come from sustainable

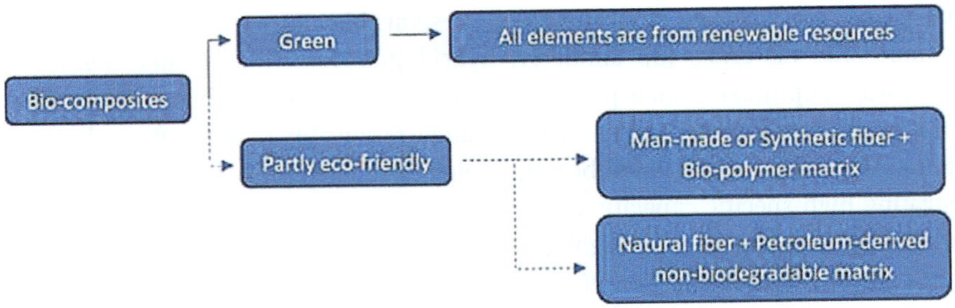

Figure 9.2: Categorization of biocomposites [15–17].

sources, and have minimal ecological impact. Unlike conventional composites that rely on fossil fuels, these materials offer a renewable and planet-friendly option [4]. Bio-based composites can generally be categorized into two types based on their environmental impact, as depicted in Figure 9.2. A fully 'green' composite material is achieved when all its constituents originate from renewable resources, thereby contributing to a decrease in carbon emissions and diminishing dependence on petroleum-derived substances. Conversely, a composite possessing only partial environmental friendliness arises when either the reinforcing fiber or the binding matrix is derived from non-renewable sources [15].

Natural fibers can be broadly categorized based on their source: plant-based, derived from minerals, and sourced from animals. The abundance and accessibility of plant-derived fibers make them the most widely utilized and researched materials in industries [18]. These plant fibers, originating from diverse sources, share a common composition of organic compounds that contribute to their unique properties.

Natural fibers are often preferred over synthetic alternatives for several reasons. They are easily accessible, cost-effective, lightweight, and have a good strength-to-weight ratio. Additionally, they offer excellent sound-dampening properties, require less energy in production, have a smaller carbon footprint, and are biodegradable [16]. Some studies highlight the significant advantages of natural fibers, noting that they are less expensive and less energy-intensive to produce compared to traditional reinforcing fibers like carbon and glass [19]. However, natural fibers present several challenges, including inherent variability in quality and mechanical properties, a tendency towards moisture absorption, diminished load-bearing capacity compared to some synthetic counterparts, and the necessity of employing lower processing temperatures [20–22].

Different plant sources yield fibers with varying properties. Leaf fibers, such as those derived from sisal and pineapple, exhibit superior yield strength compared to many other fiber types [23, 24]. Stem fibers, such as flax, hemp along with jute and kenaf, are widely utilized as fibers in composite materials [1]. Grass fibers, like bamboo and elephant grass, are renowned for their abundance, lightweight nature, and ease of extraction, while seed fibers, such as those derived from cotton and kapok, are obtained from the protective coverings of plant seeds [25]. Stalk fibers, such as those from sugarcane and corn, combine strength with low density. Among these, corn straw fiber (CSF) is particularly abundant, with global maize stover production reaching about one billion tons annually [26]. CSF comprises two primary components: the skin (CSFS), exhibiting high strength suitable for structural applications, and the core (CSFC), which provides excellent thermal performance and impact absorption. *Phragmites australis*, commonly known as common reed, plays a crucial role in sustainable industries, finding applications in ethanol production, composite materials, paper manufacturing, and soil remediation.

Elanchezhian et al. examined the mechanical performance characteristics of natural fiber composites and examined materials such as abaca, jute, and sisal. The study emphasized the multifaceted nature of these fibers, highlighting their high strength-to-weight ratio, affordability, and favorable mechanical performance, while also noting their non-abrasive nature and environmentally beneficial characteristics, such as biodegradability and renewability. Notably, they observed that abaca exhibited outstanding mechanical properties, jute exhibited excellent flexural strength, and sisal demonstrated noteworthy hardness [27].

Natural fibers, including bast fibers such as kenaf, hemp, jute, and sisal, are integral to crafting instrument panels, door panels, seat components, and load floors. Banana fibers are used to wrap paper, while coir enhances doormats, rugs, mattresses, and car seats. Cotton excels in insulation and soundproofing. Coconut fibers contribute to the comfort of back cushions, head restraints, and seat surfaces. Flax, known for its strength, is used in parcel shelves and panels, whereas kenaf reinforces inner door panels. Wood fibers offer natural touch to seat components, and wool adds elegance to seat coverings. This diverse use of natural fibers underscores their importance in creating functional and aesthetically pleasing products [28]. Figure 9.3 showcases the classification of natural fibers.

Bio-fiber composites find widespread use across diverse sectors, including construction, automotive manufacturing, aerospace, electrical equipment, and the biomedical field, due to their inherent benefits compared to synthetic counterparts [30]. These composites offer several benefits, including lightweight properties, sound insulation, favorable mechanical behavior, and eco-friendliness. While offering numerous advantages, a significant limitation of these materials is their relatively short lifespan. Pioneering efforts in 1941 by Ford Motor Company showcased the potential of hemp and flax fibers in automotive applications. Since then, the utilization of natural fibers, such as kenaf, sisal, and flax, has expanded considerably. Notably, natural fiber composites have found applications in diverse construction elements, including roof walls, structural beams, and even bridges [22], and their primary objective in industries is to minimize weight and production costs. Research indicates that incorporating natural fiber-reinforced composites can significantly reduce costs by approximately 20 % while achieving a substantial weight reduction of around 30 %. This weight savings is particularly crucial for the automotive sector, where lighter vehicle components directly contribute to improved fuel economy. Natural fiber composites are widely utilized in

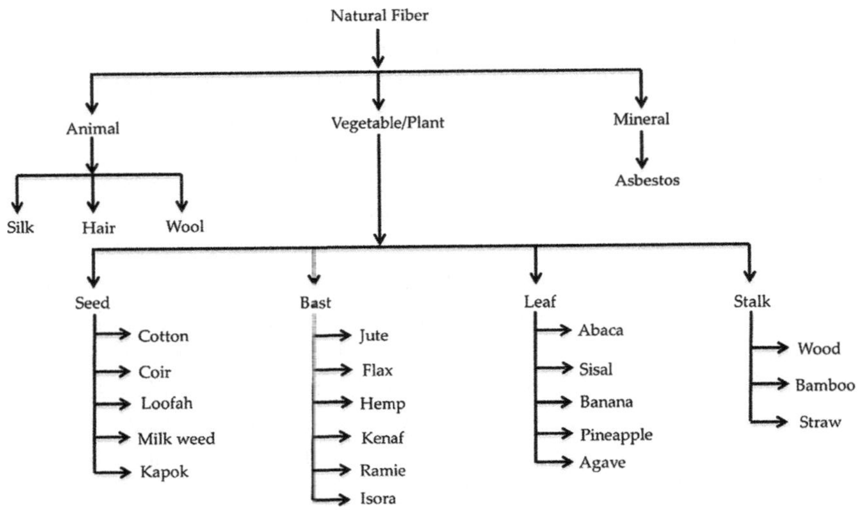

Figure 9.3: Exploring the diverse origins of natural fibers: a classification [29].

automotive interiors, finding applications in components such as side panels, rear cargo covers, seating surfaces, interior trim, and instrument panels [31].

9.3 Applications of natural fiber composites in wind turbine rotors

Batu et al. [32] conducted a comprehensive evaluation of the performance characteristics of natural fiber-reinforced composites when applied to the fabrication of wind turbine blades, highlighting significant advantages over glass/epoxy composites. Replacing glass/epoxy with natural fibers reduced blade weight by 27.9–38 %, with fibers such as Ramie, Curaua, and pineapple showing comparable mechanical performance in stress, deformation, and natural frequencies. Additionally, these materials offer renewable, recyclable, and cost-effective benefits.

Kalagi et al. [33] demonstrated that flax and sisal fiber-reinforced composites could replace traditional materials in various wind turbine components. Similarly, Temesgen et al. [34] investigated hybrid composites using E-glass, Nacha (*Hibiscus macranthus*), and sisal fibers. Using a Taguchi design and hand lay-up fabrication, the study identified optimal fiber compositions based on tensile, compressive, and flexural strengths, with Nacha fiber significantly influencing performance.

Rozikin [35] explored NFRP composites made from cane dregs fiber treated with 5 % NaOH delignification. Compression-molded specimens demonstrated optimal impact strength (0.040 J/mm^2), tensile strength (7,700 MPa), and tensile elastic modulus

(1,287.03 MPa) after 2-h treatments. These findings suggest delignification treatment enhances mechanical properties for wind turbine applications.

Balachandran et al. [36] studied hybrid natural fiber composites (HNFCs) in solar desalination systems, integrating Pharsalus vulgaris and nano-silica insulation. Experimental results showed significant increases in freshwater production (35 % and 21 %) compared to conventional systems, underscoring the thermal efficiency and insulation potential of HNFCs.

Beyond wind turbines, NFCs are being explored in hydrogen storage tanks for their lightweight and durable properties, geothermal energy systems for thermal and structural optimization, and biofuel production due to their renewable nature. NFCs are also under investigation for carbon capture and storage applications, demonstrating their versatility in addressing energy challenges. Ongoing research into these technologies highlights the multifaceted potential of NFCs in advancing sustainable energy solutions.

9.4 Life cycle assessment

The utilization of diverse products in everyday life inevitably exerts a detrimental impact on the environment throughout their entire lifecycle. Recognizing and comprehending the ensuing environmental consequences of a product, along with discerning variations in their environmental performance, is paramount. This crucial understanding has garnered significant attention from a wide range of stakeholders who are actively striving to mitigate environmental harm and foster sustainable practices. Currently, Life Cycle Assessment (LCA) stands as a leading approach for comprehensive assessment and quantification of the environmental repercussions of products, services, and materials across their entire lifecycle [2].

The versatility of the LCA methodology has significantly increased since its inception. While initially centered on evaluating final products, its application now encompasses a broader spectrum of activities, including those related to resource extraction, manufacturing processes, and end-of-life disposal [37].

Life Cycle Assessment serves as a tool for evaluating prospective environmental consequences and resource utilization across the entire life cycle of a product. This encompasses stages, such as raw material acquisition, production, utilization, and waste management. The waste management stage comprises both disposal and recycling. The term 'product' encompasses both tangible goods and intangible services and is a thorough evaluation that takes into account all characteristics and aspects of the natural environment, human health, and resources [38].

Various types of LCA exist, and in general a more thorough analysis requires collecting as much comprehensive information as possible. The level of analytical rigor required for an LCA study varies significantly depending on its intended purpose. For instance, an internal assessment may necessitate less in-depth analysis compared to a report intended for public dissemination or marketing purposes. Furthermore, the scope

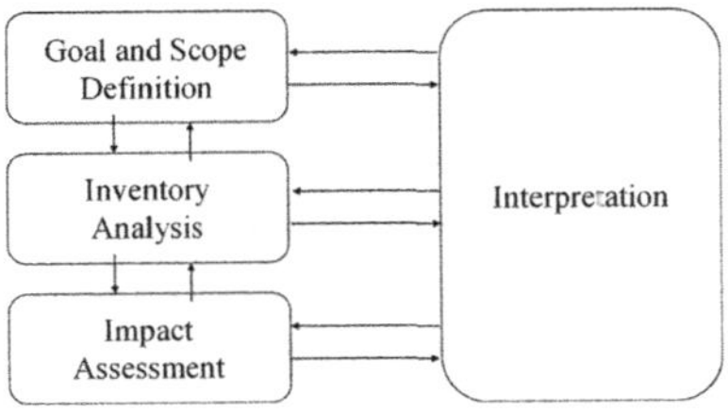

Figure 9.4: Life cycle assessment methodology [40].

of LCA assessments encompasses a wide spectrum, including the development of Environmental Product Declarations (EPDs) adhering to specific product or sector standards, focused evaluations of single environmental impacts such as carbon or water footprints, Social LCAs, and long-term monitoring studies. All LCA studies, regardless of their specific scope, should adhere to the standardized framework outlined in ISO 14040:2006 and ISO 14044:2006 [38, 39], according to which LCA was characterized by four sequential stages illustrated in Figure 9.4.

Life Cycle Assessment consists of four main phases. The initial phase, Goal and Scope Definition, establishes the study's objectives and framework. During this phase, researchers define the project's intended use and research questions. They also determine key parameters like system boundaries, which may extend from cradle-to-gate or cradle-to-grave, and establish functional units that serve as quantifiable reference points for analysis. These parameters enable effective comparison between different systems and products [41].

Following this, the Life Cycle Inventory (LCI) phase focuses on collecting comprehensive data about resource flows throughout a product's lifecycle based on the defined functional unit. This involves detailed documentation of inputs, such as raw materials and energy consumption, as well as outputs like emissions to air and water. The LCI produces a detailed overview of environmental interactions within the studied system [2].

The third phase, Life Cycle Impact Assessment (LCIA), evaluates the environmental significance of the collected inventory data. Researchers analyze various environmental impact categories, such as carbon and water footprints, along with effects like eutrophication, acidification, and human toxicity, to understand the environmental implications of the studied product or material [42].

The final phase involves interpreting the analyzed data to reach meaningful conclusions. During this interpretation stage, researchers evaluate environmental performance metrics, conduct comparative analyses between different products or materials, and identify potential areas for environmental improvement while considering trade-offs between various impact categories [43].

9.5 Life cycle assessment of natural fiber composites

Utilizing natural fiber composites in various applications is recognized for their environmentally friendly attributes, which are attributed to the extraction of raw materials that have minimal adverse effects on the environment, including low CO_2 release, minimal energy consumption, and the absence of soil contamination. Particularly in the automotive industry, components constructed from natural fiber composites tend to have a lower overall weight. Consequently, vehicles incorporating these components experience reduced weight, which leads to lower fuel consumption. This, in turn, results in decreased carbon dioxide emissions, aligning with environmental preservation efforts. Therefore, the adoption of natural fiber composites is a more environmentally favorable choice, particularly for addressing concerns related to environmental impact [2].

Saraiva et al. [44] examined the environmental sustainability of two mango packaging options: one crafted from a composite material of polyethylene and natural fibers and the other from cardboard. The findings suggest that composite packaging exhibits a reduced environmental footprint compared to the cardboard alternative, especially after four uses in Brazil and after 29 or 35 reuses in Europe, depending on the fiber content. The composite packaging comprises a reusable frame composed of high-density polyethylene reinforced with residual natural sponge fibers paired with a recyclable tray made from high-impact polystyrene. The study also identified potential enhancements and underscored the significance of considering context-specific evaluations in assessing environmental impacts.

A research investigation by Pegoretti et al. [45] examined three different acoustic automotive components manufactured in Brazil's automotive industry. Their analysis compared dual-layer polyurethane (DL-PU) panels, recycled textile absorption (ABA-cotton) panels, and recycled textile DL (DL-cotton) panels. Using Life Cycle Assessment methodology, they evaluated each panel type across production, usage, and end-of-life stages. Their findings revealed that dual-layer recycled fiber panels (DL-cotton), which incorporate two layers of varying densities, demonstrated superior environmental performance. This advantage persisted regardless of end-of-life scenarios, primarily due to several factors: the component's lighter weight which benefits transportation efficiency, reduced fossil resource requirements, lower energy consumption during manufacturing, and the prevention of textile waste disposal. The research also identified distinctive characteristics of recycled fibers compared to virgin materials, particularly noting their dual environmental impact through agricultural production contributions and landfill avoidance [45].

Luz et al. [46] investigated the environmental benefits of replacing talc with sugarcane bagasse fibers as reinforcement in polypropylene composites for automotive components. Using Life Cycle Assessment (LCA), the study compared the environmental performance of two polypropylene composites: one reinforced with talc and the other with sugarcane bagasse fibers. The results showed that the sugarcane bagasse fiber-reinforced composite had a lower environmental impact than the talc-reinforced composite, with lower global warming potential, acidification potential, and eutrophication potential.

Akhshik et al. [47] compared the environmental performance of a hybrid fiber-reinforced engine beauty cover to that of a glass fiber-reinforced version. The study concluded that the hybrid fiber-reinforced engine beauty cover has a lower environmental impact than the glass fiber-reinforced counterpart. It not only reduces emissions during car operation due to fuel savings from lightweighting but also has a more environmentally friendly production and end-of-life process. The study noted that the hybrid composite, despite containing 10 % carbon fiber, outperforms the glass fiber composite. Additionally, the cost of producing the hybrid fiber-reinforced engine beauty cover was lower than the glass fiber-reinforced alternative.

An LCA study was conducted to evaluate the environmental impact of using natural fibers as alternative materials in paper pulp production. The investigation focused on two natural fibers, flax and hemp, and performed LCA analysis to compare the environmental effects of both. The study adopted a cradle-to-gate approach, covering the entire life cycle from raw material cultivation to paper pulp production (pulp mill gate). The LCA was modeled using SimaPro software based on the CML baseline 2000 LCIA method. The environmental impacts were assessed in terms of global warming potential, acidification, eutrophication, photochemical oxidant formation, energy use, and pesticide application. The results showed that flax fiber is more environmentally friendly than hemp fiber for paper pulp production. However, flax fiber had higher irrigation and pesticide use compared to hemp. The study suggested improvements in these areas to make hemp fiber a more viable raw material for paper pulp production [48].

9.6 Life cycle assessment for energy applications

Nan Cong et al. [49] investigated the carbon reduction potential of various end-of-life (EoL) wind turbine blade disposal methods in China. The study aimed to assess the carbon reduction potential of different disposal scenarios using life cycle assessment (LCA) methodology. The authors evaluated the environmental impacts of four disposal methods: landfill, incineration, mechanical recycling, and pyrolysis. The results revealed that pyrolysis was the most environmentally friendly method, followed by mechanical recycling, incineration, and landfill. The study concluded that EoL wind turbine blade disposal in China has considerable carbon reduction potential, and the adoption of pyrolysis and mechanical recycling can significantly reduce carbon emissions.

The study conducted by Angela J. Nagle et al. [50] examined three disposal options for composite waste from decommissioned wind turbine blades in Ireland: co-processing in cement kilns in Germany, co-processing in Ireland, and landfill in Ireland. The study aimed to determine the most sustainable disposal method for Irish blade waste over the next decade using life cycle assessment. The results showed that co-processing in Ireland had the lowest environmental impact due to material substitution and reduced transportation. Material substitution was found to have a greater effect than the added transport distance between Ireland and Germany. However, the authors warned that the convenience of co-processing could discourage repurposing efforts, stressing the need for further research comparing co-processing costs with alternative repurposing methods. The study also established a baseline impact scenario for future comparison with repurposing solutions, aligning with higher priorities in the European Waste Hierarchy.

The research, conducted by Casas-Ledón et al. [51] examined the environmental effects of producing a new insulation material (panel) derived from Eucalyptus bark fibers. The study focused on four different variations of eucalyptus bark panels with varying bulk densities (25, 50, 75, and 100 kg/m^3). Life cycle assessments (LCA) were performed for each panel type following the ISO 14040 standards. The results revealed that the embodied energy and carbon emissions of eucalyptus bark panels were lower compared to traditional insulation materials such as expanded polyurethane, polystyrene, glass fibers, and glass wool.

In a study conducted by Comodi et al. [52], they evaluated two different residential solar water heating configurations using life cycle assessment methodology. They compared a system using unglazed polypropylene solar collectors connected to a 300-L storage tank against a conventional design utilizing glazed panels with an equivalent storage capacity. Their analysis, performed with the Eco-Indicator 99 method using the Egalitarian Approach, revealed scores of 49.7 and 18.3 eco-indicator points for the glazed and unglazed systems respectively. The researchers analyzed various performance indicators including energy efficiency, carbon dioxide emissions, and economic returns for both configurations. They evaluated payback periods considering both gas and electric heating alternatives. The unglazed panel configuration achieved energy payback within 2–5 months, while the glazed system required 5–12 months. Regarding carbon dioxide emissions, the unglazed panels reached payback in one to two months, whereas the glazed system needed 12–30 months.

The collective findings from multiple research studies demonstrate the essential function of Life Cycle Assessment in evaluating environmental effects across different energy applications. These investigations have established LCA's effectiveness as a quantitative tool for measuring environmental impacts of energy systems. This quantitative insight becomes instrumental in optimizing the design and operation of energy systems and identifying areas that warrant improvement. Through the judicious application of LCA, the overarching goal is to guarantee that energy systems are not only effective in meeting their intended purposes, but are also crafted and managed in a manner that aligns with the principles of environmental sustainability. Thus, the synthesis of these studies reinforces the imperative of integrating LCA as a foundational tool for the ongoing pursuit of environmentally conscious energy solutions.

9.7 Conclusions and future recommendations

Studies utilizing Life Cycle Assessment methods demonstrate the significant advantages of natural fiber composites (NFCs), especially in energy applications like wind turbine blade construction. Natural materials including Ramie, Curaua, flax, and sisal demonstrate mechanical capabilities similar to conventional glass/epoxy composites while offering substantial weight reductions between 27.9 % and 38 %. These weight savings enhance wind turbine performance while supporting more environmentally sustainable and economical energy production methods. Wind turbine blades manufactured with NFCs demonstrate improved environmental performance compared to traditional materials, which typically generate considerable carbon emissions. Life Cycle Assessment serves as an essential analytical framework for measuring environmental impacts across a product's complete lifecycle, spanning from initial material extraction through final disposal or recycling processes. By assessing environmental, health, and resource impacts, LCA helps identify areas for improvement. Studies consistently show that NFCs offer significant environmental benefits across various energy application. These benefits include reduced CO_2 emissions, lower energy consumption, and weight reduction, all of which contribute to more efficient and less resource-intensive production processes. As industries seek to reduce their carbon footprints, NFCs offer a promising alternative to conventional, energy-intensive materials. In the energy sector, LCA has been crucial for evaluating material sustainability. For example, LCA studies on wind turbine blade disposal have identified sustainable methods, such as mechanical recycling and pyrolysis. Additionally, LCA is used to analyze composite waste management and insulation material production, helping to minimize waste and carbon emissions. These studies emphasize the role of LCA in optimizing energy systems by reducing energy consumption and mitigating carbon emissions throughout a product's life cycle.

While NFCs hold great promise, challenges remain in fully optimizing their sustainability. Key areas for improvement include raw material extraction, production processes, durability during use, and end-of-life management. Enhancing the mechanical properties of NFCs and their performance in extreme conditions is critical for their wider adoption. Future research should focus on refining LCA models for NFCs in energy applications, as well as exploring their potential in energy storage, where their lightweight and sustainable properties could provide substantial benefits. Expanding NFC applications within the energy sector is also essential. Exploring new uses, such as energy-efficient insulation or components for solar energy systems, could lead to innovative, environmentally friendly solutions. These advancements would contribute to the broader goal of developing more sustainable energy systems. Ultimately, the integration of NFCs into energy applications will help reduce the environmental impact of energy production and waste management, advancing sustainability in the energy industry.

Acknowledgments: The authors would like to thank the editors S.M. Sapuan, Mohd Roshdi Hassan, Eris Elianddy Supeni and Azizan As'arry for their guidance and review of this article before its publication.

References

1. Peças P, Carvalho H, Salman H, Leite M. Natural fibre composites and their applications: a review. J Compos Sci 2018;2:66.
2. Mansor MR, Salit MS, Zainudin ES, Aziz NA, Ariff H. Life cycle assessment of natural fiber polymer composites. Agric Biomass Potential Mater 2015:121–41. https://doi.org/10.1007/978-3-319-13847-3_6/FIGURES/2.
3. Mansor MR, Mastura MT, Sapuan SM, Zainudin AZ. The environmental impact of natural fiber composites through life cycle assessment analysis. Durability Life Predict Biocompos, Fibre-Reinf Compos Hybrid Compos 2018;257–85.
4. Gholampour A, Ozbakkaloglu T. A review of natural fiber composites: properties, modification and processing techniques, characterization, applications. J Mater Sci 2020;55:829–92.
5. Mohammed L, Ansari MNM, Pua G, Jawaid M, Islam MS. A review on natural fiber reinforced polymer composite and its applications. Int J Polym Sci 2015;2015. https://doi.org/10.1155/2015/243947.
6. Kamarudin SH, Mohd Basri MS, Rayung M, Abu F, Ahmad S, Norizan MN, et al. A review on natural fiber reinforced polymer composites (NFRPC) for sustainable industrial applications. Polymers 2022;14:3698.
7. CompositesWorld. Composites end markets: energy. 2023. https://www.compositesworld.com/articles/composites-end-markets-energy-2023 [Accessed 04 Nov 2023].
8. Kalagi GR, Patil R, Nayak N. Experimental study on mechanical properties of natural fiber reinforced polymer composite materials for wind turbine blades. Mater Today Proc 2018;5:2588–96.
9. Reddy KS, Singla H. Optimization of woven jute/glass fibre-reinforced polyester hybrid composite solar parabolic trough collector. In: IOP conference series: materials science and engineering. Hiroshima, Japan: IOP Publishing; 2017:012016 p.
10. Yuan H, Wang F, Li S, Lin Z, Huang J. A cellulose substance derived nanofibrous CoS–nanoparticle/carbon composite as a high-performance anodic material for lithium-ion batteries. New J Chem 2020;44:1846–57.
11. Syduzzaman M, Al Faruque MA, Bilisik K, Naebe M. Plant-based natural fibre reinforced composites: a review on fabrication, properties and applications. Coatings 2020;10:973.
12. Life cycle assessment (LCA) – complete beginner's guide. https://ecochain.com/blog/life-cycle-assessment-lca-guide/ [Accessed 04 Nov 2023].
13. 2.4. Life cycle assessment | EME 807: technologies for sustainability systems. https://www.e-education.psu.edu/eme807/node/690 [Accessed 04 Nov 2023].
14. Ebrahimnezhad-Khaljiri H, Eslami-Farsani R, Najafi M, Saeedi A. Introduction to epoxy/natural fiber composites. In: Handbook of epoxy/fiber composites. Singapore: Springer; 2022:1–31 pp.
15. Mitra BC. Environment friendly composite materials: biocomposites and green composites. Def Sci J 2014;64:244–61.
16. Natural fibers, biopolymers, and biocomposites – Google Books. https://books.google.com.my/books?hl=en&lr=&id=AwXugfY2oc4C&oi=fnd&pg=PP1&ots=gNQeMC37TO&sig=9cHOf_7AGaQJPJwwRoAO_95BLAE&redir_esc=y#v=onepage&q&f=false [Accessed 04 Nov 2023].
17. Dicker MPM, Duckworth PF, Baker AB, Francois G, Hazzard MK, Weaver PM. Green composites: a review of material attributes and complementary applications. Compos Part A Appl Sci Manuf 2014;56:280–9.
18. Cristaldi G, Latteri A, Recca G, Cicala G. Composites based on natural fibre fabrics. In: Woven fabric engineering, Chapter 17, London, UK: Intech Open; 2010

19. Huda MS, Drzal LT, Ray D, Mohanty AK, Mishra M. Natural-fiber composites in the automotive sector. In: Properties and performance of natural-fibre composites. UK: Woodhead Publishing Limited; 2008:221–68 pp.
20. Pickering KL, Aruan Efendy MG, Le TM. A review of recent developments in natural fibre composites and their mechanical performance. Compos Part A Appl Sci Manuf 2016;83:98–112.
21. Rohan T, Tushar B, Mahesha GT. Review of natural fiber composites. IOP Conf Ser Mater Sci Eng 2018;314: 012020.
22. yan Cheung H, po Ho M, Tak Lau K, Cardona F, Hui D. Natural fibre-reinforced composites for bioengineering and environmental engineering applications. Compos B Eng 2009;40:655–63.
23. Cesarino I, Carnietto MB, Bronzato GRF, Leao AL. Fabrication of pineapple leaf fibers reinforced composites. Green Energy Technol 2020:265–77.
24. Radoor S, Karayil J, Rangappa SM, Siengchin S, Parameswaranpillai J. A review on the extraction of pineapple, sisal and abaca fibers and their use as reinforcement in polymer matrix. Express Polym Lett 2020;14:309–35.
25. Sanjay MR, Arpitha GR, Yogesha B. Study on mechanical properties of natural – glass fibre reinforced polymer hybrid composites: a review. Mater Today Proc 2015;2:2959–67.
26. Qi Z, Wang B, Sun C, Yang M, Chen X, Zheng D, et al. Comparison of properties of poly(lactic acid) composites prepared from different components of corn straw fiber. Int J Mol Sci 2022;23:6746.
27. Ortega Z, Bolaji I, Suárez L, Cunningham E. A review of the use of giant reed (Arundo donax L.) in the biorefineries context. Rev Chem Eng 2024;40:305–28.
28. Kumar S, Manna A, Dang R. A review on applications of natural Fiber-Reinforced composites (NFRCs). Mater Today Proc 2022;50:1632–6.
29. Hasan KMF, Horváth PG, Alpár T. Potential natural fiber polymeric nanobiocomposites: a review. Polymers 2020;12:1072.
30. Li Y, Mai YW, Ye L. Sisal fibre and its composites: a review of recent developments. Compos Sci Technol 2000;60:2037–55.
31. Gurunathan T, Mohanty S, Nayak SK. A review of the recent developments in biocomposites based on natural fibres and their application perspectives. Compos Part A Appl Sci Manuf 2015;77:1–25.
32. Batu T, Lemu HG, Sirhabizuh B. Study of the performance of natural fiber reinforced composites for wind turbine blade applications. Adv Sci Technol 2020;14:67–75.
33. Kalagi GR, Patil R, Nayak N. Experimental study on mechanical properties of natural fiber reinforced polymer composite materials for wind turbine blades. Mater Today Proc 2018;5:2588–96.
34. Miliket TA, Ageze MB, Temesgen Tigabu M, Zeleke A. Experimental characterizations of hybrid natural fiber-reinforced composite for wind turbine blades. Heliyon 2022;8:e09092.
35. Rozikin M, Kalimantan IT, Rozikin MN, Laksono AD, Zulfikar A. Composite Nfrp (natural fiber reinforced plastic) cane dregs fiber with 5% NaOH delignification treatment using compression molding techniques in wind turbine blades applications. 2023. https://doi.org/10.21203/RS.3.RS-3023271/V1.
36. Balachandran GB, David PW, Vijayakumar ABP, Kabeel AE, Athikesavan MM, Sathyamurthy R. Enhancement of PV/T-integrated single slope solar desalination still productivity using water film cooling and hybrid composite insulation. Environ Sci Pollut Control Ser 2020;27:32179–90.
37. (16) (PDF) ReCiPE 2008: a life cycle impact assessment method which comprises harmonised category indicators at the midpoint and the endpoint level. https://www.researchgate.net/publication/302559709_ReCiPE_2008_A_life_cycle_impact_assessment_method_which_comprises_harmonised_category_indicators_at_the_midpoint_and_the_endpoint_level [Accessed 29 Nov 2023].
38. ISO. ISO 14040 international standard, environmental management-life cycle assessment-principles and framework. Geneva, Switzerland: International Organisation for Standardization; 2006. https://www.iso.org/standard/37456.html.
39. ISO 14044:2006 – environmental management – life cycle assessment – requirements and guidelines. https://www.iso.org/standard/38498.html [Accessed 29 Nov 2023].

40. Fonseca A, Ramalho E, Gouveia A, Figueiredo F, Nunes J. Life cycle assessment of PLA products: a systematic literature review. Sustainability 2023;15:12470.
41. Muralikrishna IV, Manickam V. Environmental impact assessment and audit. Environ Manag 2017:77–111. https://doi.org/10.1016/B978-0-12-811989-1.00006-3.
42. Rebitzer G, Ekvall T, Frischknecht R, Hunkeler D, Norris G, Rydberg T, et al. Life cycle assessment: Part 1: framework, goal and scope definition, inventory analysis, and applications. Environ Int 2004;30:701–20.
43. Curran MA. Life cycle assessment handbook: a guide for environmentally sustainable products. United States: John Wiley & Sons; 2012.
44. Saraiva AB, Pacheco EB, Gomes GM, Visconte LL, Bernardo C, Simões CL, et al. Comparative lifecycle assessment of mango packaging made from a polyethylene/natural fiber-composite and from cardboard material. J Clean Prod 2016. https://doi.org/10.1016/j.jclepro.2016.08.135.
45. dos Santos Pegoretti T, Mathieux F, Evrard D, Brissaud D, Roberto de Franç Arruda J. Use of recycled natural fibres in industrial products: a comparative LCA case study on acoustic components in the Brazilian automotive sector. Resour Conserv Recycl 2014;84:1–14.
46. Luz SM, Caldeira-Pires A, Ferrão PMC. Environmental benefits of substituting talc by sugarcane bagasse fibers as reinforcement in polypropylene composites: ecodesign and LCA as strategy for automotive components. Resour Conserv Recycl 2010;54:1135–44.
47. Akhshik M, Panthapulakkal S, Tjong J, Sain M. Life cycle assessment and cost analysis of hybrid fiber-reinforced engine beauty cover in comparison with glass fiber-reinforced counterpart. Environ Impact Assess Rev 2017;65:111–17.
48. González-García S, Hospido A, Feijoo G, Moreira MT. Life cycle assessment of raw materials for non-wood pulp mills: hemp and flax. Resour Conserv Recycl 2010;54:923–30.
49. Cong N, Song Y, Zhang M, Wu W. Life cycle assessment of carbon reduction potential of EoL wind turbine blades disposal scenarios in China. Environ Impact Assess Rev 2023;100:107072.
50. Nagle AJ, Delaney EL, Bank LC, Leahy PG. A Comparative Life Cycle Assessment between landfilling and Co-Processing of waste from decommissioned Irish wind turbine blades. J Clean Prod 2020;277:123321.
51. Casas-Ledón Y, Daza Salgado K, Cea J, Arteaga-Pérez LE, Fuentealba C. Life cycle assessment of innovative insulation panels based on eucalyptus bark fibers. J Clean Prod 2020;249:119356.
52. Comodi G, Bevilacqua M, Caresana F, Paciarotti C, Pelagalli L, Venella P. Life cycle assessment and energy-CO2-economic payback analyses of renewable domestic hot water systems with unglazed and glazed solar thermal panels. Appl Energy 2016;164:944–55.

Yusuf Jameel, Abdul Halim Muhammad Firdaus and Mohd Sapuan Salit*

10 Nanocomposites in energy conversion

Abstract: Nanocomposites, particularly those based on graphene, exhibit considerable ability in energy storage and conversion due to their unique characteristics and simplicity of manufacturing and functionalization. High carrier mobility, rapid charge recombination rate, along with long-term stability are only a few of these materials exceptional qualities. This chapter provides an outline of the recent advances in the nanocomposites and uses of nanocomposites in the disciplines of energy conversion (solar cells) and storage (super-capacitors). This chapter emphasizes the difficulties in using graphene-based materials in solar cells and super-capacitors. Future studies are also suggested to create innovative methods for the design and synthesis of graphene-based nanocomposites, focussing on functionalization strategies, dispersion techniques, and structural optimization.

Keywords: bio composites; energy conversion; metal organic framework; supercapacitor

10.1 Introduction

Nanocomposites are multiphase solid materials with minimum one nano scaled element, have been attracting attention due to their interesting properties. It is composed of two or more separate components, each of which has unique features. Energy conversion and storage have advanced significantly because of the development of nanostructured and composite materials. Low-dimensional materials and composites are being designed and developed for a variety of applications, such as photocatalysts for photoelectrochemical devices that produce solar fuel, semiconductor nanomaterials for next-generation solar cells, high specific surface area electrodes for effective energy storage systems like supercapacitors and batteries, and nano catalysts for increased triple phase boundaries in fuel cells [1]. Combining them results in a material that is specifically designed to accomplish a specific purpose, such as becoming lighter, or stronger. It also increases the material's stiffness and strength. There are several ways that nanocomposites can be

***Corresponding author: Mohd Sapuan Salit**, Advanced Engineering Materials and Composites Research Centre (AEMC), Department of Mechanical and Manufacturing Engineering, Universiti Putra Malaysia, 43400 UPM Serdang, Selangor, Malaysia, E-mail: sapuan@upm.edu.my. https://orcid.org/0000-0002-4598-6358
Yusuf Jameel and Abdul Halim Muhammad Firdaus, Advanced Engineering Materials and Composites Research Centre (AEMC), Department of Mechanical and Manufacturing Engineering, Universiti Putra Malaysia, 43400 UPM Serdang, Selangor, Malaysia, E-mail: yusufjameel111@gmail.com (Y. Jameel), muhdfirdaus0701@gmail.com (A.H. Muhammad Firdaus). https://orcid.org/0000-0003-3583-8383 (Y. Jameel). https://orcid.org/0009-0004-4853-1449 (A.H. Muhammad Firdaus)

As per De Gruyter's policy this article has previously been published in the journal Physical Sciences Reviews. Please cite as: Y. Jameel, A. H. Muhammad Firdaus and M. S. Salit "Nanocomposites in energy conversion" *Physical Sciences Reviews* [Online] 2024. DOI: 10.1515/psr-2024-0028 | https://doi.org/10.1515/9783111234762-010

used for energy conversion, including decreasing the weight of the devices or structures, increasing electrical conductivity, and improving thermal insulation [2]. A new field of study is emerging for the development of green energy alternatives and its effective storage technologies considering the world's growing consumption of fossil fuels and environmental concerns. Nanocomposites are finding more and more applications in energy conversion technologies like lithium-ion batteries, solar cells, and piezoelectric materials. When the matrix in nanocomposites is reinforced with nanocellulose, graphene the power conversion efficiency of solar cells and the piezoelectric performance of piezoelectric materials can both be enhanced [3]. Graphene is a two-dimensional sheet of carbon atoms with a high surface area, mechanical strength, and electrical and thermal conductivity [4]. To enhance the qualities and capabilities of nanocomposites for solar cells, graphene can be used as a coating or filler. Graphene-based nanocomposites can be used to silicon, perovskite, or quantum dot solar cells, among other solar cell types and technologies [5]. Some of the advantages of graphene-based nanocomposites for solar cells are:

10.2 Properties of graphene

10.2.1 High light absorption

Graphene can function as a sensitizer to boost the solar cells' capacity to absorb light and produce charge. Additionally, graphene can form a heterojunction with silicon or perovskites to improve the solar cells' ability to absorb light and transfer charges [6].

10.2.2 High charge transport

By acting as a conductor, graphene can improve the solar cells' electrical conductivity and power factor. Additionally, graphene has the potential to function as a buffer layer, enhancing the stability and quality of the solar cells' interface. Additionally, graphene has the potential to function as an electrode to enhance solar cells' ability to transfer and collect electrons [7].

10.2.3 High stability

Graphene has the capacity to shield solar cells from oxidation and deterioration. Additionally, graphene can function as a passivation layer to lower surface recombination and boost solar cell efficiency. Additionally, graphene can be used as a modifier to adjust the solar cells' band gap and operational characteristics [8].

10.3 Background of nanocomposite-based materials

Nanocomposites are materials made of different components with multiple discrete phase domains, at least one of which is a continuous phase, and at least one phase having at least one dimension of the order of nanometres. The presence of polymer is used to further categorize the nanocomposites (polymer-based or non-polymer-based nanocomposites). There are various types of polymer-based nanocomposites that we may identify, such as polymer/ceramic, polymer/polymer, polymer/layered silicate, biocomposite, inorganic/organic polymer, and inorganic/organic hybrid nanocomposites. Metal/metal, ceramic/ceramic, and metal/ceramic nanocomposites are three categories for non-polymer based nanocomposites [9]. They can be divided into four categories: metal oxide-based, polymer-based, carbon nanotube-based, and noble metal-based nanocomposite, depending on the type of reinforcing fabric used. A polymer matrix including evenly dispersed filler particles (silicas, for example) with at least one nanometric dimension makes up a polymeric nanocomposite (PNC), a bi-phase material. The polymer matrix may consist of thermosets and thermoplasts, sometimes known as plastomers or elastomers. Good optical qualities, size stability, decreased surface defect presence, enhanced thermal stability, and flame resistance are just a few of their many positive attributes. Regretfully, they are expensive, the filler tends to clump together, and they cause dispersion issues in polymers [10].

Depending on the intended use, the electrode materials' characteristics can be adjusted. The 2D material could also be integrated with various oxides, conductive polymers, and/or carbonaceous materials to improve its mechanical strength, stability, physical, and electrochemical properties of nanocompoites. The materials used in nanocomposite systems show promise for a number of uses, including energy conversion, hydrogen storage, and energy storage (as electrode material in supercapacitors and battery devices) [11]. Numerous synthesis techniques, including chemical vapour deposition, hydrothermal method, microwave procedure, and simple spray coating approach, can be used to fabricate the nanocomposites [12].

Nanocomposites have become vital materials in energy conversion systems due to their ability to enhance performance and efficiency. Their high surface area, excellent conductivity, and chemical tunability make them suitable for applications in supercapacitors, batteries, and fuel cells.

In supercapacitors, nanocomposites are used to increase energy density and improve charge-discharge rates. Carbon-based nanocomposites, such as graphene and carbon nanotubes, provide a high surface area for ion adsorption and rapid electron transport. Incorporating metal oxides like MnO_2 or conducting polymers into nanocomposites further boosts capacitance and stability, enabling high-performance energy storage systems.

For batteries, nanocomposites enhance the electrochemical properties of electrodes. In lithium-ion batteries, silicon or graphene-based nanocomposites are widely used to improve capacity and cycling stability. Solid-state batteries also benefit from nanocomposite electrolytes, which ensure high ionic conductivity and mechanical strength, addressing issues like dendrite formation.

In fuel cells, nanocomposites improve the catalytic activity and durability of electrodes. Platinum-based nanocomposites are commonly used in proton exchange membrane (PEM) fuel cells for their high efficiency and resistance to catalyst poisoning. Carbon nanocomposite supports ensure uniform dispersion of catalyst particles, enhancing reaction kinetics and overall performance. Usage of nanocomposites in energy conversion is illustrated in Figure 10.1.

10.4 Composites formed of metal-organic framework

The composites formed from metal-organic framework (MOF) exhibit great potential as a material class for energy conversion systems because of their huge surface area, strong catalytic performance, and adjustable pore size (Figure 10.2). Several synthesis procedures, including calcination, hydrothermal reaction, solvothermal, electrodeposition, sonication, self-assembly, template, etc., are used for making MOF-derived composites employing MOFs or MOF-based composites as the precursor [14].

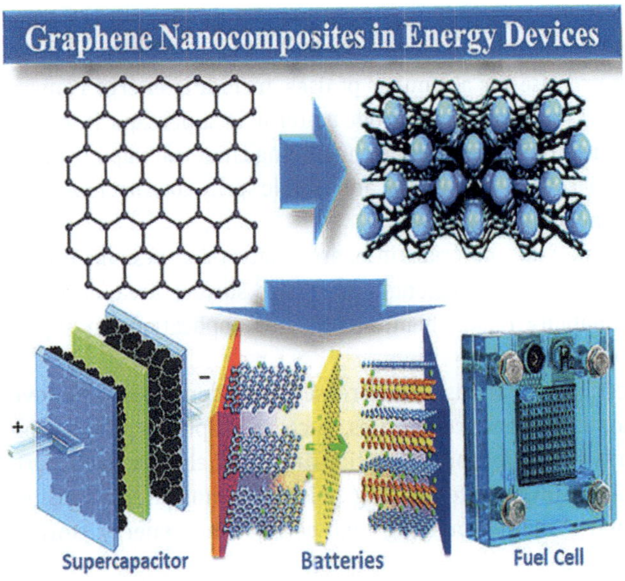

Figure 10.1: Usage of nanocomposites in energy conversion [13].

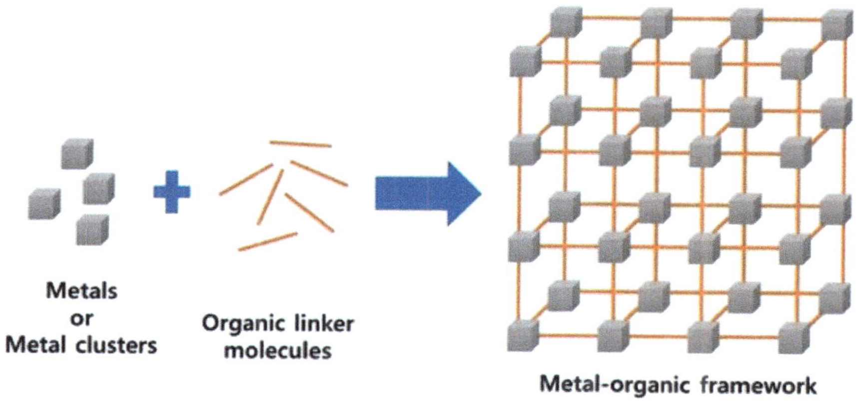

Figure 10.2: Metal-organic framework (MOF) structure [15].

10.4.1 Calcination method

Currently, the majority agree that the calcination process is an effective way to transform MOFs or MOF-based composites into MOF-derived composites [16]. Various atmospheres, including air, H_2, PH_3, Ar, sulphur/selenium vapour, and PH_3, can be used during the calcination process to prepare various MOF-derived composites, such as MOF-derived metal oxides, MOF-derived metal chalcogenides, MOF-derived metal phosphides, and MOF-derived C materials [17].

10.4.2 MOF-derived metal oxides

Using Ni-MOF as a template, Yu et al. synthesised the porous In_2O_3–NiO composites by calcination [18]. According to Wei et al., the CoCu-oxides composites were created by annealing the CoCu-MOF precursor at different temperatures – namely, 300, 350, and 400 °C, respectively [19]. Zhang et al. calcined the MIL-125/Cu-BDC composites for 6 h at 400 °C to create a TiO_2/CuO heterostructure [20].

10.4.3 MOF-derived carbon materials

In terms of the development of MOF-derived C materials, carbonization at a high temperature has attracted more interest lately than metal oxide/chalcogenide/phosphide composites made from MOFs or MOF-based composites [21]. For example, in 2021, Prof. Xu's group manufactured hollow spherical C nanosheets from MOF using a three-step procedure [17]. The FeCoNi–OH/Zn composites in this study were first made using a simple method; they were then transformed into FeCoNi-ZIF/Zn composites by soaking them in a 2-methylimidazole solution for an hour at room temperature; and finally, the FeCoNi/- hollow C spheres composites were created by a two-hour pyrolysis process at 950 °C. FeCoNi/hollow C sphere composites that were the end product displayed unique mesoporous structures that provided more response zones. Additionally, their hollow morphology encouraged gas escape, and the metal particles supplied a large number of surface-active sites. The exceptional catalytic activity of the FeCoNi/hollow C spheres composites for oxygen evolution and reduction processes was facilitated by these characteristics [14].

10.5 Properties of nanocomposites

10.5.1 Large surface area

The large surface area of nanocomposites is a defining feature that significantly enhances their performance across a range of applications. This characteristic is attributed to the nanoscale dimensions of the reinforcing fillers, such as graphene, carbon nanotubes, or nanocellulose, which provide a vastly higher surface area compared to their bulk counterparts. The increased surface area enables stronger interfacial interactions between the filler and the polymer matrix, facilitating better stress transfer. As a result, the nanocomposite exhibits improved mechanical properties, including higher tensile strength, Young's modulus, and impact resistance.

The development of dye-sensitized solar cells (DSCs) is a prime example of how using nanostructures to produce a film can considerably benefit solar cells and potentially provide a far larger internal surface area than bulk material. Which is used for DSC that converts solar energy into electricity. It usually consists of a counter electrode (also known as a cathode) that transports electrons from an external circuit, a working electrode (also known as a photoanode) for light harvesting and charge separation, and an electrolyte filled between the two electrodes that creates redox couples that allow the photoanode to be reduced. Numerous additional nanostructures, including as nanoparticles, nanowires/nanorods, nanotubes, and core–shell structures of oxides like ZnO, SnO_2, Nb_2O_5, and so on, have been used in a considerable deal of work on DSCs. The objective of recent researches are to enable additional functionalities to enhance optical absorption or electron transport while simultaneously giving the photoelectrode film a wide interior surface area [22].

10.5.2 Unique optical effects

An important aspect of employing nanostructures in solar cells is utilising optical phenomena. Nanostructures have been used to increase solar cell efficiency in a number of ways, such as the following: acting as an antireflection layer to lessen incident light loss at the solar cells front interface, (2) producing surface plasmon resonance, or (3) producing light scattering to improve the active layer's optical absorption in DSCs. These methods are based on the fundamental design principles of solar cells, which include potentially enhancing optical absorption and decreasing light loss [23].

10.6 Applications of nanocomposites

10.6.1 Nanostructured materials for solar cell

The primary goal of solar cell design is to decrease electron loss during transport and/or boost the active layer's optical absorption. Solar cell is also called as PV (photovoltaic) cell which converts sunlight energy into the renewable nature of electrical energy by the PV effect [10] (refer Figure 10.3). These materials have different functions such as (1) introduce new mechanisms, such as the multiple exciton generation effect in quantum dots, (2) provide a large surface area, such as the nanocrystalline films used for dye-sensitive solar cells, (3) generate unique optical effects to either reduce the light loss or enhance the optical absorption, based on the methods of antireflection, surface plasmon resonance, or light scattering, and (4) improve electron transport and/or collection by using one-dimensional nanostructures or specifically designed three-dimensional structures, such as host-passivation-guest hollow core–shell spheres. There are different ways from which materials can be modified to maximize light harvesting and electron transport in order to construct high-performance solar cells [24]. Using graphene-based nanocomposites is one approach to tailor nanocomposites to optimise light gathering and electron transport for the making of high-performance solar cells. Commonly employed in conventional photovoltaic cells, inorganic semiconductor materials including gallium arsenide, amorphous silicon, and sulphide salts directly produce free electrons and holes upon photon

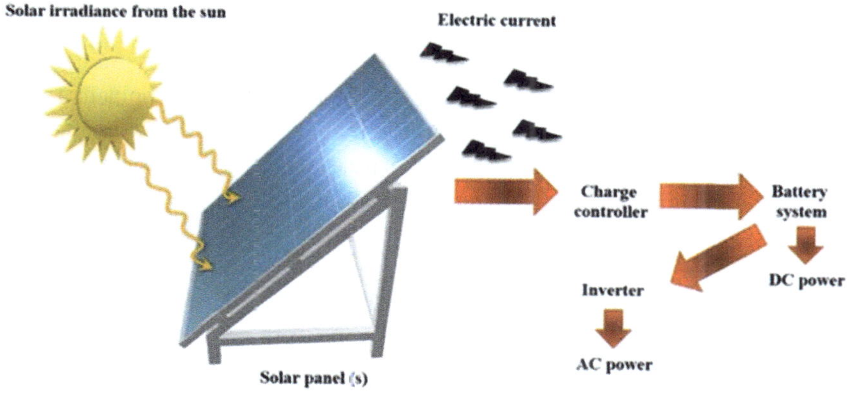

Figure 10.3: Solar cell structure and current production [10].

absorption. Even though using inorganic semiconductor materials in solar cells can achieve a power conversion efficiency (PCE) of more than 40 %, the widespread use of inorganic solar cells is still restricted because it can be challenging to modify the bandgap of inorganic semiconductors, and because the high vacuum and elevated temperature required for the elaborate fabrication processes are expensive [25]. The cost of these inorganic solar cells is significantly higher than that of traditional grid electricity. Alternative methods utilizing organic or polymer materials have drawn a lot of interest because of their affordability, portability, flexibility, and ability to be processed as solutions [26].

The following are a few benefits of nanocomposites for solar cells:

- High light absorption
  By acting as sensitizers, nanocomposites can boost the photovoltaic cells' capacity to absorb light and produce electricity. Additionally, heterojunctions between nanocomposites and other materials, like silicon or perovskites, can improve the photovoltaic cells' ability to absorb light and carry charges [27]. By incorporating nanoparticles with high refractive indices, the path length of light within the solar cell is increased, leading to better absorption [28].
- High charge transport
  By acting as conductors, nanocomposites can improve the solar cells' electrical conductivity and power factor. Additionally, buffer layers made of nanocomposites can enhance the durability and quality of the solar cells' interface. Additionally, nanocomposites can serve as electrodes to enhance solar cells' electron collection and transport [27]. By incorporating materials such as carbon nanotubes or graphene, nanocomposites provide pathways with superior electrical conductivity. This results in faster and more efficient movement of charge carriers (electrons and holes) within the solar cell [28].
- High stability
  The solar cells can be shielded from oxidation and deterioration using nanocomposites. Additionally, nanocomposites can function as passivation layers to lower surface recombination and boost solar cell efficiency. Additionally, nanocomposites can be used as adjusters to adjust the solar cells' band gap and operational efficiency [29]. Nanocomposites are composed of nanoscale materials that provide a robust framework, helping to maintain the integrity of the solar cell under various environmental conditions. This enhanced structural stability ensures that the solar cells remain effective and functional over longer periods, even when exposed to factors such as temperature fluctuations, mechanical stress, and humidity [30].

10.6.2 Nanocomposite materials for supercapacitor applications

When compared to traditional batteries, supercapacitors have a higher energy storage capacity and a higher energy delivery capability. They are electrochemical energy storage devices. These devices also have a long cycle life and excellent stability. Compared to ordinary

Table 10.1: Advantages and disadvantages of nanocomposite in energy Storage.

Advantages	Disadvantages	Materials used	Ref.
– Enhanced energy density and cycling stability in batteries. – Improved energy and power densities in supercapacitors.	– Large-scale synthesis of nanoparticles remains challenging. – Cost-effectiveness and long-term stability need addressing.	– Nanostructured particles for batteries and supercapacitors. – Tailored nanoparticles	[36]
– High energy density in lithium-sulfur batteries. – Efficient solar energy conversion and storage capabilities.	– Low overall efficiency of two-electrode photovoltaic cells (maximum 9 %). – Need for advanced, low-cost photovoltaic electrode materials.	– Organic-based materials, small molecules, polymers, MOFs, COFs. – Lithium, zinc, sodium, titanium dioxide, cadmium sulphide.	[37]
– Nanomaterials exhibit intrinsic strength and flexibility. – Low-dimensional materials enable structural energy storage solutions.	– Poor electrical conductivity and low Li^+ diffusion rate – Generation and fracture of solid-electrolyte interphase (SEI)	– Metal oxide nanostructures: SnO_2 and TiO_2 nanoparticles. – Si quantum dots below 5 nm for Si/C nanocomposites.	[38]

capacitors, these capacitors have a higher power density, a longer service life, and a faster charging speed [23]. Nanocomposite in energy storage applications shows plenty advantages and disadvantages. Table 10.1 shows advantages and disadvantages of nanocomposite in energy storage.

Supercapacitors also have the advantage of being safe, environmentally friendly, and able to function at various temperatures. The demand for wearable technology, medical equipment, industrial equipment, hybrid or electric cars, portable battery backups, and other electronic gadgets necessitates the need for high storage energy devices [31]. However, there are also some challenges in supercapacitors, such as low energy density, high cost, and environmental impact of the materials and devices [32]. By employing various materials and structures to improve the performance and efficiency of supercapacitors, nanocomposites can overcome these obstacles.

Different types of nanocomposites for supercapacitors include the following:

– Metal oxide/carbon nanocomposites
  These are carbon-based materials, such graphene, carbon nanotubes, or carbon nanofibers, combined into nanocomposites. High capacitance and pseudo capacitance can be obtained from metal oxides for the storage of electrical energy. High conductivity and stability can be obtained from carbon-based materials for transmitting electrical energy. For supercapacitors, metal oxide/carbon nanocomposites can offer good cycling stability, power density, and energy density. $MnCo_2O_4/NiCo_2O_4$ nanocomposites, for instance, have a high activity and are inexpensive supercapacitors [33].

- Polymer nanocomposites
  These are composites of polymers and other substances, like organic dyes, metal oxides, or metal nanoparticles. For the purpose of making supercapacitors, polymers can offer flexibility and processability. Other materials have the potential to improve supercapacitors by offering electroactivity and functionality. For supercapacitors, polymer nanocomposites can have high capacitance, a high power factor, and durability. For instance, excellent conductivity and stable supercapacitors can be made using polyaniline/silver nanocomposites [34].
- Carbon-based nanocomposites
  These include graphene, carbon nanotubes, carbon nanofibers, and carbon spheres, among other carbon-based materials. For supercapacitors, carbon materials can offer high conductivity, porosity, and surface area. For supercapacitors, carbon-based nanocomposites can offer good cycling stability, power density, and capacitance. For instance, spherical carbon nanoarchitectures can be employed to create inexpensive, highly effective supercapacitors [35].

10.7 Conclusions

An overview of recent developments in nanocomposites, particularly MOF-derived composite materials for energy conversion systems, is given in this chapter. The many synthetic methods – such as calcination, among others – for creating composites produced from MOFs are summarized. Additionally, a few representative uses of nanostructured materials for solar cells and supercapacitors are outlined in this chapter. Because of their distinct optical and electrical properties as well as their novel geometrical features, nanocomposites materials have demonstrated efficacy in improving the performance of these devices, as this chapter has demonstrated. In summary, nanocomposites have the following advantages: (i) they offer a large surface area that enhances the molecular adsorption or electrochemical reaction that occurs at the solid–liquid or solid–gas interface; (ii) they generate optical effects that improve the optical absorption in solar cells; (iii) they give rise to high crystallinity and/or porous structure to facilitate both electron and ion transport and electrolyte diffusion, so as to ensure the electrochemical process occurs highly.

From a scientific and technical perspective, nanocomposites still present certain difficulties. A thorough research is necessary into certain scientific conflicts or balances in nanocomposite materials. Strength versus toughness in bulk nanocomposites; thermal conductance versus interfacial thermal resistance in thermal management nanocomposites; reflection versus absorption in electromagnetic interference (EMI) nanocomposites; and conversion efficiency versus thermal conductance in photothermal conversion nanocomposites are some examples of these contrasts. One key area of focus is enhancing energy storage systems, such as lithium-ion batteries, supercapacitors, and fuel cells. By incorporating nanoparticles like carbon nanotubes, graphene, and metal oxides, nanocomposites can significantly increase energy density, improve charge/discharge rates, and enhance the overall durability of these devices. Actuators, for instance, need quick heat transfer to achieve excellent optical response, but solar steam generators need to lessen thermal conduction to maintain their high temperature and speed up water evaporation.

Acknowledgements: The authors would like to thank the editors S.M. Sapuan, Mohd Roshdi Hassan, Eris Elianddy Supeni and Azizan As'arry for their guidance and review of this article before its publication.

References

1. Zhong YL, Basu SN, Sun Z. Nanomaterials and composites for energy conversion and storage. J Miner Met Mater Soc 2021;73:2752–3.
2. Sun X, Huang C, Wang L, Liang L, Cheng Y, Fei W, et al. Recent progress in graphene/polymer nanocomposites. Adv Mater 2020;2001105:1–28.
3. Lasrado D, Ahankari S, Kar K. Nanocellulose-based polymer composites for energy applications – a review. J Appl Polym Sci 2020;137:48959.
4. Pal K, Asthana N, Aljabali AA, Bhardwaj SK, Kralj S, Penkova A, et al. A critical review on multifunctional smart materials 'nanographene' emerging avenue: nano-imaging and biosensor applications. Crit Rev Solid State Mater Sci 2022;47:691–707.
5. Ghaleb ZA, Mariatti M, Ariff ZM. Graphene nanoparticle dispersion in epoxy thin film composites for electronic applications: effect on tensile, electrical and thermal properties. J Mater Sci Mater Electron 2017; 28:808–17.
6. Sang T, Dereshgi SA, Hadibrata W, Tanriover I, Aydin K. Highly efficient light absorption of monolayer graphene by quasi-bound state in the continuum. Nanomater (Basel) 2021;11:484.
7. Mics Z, Tielrooij KJ, Parvez K, Jensen SA, Ivanov I, Feng X, et al. Thermodynamic picture of ultrafast charge transport in graphene. Nat Commun 2015;6:7655.
8. Kwon YJ, Kim Y, Jeon H, Cho S, Lee W, Lee JU. Graphene/carbon nanotube hybrid as a multi-functional interfacial reinforcement for carbon fiber-reinforced composites. Composites Part B 2017;122:23–30.
9. Sandri G, Bonferoni M, Rossi S, Ferrari F, Aguzzi C, Viseras C, et al. Clay minerals for tissue regeneration, repair, and engineering. In: Wound healing biomaterials. UK: Woodhead Publishing; 2016.
10. Kurc B, Pigłowska M, Rymaniak Ł, Fu F. Modern nanocomposites and hybrids as electrode materials used in energy carriers. Nanomaterials (Basel) 2021;11:538.
11. Peymanfar R, Mohammadi A, Javanshir S. Preparation of graphite-like carbon nitride/polythiophene nanocomposite and investigation of its optical and microwave absorbing characteristics. Compos Commun 2020;21:100421.
12. Forouzandeh P. 2D Nanomaterials and composites for energy storage and conversion. In: 2D Materials for energy storage and conversion. UK: IOP Publishing Ltd; 2021:1–22 pp.
13. Kausar A, Ahmad I, Zhao T, Aldaghri O, Ibnaouf KH, Eisa MH. Graphene nanocomposites as innovative materials for energy storage and conversion — design and headways. Int J Mol Sci 2023;24:11593.
14. Qian Y, Zhang F, Zhao S, Bian C, Mao H, Kang DJ, et al. Recent progress of metal-organic framework-derived composites: synthesis and their energy conversion applications. Nano Energy 2023;111:108415.

15. Heo DY, Do HH, Ahn SH, Kim SY. Metal-organic framework materials for perovskite solar cells. Polymers 2020;12:2061.
16. Sui X, Huang X, Pu H, Wang Y, Chen J. Tailoring MOF-derived porous carbon nanorods confined red phosphorous for superior potassium-ion storage. Nano Energy 2021;83:105797.
17. Lin J, Zeng C, Lin X, Xu C, Su C-Y. CNT-assembled octahedron carbon-encapsulated Cu3P/Cu heterostructure by in situ MOF-derived engineering for superior lithium storage: investigations by experimental implementation and first-principles calculation. Adv Sci 2020;7:2000736.
18. Yu Q, Jin R, Zhao L, Wang T, Liu F, Yan X, et al. MOF-derived mesoporous and hierarchical hollow-structured in(2)O(3)-NiO composites for enhanced triethylamine sensing. ACS Sens 2021;6:3451–61.
19. Wei C, Li X, Xiang W, Yu Z, Liu Q. MOF derived seaweed-like CoCu oxides nanorod arrays for electrochemical non-enzymatic glucose sensing with ultrahigh sensitivity. Sensor Actuators B Chem 2020;324:128773.
20. Zhang Y, Hu W, Wang D, Reinhart B, Huang J. Electron shuttle in MOF derived TiO 2/CuO heterojuction boosts light driven hydrogen evolution. J Mater Chem A 2021;9:6180–7.
21. Wang C, Kim J, Tang J, Na J, Kang Y, Kim M, et al. Large-scale synthesis of MOF-derived superporous carbon aerogels with extraordinary adsorption capacity for organic solvents. Angew Chem Int Ed Engl 2020;59: 2066–70.
22. Zhang Q, Yodyingyong S, Xi J, Myers D, Cao G. Oxide nanowires for solar cell applications. Nanoscale 2012;4: 1436–45.
23. Zhang Q, Uchaker E, Cao G. Nanomaterials for energy conversion and storage. Chem Soc Rev 2013;42: 3127–71.
24. Aly AM. Performance evaluation of nanostructured solar cells. Seehra MS, editor. Rijeka: IntechOpen; 2017. Ch. 10.
25. Liu J, Xue Y, Zhang M, Dai L. Graphene-based materials for energy applications. MRS Bull 2012;37:1265–72.
26. Tale B, Nemade KR, Tekade PV. Graphene based nano-composites for efficient energy conversion and storage in Solar cells and supercapacitors: a review. Polym Plast Technol Mater 2021;60:784–97.
27. Nongthombam S, Swain BP. Nanowires/graphene nanocomposites for photovoltaic applications BT – nanostructured materials and their applications. Swain BP, editor. Singapore: Springer Singapore; 2021: 131–42 pp.
28. K.V GM, George J, Balachandran M. Polymer-nanocarbon composites: a promising strategy for enhanced performance of organic solar cells. Emerg Mater 2024;7:17–33.
29. Ntanovasilis G, Zaverdas I, Ahmed T, Markoulidis F, Lekakou C. Polymer blends and polymer nanocomposites for photovoltaic (PV) cells and an investigation of the material deposition techniques in PV cell fabrication. J Compos Sci 2021;5:263.
30. Nuriyeva S, Karimova A, Shirinova H, Jafarova S, Abbas G, Zamchiy A, et al. Novel structures for PV solar cells: fabrication of Cu/Cu2S-MWCNTs 1D-hybrid nanocomposite. Micromachines 2024;15:1318.
31. Nayak A, Bhushan B, Kotnala S, Kukretee N, Chaudhary P, Tripathy AR, et al. Materials today: proceedings nanomaterials for supercapacitors as energy storage application: focus on its characteristics and limitations. Mater Today Proc 2023;73:227–32.
32. Bhoyate S, Kahol PK, Gupta RK. Nanostructured materials for supercapacitor applications. In: Nanoscience. London: The Royal Society of Chemistry; 2018, 5.
33. Lee H-M, Muralee Gopi CVV, Rana PJS, Vinodh R, Kim S, Padma R, et al. Hierarchical nanostructured $MnCo_2O_4$–$NiCo_2O_4$ composites as innovative electrodes for supercapacitor applications. New J Chem 2018; 42:17190–4.
34. Li L, Gan L, Liu M. Recent advances in carbon-based supercapacitors. Mater Adv 2020;1:945–66.
35. Wu D, Xie X, Zhang Y, Zhang D, Du W, Zhang X, et al. MnO 2/carbon composites for supercapacitor: synthesis and electrochemical performance 2020;7:1–16.
36. Al Mahmud MZ. A concise review of nanoparticles utilized energy storage and conservation. J Nanomater 2023;2023:5432099.

37. You L. Integrated photo – rechargeable batteries: photoactive nanomaterials and opportunities. In: 8th international conference on energy science and applied technology (ESAT 2023). E3S Web of Conferences; 2023, vol 375. [Online].
38. Das H, Pathak B, Khanam S, Kalita PK, Datta P. Nanomaterials for next generation energy storage applications. MRS Commun 2022;12:285–94.

Anitha R. Vaddinuri*, Satheeshkumar Palanisamy, Anil Kumar Dussa, Ganesan Shyni, Joseph Selvi Binoj, Shukur Abu Hassan and Rushdan Ahmad Ilyas

11 Performance of 6 × 6 CNT transistor array using composite nanomaterials for biomedical applications

Abstract: This study is the simulation of several forms of the most recent Field Effect Transistor (FET) and its electrical characteristics and designed using nanomaterials. The best FET to employ for sensor design in biomedical applications has been determined based on many examples and investigations. In order to overcome the difficulties of detecting disease-causing antigens, FET-based biosensors have been developed to increase sensitivity to disease-causing antigens at the micro- and nanoscale. To find the best transistor, various transistor types with various material layers are introduced. The sensor array consists of 64 cells, arranged into three categories – core, periphery, and middle – designed to enhance the sensitivity of cell detection. Three reference voltages are produced by a current reference circuit to bias the bottom gates of the CNTFETs, which generate maximum current in response to electrostatic changes at the top gate. These voltages are then processed to determine the concentration of the analyte, with the sensor array achieving a sensitivity of 130 μA/V. In NANOHUB.ORG (online Tool), all simulations are run, and performance is displayed in the form of simulation results, graphs, and tables.

Keywords: biosensor; antigens; antibodies; field effect transistors; composite nanomaterials; biomedical applications

---

***Corresponding author: Anitha R. Vaddinuri**, Department of Electronics and Communication Engineering, BMS Institute of Technology & Management, Yelahanka, 560064, Bengaluru, India, E-mail: anithavr@bmsit.in
Satheeshkumar Palanisamy and Anil Kumar Dussa, Department of Electronics and Communication Engineering, BMS Institute of Technology & Management, Yelahanka, 560064, Bengaluru, India
Ganesan Shyni, Department of Electrical and Electronics Engineering, University College of Engineering, Konam, Nagercoil, 629004, Tamil Nadu, India
Joseph Selvi Binoj, Institute of Mechanical Engineering, Saveetha School of Engineering, Saveetha Institute of Medical and Technical Sciences (SIMATS), Chennai, 602105, Tamilnadu, India
Shukur Abu Hassan, Centre for Advanced Composite Materials, Office of Deputy Vice-Chancellor (Research & Innovation), UniversitiTeknologi Malaysia, Johor Bahru, 81310, Johor, Malaysia
Rushdan Ahmad Ilyas, School of Chemical and Energy Engineering, Faculty of Engineering, UniversitiTeknologi Malaysia, Johor Bahru, 81310, Johor, Malaysia

As per De Gruyter's policy this article has previously been published in the journal Physical Sciences Reviews. Please cite as: A. R. Vaddinuri, S. Palanisamy, A. K. Dussa, G. Shyni, J. S. Binoj, S. A. Hassan and R. A. Ilyas "Performance of 6 × 6 CNT transistor array using composite nanomaterials for biomedical applications" *Physical Sciences Reviews* [Online] 2024. DOI: 10.1515/psr-2024-0029 | https://doi.org/10.1515/9783111234762-011

11.1 Introduction

Devices based on Silicon-On-Insulator (SOI) FETs were compared in terms of their electrical properties. The wet etching process, passivation, and patterning to the gate areas are used to define the ISFET gates and contacts. Si3N4 gate sensing layers on SiO_2 are 10 nm thick. They demonstrate how Ids for DU145 tumor cells change over time [1]. The confinement of electrons is what gives single wall CNTs their distinctive features. It has high current densities of about 109 A/Cm2. Here are provided the output electrical characteristics of a 300 nm long CNTFET and the transfer electrical characteristics of an 800 nm long CNTFET under room temperature in order to explore the electrical qualities. The subsequent discussion included CNTFET's present relationship. The concentration of the chemical analyte in the mixture was displayed. The **chemical analyte concentration** refers to the amount of a specific substance (the analyte) present in a given sample, typically expressed in units like molarity (M), micrograms per milliliter (µg/mL), or parts per million (ppm). In the context of biosensors, including those based on **Carbon Nanotube Field-Effect Transistors (CNTFETs)**, analyte concentration is often measured in terms of the **change in electrical properties** of the sensor due to the interaction between the analyte and the sensing material (such as the antibody or functionalized surface of the CNT) [2].

This sensor uses an electronic and digital signal to translate chemical reactions into biological discoveries. They emphasized how an antibody covering can help a single virus particle bind to a silicon nanowire [3]. Due to their exceptionally high Young's moduli and tensile strengths, silicon nanowire and CNTs are incredibly robust materials. In order to comprehend the characteristics and functions of the nano-biosensor, the benefits and drawbacks of various manufacture, fictionalization, and assembly techniques of the nano sensors are thoroughly examined [4]. A network of CNTFETs functionalized with anti-Salmonella antibodies was used to identify *Salmonella infantis*. These analyte interact with a functionalized SWCNT that has been treated with Tween 20 and functionalized with anti-Salmonella antibodies. On the carbon nanotubes, anti-Salmonella antibodies were adsorbed [5]. By changing the current measurement, noise suppression techniques in CNTFETs are used to explain the detection of biomolecules utilizing CNTFETs. Since they exhibit good performance as sensors and in electrical properties, they came to the unavoidable conclusion that CNTFET-based biosensors are a potential technology for future biosensors [6]. Optical and SEM images of arrayed CNTFETs for label-free biosensors have shown the detection of immunoglobulin E (IgE) using aptamer-based immunosensors. The aptamer-modified CNTFETs' CNT channel conductance modulation is used to detect biomolecules [7]. To enable nano-biosensing based on cytochromes, a biosensor array made of carbon nanotube field effect transistors was constructed. The suggested CNT-based CNFET biosensor array achieves a high degree of sensitivity thanks to a greater current flow. The design demonstrates enhanced specificity at the nanoscale using multigate controller [8], using graphene sheets set up as an FET for in-the-moment chemical sensing. As a specialized redox enzyme oxidizes glucose molecules, the conductance of the graphene transistor changes. It demonstrated a graphene FET with functionalized glucose. A change in current increases while taking into account a glucose concentration [9]. Small biomolecule detection using graphene-based biosensors was proved to be sensitive and selective. Graphene-based biosensors have shown good selectivity and sensitivity for small bimolecular detection. A convincing comparison of the sensing performance of graphene- and CNT-based biosensors is reported [10]. To identify Mycobacterium TB isolates, a CNTFET sensor in a biosensor device with CNT channels that have been altered with aptamers is employed. A promising approach for creating an integrated, highly selective, multiplexed real-time biosensor is the aptamer-based CNTFET device [11]. The physics of carbon nanotubes and the computation of the sub-band for zigzag CNT are used to propose the chirality CNT with tiny band-gap that actively participates in the modeling of nano-biosensor [12].

11.2 Properties of silicon nanowire MOSFET and carbon nanotube MOSFET

Numerous nonideal features that are significantly different from MOSFET are emerging when MOS infrastructure scales down more into the nanoscale. The rolled shape of graphene called a carbon nanotube (CNT) lends itself to a variety of uses in nanoelectronic circuits and eliminates those effects. Due to their unique structural, electrical, and mechanical characteristics, CNT field effect transistors, also known as CNTFETs, have recently received interest as potential building blocks for future nanoelectronics. These offer unique benefits over MOSFETs, which also benefit from the semiconducting properties of CNT, including high carrier mobility, a large MFP (mean free path), ballistic conduction, acceptable contact resistance, quick switching speed, and low heat dissipation. Despite being expensive and unreliable, CNT is essential for nanoscale production because of its structural characteristics. There are three distinct terminal types: a source, a gate, and a drain. Top and bottom gates are two different sorts of gates; the top gate serves as a sensor. In CNTFET devices, the gate region receives an antibody coating. The next step involves placing an antibody sample over the gate area, where an antibody and antigen response alters the conduction current. The concentration of disease-causing antigens is determined using this concept.

Carbon nanotubes have remarkable qualities for use in nanoelectronics. A CNTFET constructed from semiconducting nanotube is the best structure and has the best characteristics. The performance of the CNTFET in terms of electrical characteristics in comparison to silicon nanowire FETs is assessed using the tool provided by nanohub.org. Table 11.1 displays the CNTFET's input settings.

Figure 11.1 depicts the properties of the conductor's diameter with drain current, and the model is simulated using the web tool nanohub.org. Output drain current is one of electricity's characteristics. It is an essential component of the device's design and modeling and conclusively shows that the CNTFET has better value characteristics than the silicon nanowire FET.

Table 11.1: Input parameters of CNTFET.

S. No	Input parameters	Value
1.	Insulator thickness	1.50 nm
2.	Nanotube diameter	2.00 nm
3.	Insulator relative dielectric constant	3.90
4.	Temperature	300.00 K
5.	Initial source Fermi level	−0.320eV
6.	Gate control parameter	0.880
7.	Drain control parameter	0.035

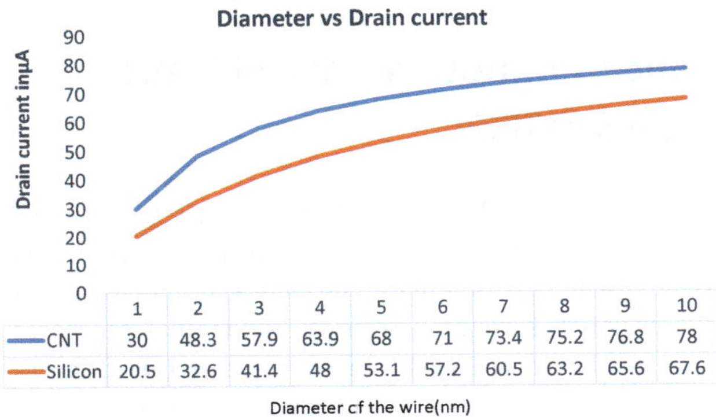

Figure 11.1: Drain current versus diameter of the wire.

11.3 Results and discussion

11.3.1 Design of 6 × 6 biosensor array

Carbon nanotubes are the perfect material for creating future electronics because of their numerous distinctive characteristics that lead to extraordinarily high current conduction. CNTs are used to create electronic circuits that operate more efficiently than those without them. By using chemical vapor deposition (CVD), the CNT is produced. Nanotubes of all types, including metallic, semiconducting, single-walled, and multi-walled ones, cannot be controlled, even when placed exactly [13–17].

Each node of the sensor array contains two gates, the top gate (the detecting gate) and the rear gate (the biasing gate), in accordance with the geometrical arrangement of CNTFETs depicted in Figure 11.2. Depending on where the sensors are placed inside the sensor array, the back gate is fixed biased with defined values of 0 V, 0.2 V, and 0.4 V. One of the antigens present in blood samples, cervical specific antigen, can combine with CSA antibody found in the top gate.

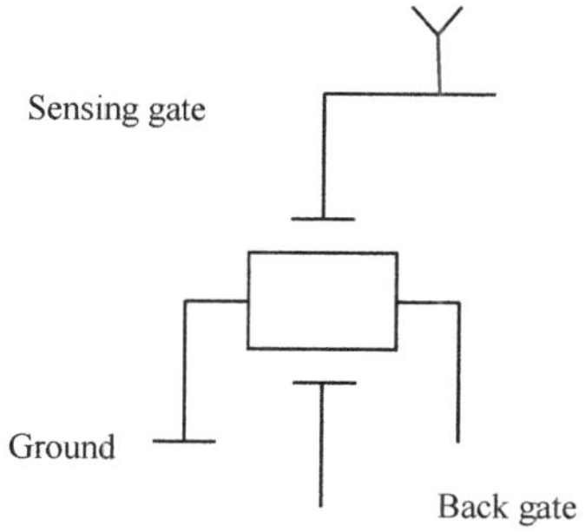

Figure 11.2: Structure of single biosensor.

The power supply affects the saturation current of a CNTFET sensor. The 6 × 6 CNTFET sensor array's power source changes, which affects the propagation delay in the sensor cells. The faster operation of a sensor array cell at a higher voltage reduces propagation delay. The most popular technique for evaluating stability is static Noise Margin analysis, which measures the maximum permissible DC noise voltage without impairing the output. Different supply voltages are discovered in order to ensure that the sensor cells have comparable propagation delays across the board for all sensors in the perimeter, center, and core. The two gates on the CNTFET that acts as the sensor array cell are the top gate and the back gate. To make sure that the process corners are being addressed to obtain consistent propagation delay, various voltages are supplied to the back gate of CNTFETs. The drain voltage of every sensor in the sensor array is predetermined by a bias, which ranges from 0.2 V to 1.8 V. Three groups make up the suggested sensor array in Figure 11.3. Core sensors, middle sensors, and periphery sensors make up the three categories. Figure 11.3 above depicts a 6 × 6 array of biosensors. Using a 6 × 6 array, the biosensor is optimized. A sample of the analyte being tested is placed over the sensor array; each CNTFET biosensor has an antibody that binds to the analyte's antigen. The sensor nodes will draw current as the sample analyte is placed over the array because of bio-potential, which is proportional to the concentration of the antigen present in the sample.

Table 11.2 shows that the 6 × 6 array biosensor produced maximum current of 2.203 mA due to the different gate bias. By comparing silicon nanowire and CNTFET transistor, carbon nanotube array have nearly 28 % more effective than the silicon nanowire-based sensor array.

11.4 Fabrication process of CNTFET

A Carbon Nanotube Field Effect Transistor (CNTFET) is a type of transistor that uses carbon nanotubes (CNTs) as the channel material. CNTFETs are promising for future electronic devices due to their superior electrical, thermal, and mechanical properties.

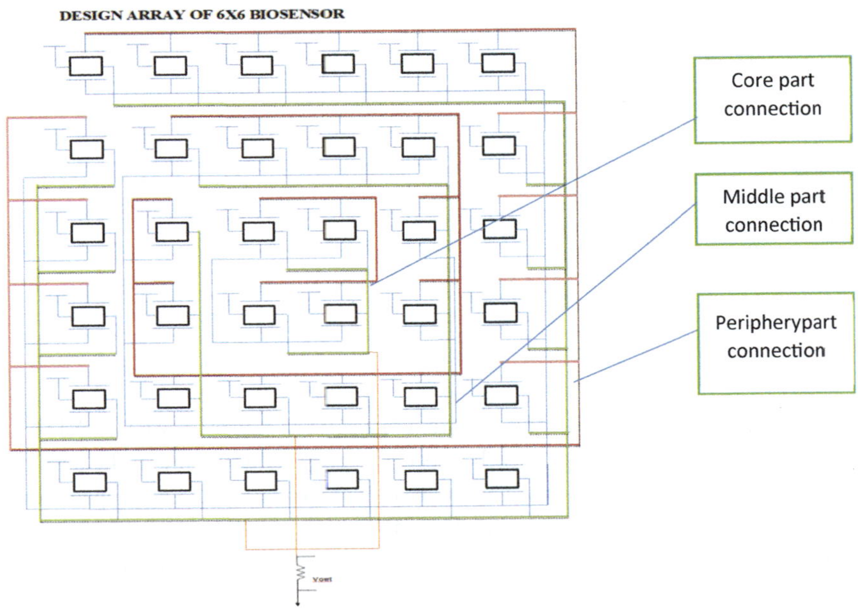

Figure 11.3: Design of a 6 × 6 array biosensor.

Table 11.2: Comparison of performance analysis of sensor models.

Sensor model	Drain current of 6 × 6 Array of sensor with constant gate voltage	Drain current of Proposed 6 × 6 Array of sensor with variable gate voltage
Silicon nanowire FET	1.1736 mA	1.578 mA
CNTFET	1.738 mA	2.203 mA

The fabrication process involves several critical steps to ensure proper alignment, integration, and functionality of CNTs. Here is an overview:

I. **Substrate Preparation**
   - **Material Selection**: Use a silicon wafer coated with a thin layer of silicon dioxide (SiO_2) as the substrate.
   - **Surface Cleaning**: Clean the substrate to remove impurities using standard chemical cleaning processes, such as RCA cleaning.

II. **Gate Fabrication**
   - **Gate Oxide Deposition:** Deposit a high-quality insulating layer (e.g., SiO_2 or high-k dielectric) over the substrate.
   - **Gate Patterning:** Use photolithography or electron beam lithography (EBL) to define the gate regions.

- **Metal Deposition:** Deposit the gate metal (e.g., gold, aluminum, or titanium).
- **Lift-Off Process:** Remove excess metal to leave the desired gate structure.

III. **Carbon Nanotube Placement**
- **CNT Synthesis:**
  - Synthesize CNTs using methods like chemical vapor deposition (CVD), arc discharge, or laser ablation.
- **Dispersion and Alignment:**
  - Disperse CNTs in a solvent and apply them to the substrate.
  - Align CNTs using methods like dielectrophoresis, fluid flow alignment, or lithographic techniques.
- **Selective Placement:**
  - Remove unwanted CNTs through patterning and etching, leaving the desired CNTs on the gate region.

IV. **Source and Drain Electrode Fabrication**
- **Electrode Definition:**
  - Use photolithography or EBL to define source and drain regions on either side of the CNT channel.
- **Metal Deposition:**
  - Deposit source and drain contact metals (e.g., gold, palladium, or platinum) to ensure low-resistance contact with the CNT.
- **Lift-Off Process:**
  - Remove excess metal to form the electrodes.

V. **Passivation and Packaging**
- **Passivation Layer:**
  - Deposit a passivation layer (e.g., Si_3N_4 or polymer) to protect the device from environmental damage.
- **Electrical Contacts:**
  - Make external connections to the source, drain, and gate terminals for device testing and integration.

VI. **Testing and Optimization**
- **Electrical Testing:**
  - Measure current–voltage (I–V) characteristics to verify the operation of the CNTFET.
- **Optimization:**
  - Optimize parameters such as CNT density, alignment, and contact resistance to improve device performance.

VII. **Challenges in Fabrication**
- **CNT Purity and Chirality:** Ensuring the use of semiconducting CNTs (as opposed to metallic ones) is critical for FET operation.
- **Alignment and Placement:** Precise alignment of CNTs is challenging and crucial for reproducibility.

- **Contact Resistance**: Achieving low-resistance metal-CNT contacts remains a key hurdle.
- **Scalability**: Integrating CNTFETs at a large scale with consistent properties is a complex task.

11.5 Conclusions

The biosensors are faster and can handle higher currents in the order of 70 μA and are able to operate at frequencies greater than 50 MHz, both N channel and P channel. Biosensors have been modeled using to realize inverters using H-Spice to estimate the electrical characteristics. Biosensors can be used as active device in place of nano belt to detect CSA present in blood or urine samples. The design of sensor array was with planar structure CNTFET with back gate bias voltage and top gate sensing node. The 64 sensors are grouped into three modes such as core sensors, middle sensors, and periphery sensors. The back gate bias voltage of 0 V, 0.2 V, and 0.4 V set to three groups of sensors provides maximum current flow with improved sensitivity. The series connection of three groups of sensors provides maximum current flow with any changes in top gate voltages. The CNTFET-based current mirror was with accumulated node to accumulate the current flow from sensor array. The CNTFET-based differential amplifier was with reference node voltage to detect cervical cancer. The designed biosensor array achieves 30–40 % improvement in terms of current sensitivity; hence, the designed sensor, sensor array, and the signal conditioning circuits are suitable for biosensing applications.

Practical implications and future research

The 6 × 6 CNT FET array model presents a transformative platform for next-generation technologies in sensing, computing, and flexible electronics. Future research focusing on material innovations, fabrication techniques, and application-specific optimization can amplify its impact, paving the way for breakthroughs in healthcare, IoT, and sustainable electronics.

Acknowledgments: The authors would like to thank the editors S.M. Sapuan, Mohd Roshdi Hassan, Eris Elianddy Supeni, and Azizan As'arry for their guidance and review of this article before its publication.

References

1. Pyo J. Highly sensitive CNT-based field-effect transistor sensors for biomolecular detection. Adv Funct Mater 2022;32:2108133.

2. Sahoo S. Recent advancements in carbon nanotube-based field-effect transistor biosensors for point-of-care diagnostics. Bioelectronics 2021;37:100048.
3. Huang X. Functionalized CNTs for biomedical applications: a review of the recent advances in drug delivery and sensors. J Nanosci Nanotechnol 2023;23:1222–40.
4. Wanekaya AK, Chen W, Myung NV, Mulchandani A. Nanowire-based electrochemical biosensors. Electroanalysis 2006;18:533–50.
5. Villamizar RA, Maroto A, Xavier Rius F, Inza I, Figueras MJ. Fast detection of Salmonella Infantis with carbon nanotube field effect transistors. Biosens Bioelectron 2008;24:279–83.
6. Yamamoto Y, Maehashi K, Ohno Y, Matsumoto K. Highly sensitive biosensors based on high-performance carbon nanotube field-effect transistors. Sensor Mater 2009;21:351–61.
7. Maehashi K, Matsumoto K, Takamura Y, Tamiya E. Aptamer based label free immunosensors using carbon nanotube field effect transistors. Electroanalysis 2009;21:1285–90.
8. Bobba S, Carrara S, De Micheli G. Design of a CNFET array for sensing and control in P450 based biochips for multiple drug detection. In: International Symposium on Circuits and Systems (ISCAS 2010), Paris, France; 2010:1711–14 pp.
9. Huang Y, Dong X, Shi Y, Li CM, Li L-J, Chen P. Nanoelectronic biosensors based on CVD grown graphene. Nanoscale 2010;2:1485–8.
10. Kuila T, Bose S, Khanra P, Mishra AK, Kim NH, Lee JH. Recent advances in graphene-based biosensors. Biosens Bioelectron 2011;26:4637–48.
11. Wang H-W, Huang J-T, Lin C-C. Real-time detecting concentration of Mycobacterium tuberculosis by CNTFET biosensor. World Acad Sci Eng Technol 2013;79.
12. Farhana S, ZahirulAlam AHM, Khan S. Small band-gap-based CNT for modeling of nano sensor. Proc Comput Sci 2014;42:122–9.
13. Liu Y, Liu D, Alpuche-Aviles MA, Ma W. Carbon nanotube-based biosensors for the detection of biomolecules: applications in healthcare. Front Bioeng Biotechnol 2023;11:774501.
14. Choi Y. Carbon nanotube-based field-effect transistor biosensors for rapid and ultrasensitive detection of pathogens. Biosens Bioelectron 2022;195:113591.
15. avey A, Nam SW, Friedman RS, Lieber CM, Dai H. High-field electron transport in single-walled carbon nanotube transistors. Nature 2003;424:654–7.
16. Freitag M. Carbon nanotube electronics and devices. Carbon Nanotub: Prop Appl 2006:83–117. https://doi.org/10.1201/9781420004212.ch4.
17. Wind SJ, Appenzeller J, Martel R, Derycke VPPA, Avouris P. Vertical scaling of carbon nanotube field-effect transistors using top gate electrodes. Appl Phys Lett 2002;80:3817–19.

Munir Faraj Alkbir*, Adnan Bakri, Fatihhi Januddi, Nuha Awang, Salit Mohd Sapuan, Suhad D. Salman, Alhadi A. Abosbaia, Zulhaimi Mohammad and Mmhamad Shahrul Effendy Kosnan

12 Energy absorption capacity of biocomposites

Abstract: This paper explores the energy absorption capacity of biocomposites and its impact on sustainability and environmental benefits. Biocomposites consisting of natural fibers within a polymer matrix have attracted attention for their potential to improve impact resistance and safety while reducing environmental impact. This review critically examines the current state of research in this field, highlighting the importance of biocomposites in utilizing renewable and recyclable resources, energy-efficient processing, minimizing toxicological impacts, and promoting responsible waste management. This also highlights the importance of assessing emissions, including volatile organic compounds and nanoparticles, from an environmental and toxicological perspective. Additionally, the paper takes into account the degradation susceptibility of sustainable biocomposites, emphasizing the need to ensure their structural integrity during their service life and their ultimate biodegradability and assimilability during composting. The results presented in this review highlight the multifaceted advantages of biocomposites, making them a promising choice for sustainable materials with excellent energy absorption capabilities.

Keywords: biocomposites; energy absorption; sustainability; environmental benefits; renewable resources

***Corresponding author: Munir Faraj Alkbir**, Advanced Facilities Engineering Technology Research Cluster (AFET), Plant Engineering Technology (PETech) Section, Universiti Kuala Lumpur Malaysian Institute of Industrial Technology, 81750 Masai, Johor, Malaysia, E-mail: munir@unikl.edu.my

Adnan Bakri, Fatihhi Januddi, Nuha Awang, Zulhaimi Mohammad and Mmhamad Shahrul Effendy Kosnan, Advanced Facilities Engineering Technology Research Cluster (AFET), Plant Engineering Technology (PETech) Section, Universiti Kuala Lumpur Malaysian Institute of Industrial Technology, 81750 Masai, Johor, Malaysia

Salit Mohd Sapuan, Advanced Engineering Materials and Composites Research Centre (AEMC), Department of Mechanical and Manufacturing Engineering, Universiti Putra Malaysia, 43400 UPM Serdang, Selangor Darul Ehsan, Malaysia; and Laboratory of Biocomposite Technology, Institute of Tropical Forestry and Forest Products (INTEROP), Universiti Putra Malaysia, 43400 UPM Serdang, Selangor, Malaysia

Suhad D. Salman, Mechanical Engineering Department, College of Engineering, Mustansiriyah University, Baghdad, Iraq

Alhadi A. Abosbaia, Mechanical Engineering, Faculty of Engineering, University of Zawia, Zawia, Libya

As per De Gruyter's policy this article has previously been published in the journal Physical Sciences Reviews. Please cite as: M. F. Alkbir, A. Bakri, F. Januddi, N. Awang, S. M. Sapuan, S. D. Salman, A. A. Abosbaia, Z. Mohammad and M. S. E. Kosnan "Energy absorption capacity of biocomposites" *Physical Sciences Reviews* [Online] 2025. DOI: 10.1515/psr-2024-0030 | https://doi.org/10.1515/9783111234762-012

12.1 Introduction

In recent years, there has been growing interest in the development and utilization of biocomposite materials due to their unique combination of eco-friendliness, lightweight nature, and remarkable mechanical properties. Biocomposites are materials formed by embedding natural fibers, such as jute, flax, hemp, or bamboo, within a biodegradable polymer matrix, resulting in a novel class of sustainable materials. These biocomposites have shown significant promise in various engineering applications, particularly in areas where energy absorption is of paramount importance, such as the automotive, aerospace, and sports equipment industries [1].

The energy absorption capacity of biocomposites is a focal point in contemporary materials science, commanding attention from a diverse spectrum of stakeholders, including researchers and industry experts [2]. This focus is not incidental but a response to the urgent need for innovative materials that address environmental concerns while meeting the rigorous performance demands of modern engineering. At the core of this attention is the intrinsic nature of biocomposites, which differentiates them from conventional synthetic composites [3].

Unlike traditional synthetic composites, biocomposites offer unparalleled ecological advantages. These advantages stem from their organic components – natural fibers such as jute, flax, hemp, or bamboo – embedded within a biodegradable polymer matrix. The result is a composite material that aligns with the principles of sustainability [4]. Biocomposites exhibit a significantly lower carbon footprint during production, and their end-of-life disposal poses a substantially reduced environmental burden. As a result, they provide a compelling solution for industries aiming to minimize ecological impact while maintaining a competitive edge [5].

The appeal of biocomposites extends beyond their environmental benefits. Their ability to balance environmental responsibility with robust mechanical and energy absorption properties has driven a resurgence of interest in their research and development. This unique combination of sustainability and performance has inspired systematic investigations into the mechanisms governing the energy absorption capabilities of biocomposites [6].

Moreover, efforts to comprehend the energy absorption dynamics of biocomposites include identifying numerous factors that influence their performance. These factors encompass the choice of natural fibers, the specific polymer matrix, manufacturing methods, and the engineered composite architecture. For example, biocomposites reinforced with flax fibers in a tailored polymer matrix have shown an increase in energy absorption by up to 30 % compared to traditional composites, showcasing their potential for high-performance applications. These variables significantly modulate energy absorption characteristics, making biocomposites versatile and adaptable for applications such as automotive safety components, aerospace structures, and sporting equipment [7, 8].

The primary aim of this study is to explore the energy absorption capacity of biocomposites, focusing on the mechanisms, factors influencing energy absorption, and potential applications. By reviewing and analyzing relevant research, particularly recent advancements and emerging trends, this study seeks to contribute to a broader understanding of sustainable materials for high-impact engineering applications.

12.2 Discussion

The energy absorption capacity of biocomposites is a critical aspect of their mechanical performance, and it has gained significant interest in various industries due to the growing emphasis on sustainability and eco-friendly materials. Here is a discussion of the key points related to the energy absorption capacity of biocomposites.

12.2.1 Sustainability and Environmental Benefits

The examines environmental and resource impacts for the design of sustainable biocomposites. Criteria for sustainable biocomposites include the use of renewable or recycled resources in manufacturing, energy-efficient and benign processing methods, avoidance of harmful environmental and toxic impacts throughout their life cycle, and effective waste management practices. The future integration of biorefineries and green chemistry promises a wider range of raw materials. Evaluating emissions, such as volatile organic compounds and nanoparticles, is crucial for understanding their environmental and toxicological impacts. Furthermore, sustainable biocomposites are prone to degradation, resulting from biodegradation, biofilm formation, and abiotic factors like water absorption and thermal oxidation. Addressing these issues is essential to maintain the structural integrity of biocomposites during their use and to ensure they are biodegradable and assimilable during composting [9, 10] (Figures 12.1–12.6).

12.2.2 Energy absorption

Crushing behavior of kenaf fiber/wooden stick-reinforced epoxy hybrid composite elliptical tubes. Wooden stick reinforcements significantly enhance energy absorption and load-carrying capacity. Elliptical tubes with eight wooden sticks achieve optimal energy absorption, about 25 % higher than unreinforced tubes, demonstrating superior crashworthiness. The presence of wooden sticks transitions crash behavior from brittle, abrupt failure to a ductile mode, allowing gradual deformation and improved energy absorption. This advancement underscores the potential of these hybrid composites for safer and more efficient structural applications, particularly in scenarios requiring enhanced impact resistance and energy dissipation capabilities [11].

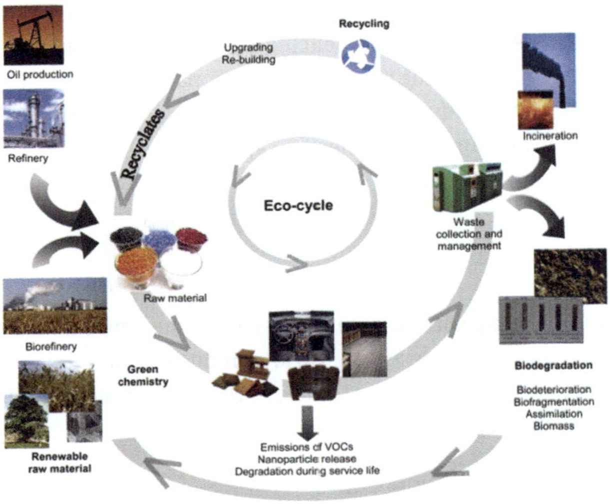

Figure 12.1: Life cycle of sustainable composites [9].

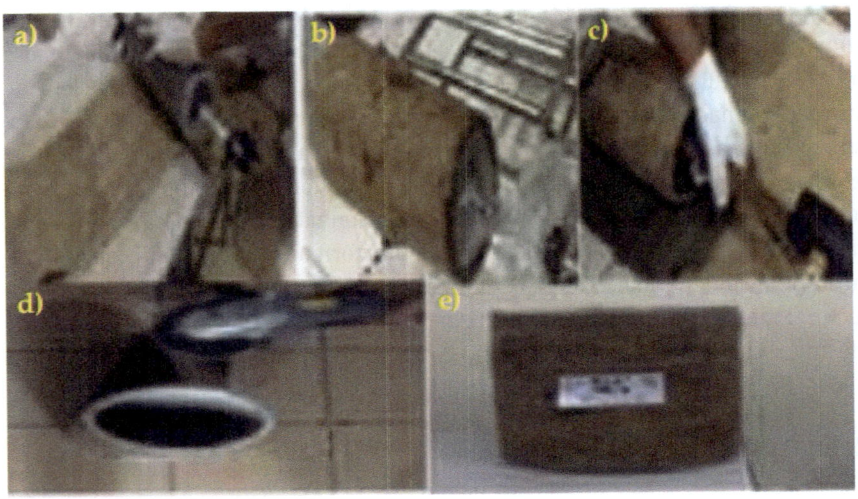

Figure 12.2: Wet and wrapped kenaf mat around the mold [11].

Edison E. Haro et al. [12] examined the impact of chonta palm wood microparticles on the mechanical behavior of high-density polyethylene (HDPE) under high strain-rate compressive and ballistic loading. The study employed an extrusion process to create biocomposites with varying chonta wood content. Results revealed improved mechanical properties in HDPE when reinforced with chonta palm wood microparticles. Notably, the

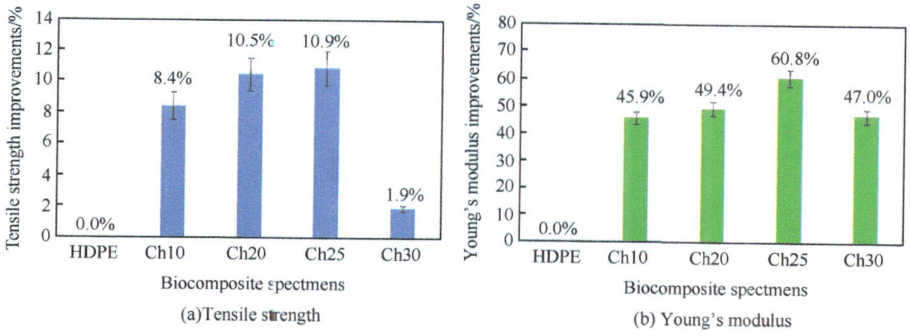

Figure 12.3: Improving tensile properties of biocomposites [12].

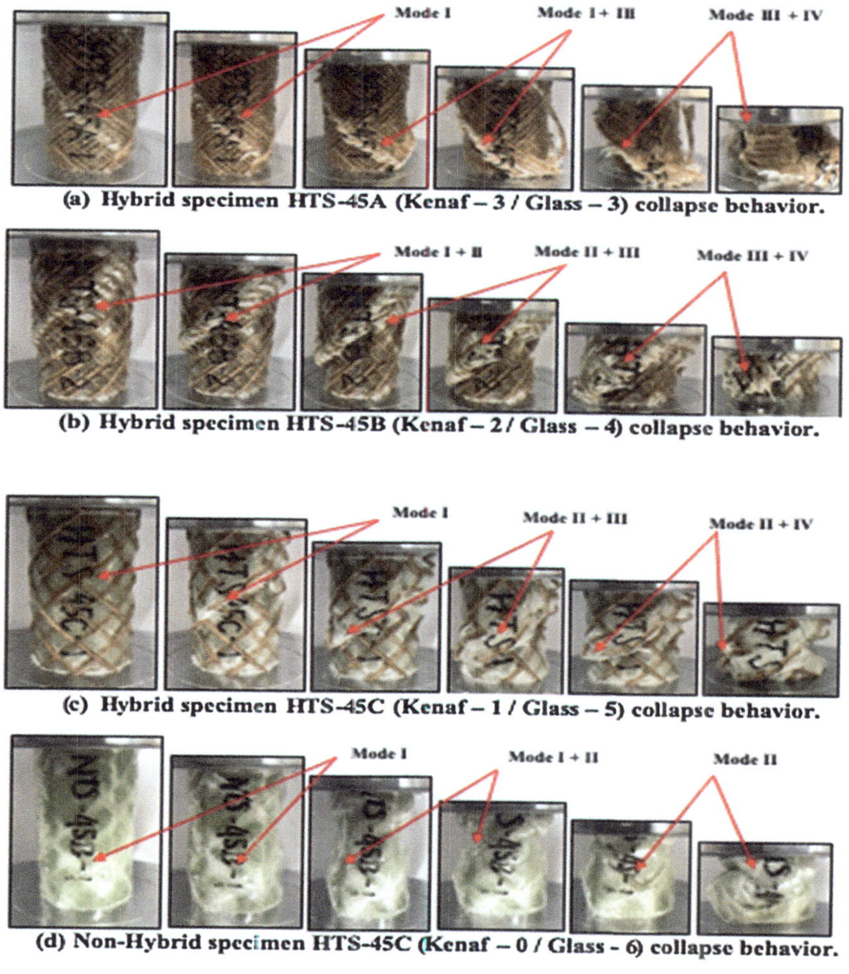

Figure 12.4: Collapse mode behavior: (a–c) hybrid kenaf/glass reinforced epoxy composite specimen tube and (d) glass fiber reinforced epoxy composite specimen tube [13].

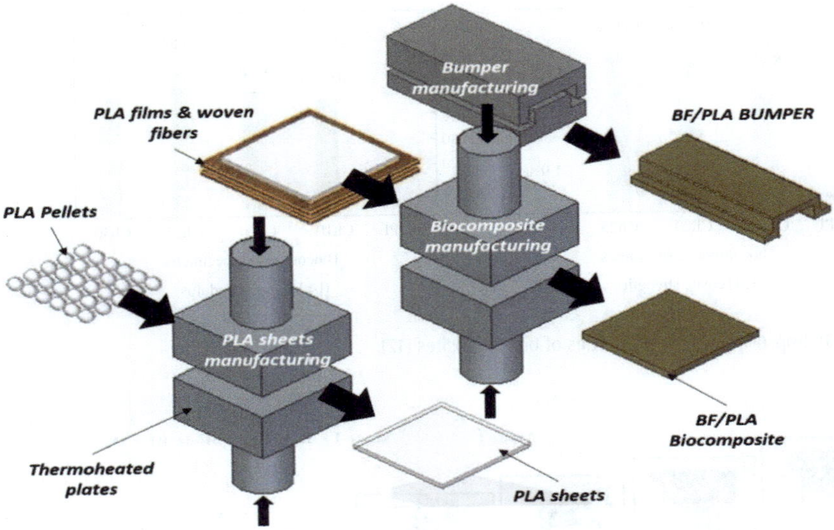

Figure 12.5: Exterior composite manufacturing method that maintains compressible form [14].

Figure 12.6: Example of extruded WPC product (provided by entec).

25 wt % chonta wood microparticles demonstrated the highest strength, stiffness, ballistic impact resistance, and impact energy absorption.

12.2.3 Energy capacity on natural fibre

The experimental work done by Alireza Taherzadeh-Fard et al. [15] investigates how different fibers affect the mechanical properties of elastomeric materials. Kevlar fibers boost ultimate strength, while carbon and glass fibers reduce strength and failure strain.

Energy absorption is generally lower in fiber-reinforced rubber, but 5 wt% carbon fibers offer a balanced improvement.

Alkbir et al. [10] discussed the use of natural and synthetic fibers in engineering applications as lightweight alternatives to metals. It explores their structural rigidity, fabrication methods, and their effects on crashworthiness. The review emphasizes that failure modes and crushing behavior are influenced by the type of materials and structure in composite tubes.

The study by Mohd Supian Abu Bakar et al. [13] explores the energy absorption of hybrid kenaf/glass fiber-reinforced epoxy composite tubes under axial quasi-static compression. Various stacking sequences are examined, and the results show that the hybrid specimen with a higher fiber volume fraction outperforms the synthetic glass fiber-reinforced epoxy in terms of energy absorption and progressive crushing behavior. This suggests that kenaf fiber can be a valuable reinforcement in composite tubes for energy absorption applications.

12.2.4 Application of Energy absorption capacity of biocomposites

A. Díaz-Álvarez et al. [14] examined the behavior of bumper beams made from PLA/flax composites subjected to low-velocity impacts. The composite's biodegradability and lack of delamination make it promise. Impact tests reveal that while damage reduces residual stiffness, residual strength remains unaffected within the tested impact energy range.

Liu Jiao et al. [16] investigated fully biodegradable flax/PLA bumper beams with different cross-sectional shapes for low-velocity impact performance. Results showed that rounder sections perform better in impact, while mid-sections excel in post-impact residual properties.

Non-food crops and renewable resources hold immense potential for biocomposites, offering sustainable raw materials. While the biocomposite market is growing, further development can unlock new applications. Various reinforcing fibers and processing options exist, and advancements in biopolymers support sustained growth. However, continued research and development investment is crucial for establishing a sustainable biocomposites industry [17].

12.3 Conclusions

In conclusion, this comprehensive review underscores the significance of biocomposites in achieving a harmonious balance between sustainability and high-performance engineering. Biocomposites, composed of natural fibers within biodegradable polymers, offer a unique and eco-friendly alternative to conventional materials. Their energy absorption capacity and impact resistance make them highly valuable in industries such as automotive, aerospace, and sports equipment. The environmental benefits of biocomposites, including lower carbon footprint and reduced end-of-life disposal impact, further underscore their importance in addressing contemporary ecological concerns.

The review also emphasizes the need for continuous research and development investment to foster the growth of a sustainable biocomposites industry. Despite the promising advantages of biocomposites, ongoing efforts are necessary to unlock their full potential and establish them as a mainstream choice for sustainable materials in various engineering applications.

Overall, this review provides a comprehensive overview of the multifaceted advantages and potential of biocomposites, reaffirming their position as a promising choice for sustainable materials with excellent energy absorption capabilities. It underscores the critical role of biocomposites in the intersection of environmental responsibility and high-impact engineering, paving the way for a more sustainable and eco-conscious future.

Acknowledgment: The authors would like to thank the editors S.M. Sapuan, Mohd Roshdi Hassan, Eris Elianddy Supeni and Azizan As'arry for their guidance and review of this article before its publication.

References

1. Deb D, Patel R, Balas VE. A review of control-oriented bioelectrochemical mathematical models of microbial fuel cells. Processes 2020;8. https://doi.org/10.3390/pr8050583.
2. Buchko C, Chen LC, Shen Y, Martin DC. Processing and microstructural characterization of porous biocompatible protein polymer thin films. Polymer (Guildf) 1999;40:7397–407.
3. Nunes-Pereira J, Sencadas V, Correia V, Rocha JG, Lanceros-Méndez S. Energy harvesting performance of piezoelectric electrospun polymer fibers and polymer/ceramic composites. Sens Actuators A Phys 2013;196: 55–62.
4. Kolb G. Review: microstructured reactors for distributed and renewable production of fuels and electrical energy. Chem Eng Process: Process Intensif 2013;65:1–44.
5. Guo KW. Green nanotechnology of trends in future energy: a review. Int J Energy Res 2012;36:1–17.
6. Lv C, Tang Y, Wang L, Ji W, Chen Y, Yang S, et al. Bioconversion of yolk cholesterol by extracellular cholesterol oxidase from Brevibacterium sp. Food Chem 2002;77:457–63.
7. Thomassin JM, Pagnoulle C, Caldarella G, Germain A, Jérôme R. Impact of acid containing montmorillonite on the properties of Nafion® membranes. Polymer (Guildf) 2005;46:11389–95.
8. Sicaire AG, Vian MA, Fine F, Carré P, Tostain S, Chemat F. Ultrasound induced green solvent extraction of oil from oleaginous seeds. Ultrason Sonochem 2016;31:319–29.
9. Thomassin JM, Pagnoulle C, Caldarella G, Germain A, Jérôme R. Impact of acid containing montmorillonite on the properties of Nafion ® membranes. Polymer (Guildf) 2005;46. https://doi.org/10.1016/j.polymer.2005.10.018.
10. Alkbir MFM, Fatihhi J, Adnan B, Sapuan SM, Kosnan MSE, Mohamed SB, et al. Crushing behaviour of plain weave composite hexagonal cellular structure. Sains Malays 2020;49. https://doi.org/10.17576/jsm-2020-4909-18.

11. Alkateb M, Sapuan SM, Leman Z, Jawaid M, Ishak MR. Crushing behavior of kenaf fiber/wooden stick reinforced epoxy hybrid 'green' composite elliptical tubes. Polimery/Polymers 2018;63. https://doi.org/10.14314/polimery.2018.6.4.
12. Haro EE, Szpunar JA, Odeshi AG. Dynamic and ballistic impact behavior of biocomposite armors made of HDPE reinforced with chonta palm wood (Bactris gasipaes) microparticles. Defence Tech 2018;14. https://doi.org/10.1016/j.dt.2018.03.005.
13. Mohd Supian AB, Mohd Sapuan S, Mohd Zuhri MY, Edi Syams Z, Hamdan HY. The crashworthiness performance of stacking sequence on filament wound hybrid composite energy absorption tube subjected to quasi-static compression load. Integr Med Res 2019;9:654–66.
14. Díaz-Álvarez A, Jiao-Wang L, Feng C, Santiuste C. Energy absorption and residual bending behavior of biocomposites bumper beams. Compos Struct 2020;245. https://doi.org/10.1016/j.compstruct.2020.112343.
15. Taherzadeh-Fard A, Khodadadi A, Liaghat G, Yao XF, Mehrizi MAZ. Mechanical properties and energy absorptionindex: energy absorption capacity of chopped fiber reinforced natural rubber. Composites Part C: Open Access 2022;7. https://doi.org/10.1016/j.jcomc.2022.100237.
16. Jiao-Wang L, Loya JA, Santiuste C. Influence of cross-section on the impact and post-impact behavior of biocompositesindex: biocomposites bumper beams. Mech Adv Mater Struct 2023;30. https://doi.org/10.1080/15376494.2022.2072029.
17. Fowler PA, Hughes JM, Elias RM. Biocomposites: technology, environmental credentials and market forces. J Sci Food Agric 2006;86. https://doi.org/10.1002/jsfa.2558.

Salit Mohd Sapuan*, Rushdan Ahmad Ilyas*, Agus Nugroho and Lakshminarasimhan Rajeshkumar

13 Energy efficiency and sustainability in composites industry

Abstract: Composite materials are tailor-made materials that constitute matrix and fiber-based reinforcements that are collectively bonded to obtain a material that possesses a combination of properties of both the precursor materials. The current article reviews the wide composite materials processes; furthermore, when it comes to manufacturing composites, cost is a very influential factor. The manufacture of composites is not cheap. The processes require resources, infrastructure, and energy. Therefore, energy-saving measures are worthwhile things to research. And said measures are also included in this work. Also present are information regarding existing conventional energy sources as well as potential alternative sources for the composite industry.

Keywords: energy efficiency; sustainability; fiber reinforced composites; industry; manufacturing

13.1 Introduction

Composite materials are tailor-made materials that constitute matrix and fiber-based reinforcements that are collectively bonded to obtain a material that possesses a combination of properties of both the precursor materials. Glass reinforcements with polyester resin systems are the most widely used composites across a range of applications. Carbon reinforcements with epoxy resin systems are used in specialist applications such as aerospace and high performance automotive [1–3].

***Corresponding authors: Salit Mohd Sapuan**, Advanced Engineering Materials and Composites Research Centre (AEMC), Department of Mechanical and Manufacturing Engineering, Universiti Putra Malaysia, Serdang, Selangor, Malaysia; and Institute of Tropical Forest and Forest Products (INTROP), Universiti Putra Malaysia, 43400 UPM, Serdang, Selangor, Malaysia. E-mail: sapuan@upm.edu.my. https://orcid.org/0000-0002-4598-6358; and **Rushdan Ahmad Ilyas**, Centre for Advanced Composite Materials, Universiti Teknologi Malaysia, 81310 UTM, Johor, Malaysia; Faculty of Chemical and Energy Engineering, Universiti Teknologi Malaysia 81310 UTM, Johor, Malaysia; and Centre of Excellence for Biomass Utilization, Universiti Malaysia Perlis, 02600, Arau, Perlis, Malaysia, E-mail: ahmadilyas@utm.my

Agus Nugroho, Research Center for Advanced Materials, National Research and Innovation Agency (BRIN), South Tangerang, Banten, 15314, Indonesia

Lakshminarasimhan Rajeshkumar, AU-Sophisticated Testing and Instrumentation Centre, CoE-Advanced Materials Synthesis and Department of Mechanical Engineering, Alliance School of Applied Engineering, Alliance University, Bangalore 562106, Karnataka, India

As per De Gruyter's policy this article has previously been published in the journal Physical Sciences Reviews. Please cite as: S. M. Sapuan, R. A. Ilyas, A. Nugroho and L. Rajeshkumar "Energy efficiency and sustainability in composites industry" *Physical Sciences Reviews* [Online] 2024. DOI: 10.1515/psr-2024-0031 | https://doi.org/10.1515/9783111234762-013

The composites industry plays a significant role in various sectors, including aerospace, automotive, construction, and renewable energy [4–6]. However, with increasing concerns about climate change and environmental sustainability, there is a growing need for the industry to address energy efficiency and sustainability challenges [7]. Energy consumption, raw material utilization, waste generation, and carbon emissions are some of the critical aspects that require attention to ensure a more sustainable future.

Sustainability has become a crucial aspect in the composites industry, driving the need for eco-friendly and resource-efficient practices [8]. The adoption of sustainable approaches in composites manufacturing has gained significant attention due to its potential to reduce environmental impact and promote long-term viability. This review provides a concise overview of sustainability in the composites industry, highlighting key strategies and advancements [9–11]. It explores sustainable materials, such as bio-based resins and recycled fibers, which offer reduced carbon footprint and enhanced recyclability. Additionally, sustainable manufacturing processes, including green composites production and waste management, are examined. The review also emphasizes the importance of life cycle assessment and eco-design principles to optimize sustainability throughout the entire product lifecycle. By embracing sustainable practices, the composites industry can contribute to a greener and more sustainable future [12, 13].

This manuscript aims to explore the importance of energy efficiency and sustainability in the composites industry. It delves into the various strategies, technologies, and best practices that can be employed to enhance energy efficiency, reduce environmental impact, and promote sustainable manufacturing processes. By examining the current state of the industry, identifying challenges, and proposing innovative solutions, this paper seeks to contribute to the advancement of sustainable practices in the composites sector.

The first section of this manuscript provides an overview of the composites industry, highlighting its significance, growth, and applications. It also presents the environmental concerns associated with traditional manufacturing processes, such as high energy consumption, material waste, and carbon emissions. Understanding these challenges is essential for identifying opportunities and driving the transition toward a more sustainable composites industry.

Next, the manuscript focuses on the concept of composites manufacturing process and its relevance in the composites sector. It explores the potential benefits and opportunities that energy-efficient practices offer, technical practice, and the competitiveness. Additionally, it highlights the importance of adopting energy-efficient technologies, equipment, and manufacturing processes to optimize resource utilization and minimize energy consumption.

The third section addresses sustainability in the composites industry, emphasizing the need to energy saving measures in composite industry through good manufacturing practice. The composite industry, with its diverse applications and growing demand, is constantly seeking ways to enhance its manufacturing processes while minimizing energy consumption and reducing environmental impact. Good Manufacturing Practice (GMP) principles provide a framework for implementing energy-saving measures that can lead to significant improvements in energy efficiency and sustainability [8]. Another significant energy-saving measure is the implementation of energy management systems and practices. This involves conducting energy audits, setting energy reduction targets, and implementing energy-efficient technologies and practices. By monitoring energy consumption, identifying areas of improvement, and implementing energy-saving initiatives, manufacturers can achieve substantial energy savings and cost reductions.

One key aspect of energy-saving measures in the composite industry is the optimization of production processes. This involves the efficient use of energy-intensive equipment, such as autoclaves, curing ovens, and resin impregnation systems. Implementing advanced control systems, automated processes, and monitoring technologies can help optimize energy usage and reduce waste during manufacturing operations.

In conclusion, this manuscript aims to provide insights into the importance of energy efficiency and sustainability in the composites industry. By examining the current challenges and proposing effective strategies, it seeks to inspire researchers, industry professionals, and policymakers to collaborate and drive the adoption of energy-efficient and sustainable practices. Ultimately, the integration of energy efficiency and sustainability principles will pave the way for a greener, more efficient, and sustainable composites industry.

13.1.1 Breakdown of energy use in composites

Since the business environment of the manufacturers around the globe is facing an acute focus on reducing cost without compromising the quality of the product. For public and privately owned businesses altogether, the energy used for making the product shoots up its cost and reduces the product value at the manufacturing unit. Maintaining the high quality for a product despite minimum production cost is of greater challenge even though economic, energy-efficient, and successful technologies are developed. In recent times, this deems very important since the modern-day energy efficient technologies offer additional advantages by increasing the company productivity through the reduction of inventory material and water consumption [14, 15].

Some composite materials are potentially applied for the manufacturing of automobile components during the past decade; however, the mass production of those components with the aid of those materials are still in the development stage owing to various difficulties like rate of automation, production cost, and low productivity. Table 13.1 enlists the energy intensity of various manufacturing techniques. The energy intensities enlisted represent the energy required only for the manufacturing but not for the processing of materials. These processes also require additional energy but not as much energy as required for initial material processing. In addition to energy, many materials use solvents and additives. Generally, heat and pressure application during processing consumes a lot of energy.

Table 13.1: Energy intensities of manufacturing processes [16].

Manufacturing methods	Energy intensity, MJ/kg
Autoclave molding technique	21.7
Spray up technique	15.2
Resin transfer molding (RTM) technique	13.1
Vacuum-assisted resin infusion (VARI) technique	10.4
Cold press technique	12.1
Preform matched die technique	10.3
Sheet molding compound (SMC) technique	3.52
Filament winding technique	2.72
Pultrusion technique	3.11
Prepreg production technique	40.3
Injection molding (hydraulic) technique	19.5
Glass fabric manufacturing technique	2.9
Iron casting (cupola) technique	13.8

13.1.2 Conventional and alternative sources of energy used in the composite industry

Petroleum: High performance fiber (except glass) and resins originate almost entirely from petroleum, and then additional petroleum is consumed in order to fuel production and transportation. This poses two significant financial hurdles for the industry, the first being the fact that the cost of materials is influenced by the oil price. Second, the combustion of petroleum releases greenhouse gases into the atmosphere, which contributes to warming and climate change on a global scale [17]. **Uranium**: Uranium dioxide UO_2 composite fuel, which has improved thermal conductivity, can increase the lifespan of UO_2 fuel used in nuclear reactors to reduce costs and promote safety. Uranium mono-nitride (UN) has a high thermal conductivity too (21.5 W/mK at 600 °C), as well as higher uranium metal atom density (13.55 g/cm^3), and a similar melting temperature (2,600 °C). These better characteristics can improve the overall performance of a water reactor. UO_2–UN composite was studied in order to come up with the suitable mixture of the UO2 and UN properties, namely the thermal conductivity and uranium density. Hot press sintering is used to produce composites of UO_2–UN [18]. Another use of uranium in the composite industry is to produce composite uranium ceramic pellets. These pellets consist of uranium oxide $U_3 O_8$ particles in a silicon carbide matrix. There is a novel processing technique, which makes use of polymer infiltration and pyrolysis (PIP) to fabricate uranium ceramic composite pellets [19].

Coal: Coal is a fossilized organic material. Compared to most raw materials and natural polymers, coal has a much higher heat resistance. In fossil carbon uses as a filler for thermoplastics, alkylated is being used. **Biomass:** By using biomass, sustainable and renewable composite materials can be manufactured. From biomass, the bio-based resin

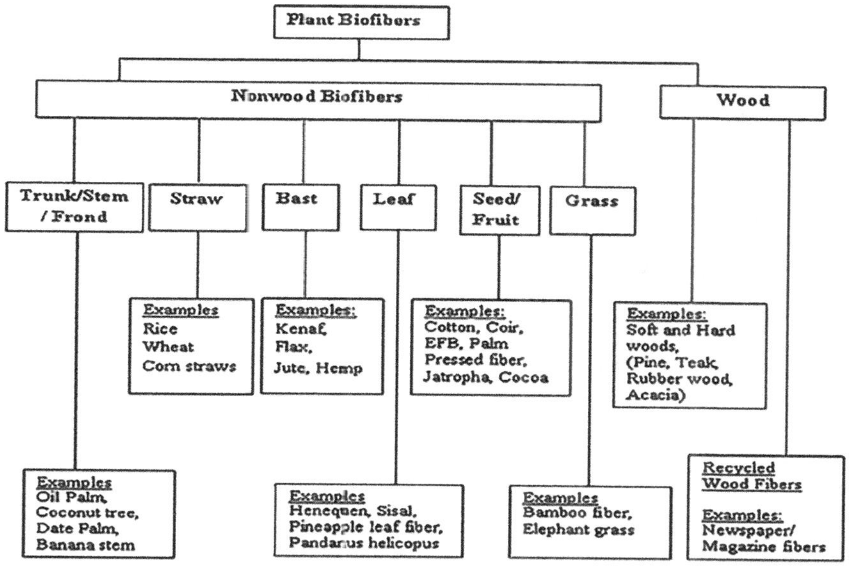

Figure 13.1: Schematic representation of reinforcing bio-fibers classification.

matrix materials and bio-based fiber reinforcements are usually derived. Major concerns regarding the use of bio-based resins as matrices. Bio-resins are resins or resin formulae sourced from a biological source and, hence, are biodegradable. Hypothetically, after usage, they can be disposed and decomposed. Traditional petroleum-based polymeric resins are being replaced by the matrix materials derived from the bio-based sources. Such biopolymers are gaining the researcher's limelight lately since these are more eco-friendly and readily available for the development of composite materials. Some of the bio-based matrix materials like starch plastic, cellulose-based plastic, corn and soy-based plastic are being reinforced with various bio-based fibers (as shown in Figure 13.1) to form biocomposites [20].

The main advantages of bio-based resources include low cost, renewable and sustainable, better specific properties, nonabrasive nature, availability, low density, higher levels of filling, and recyclability. Owing to these major points, the biodegradable reins are mostly preferred over the nonrenewable petroleum-based resins [21].

13.2 Composites manufacturing process

13.2.1 Hand lay-up

Hand lay-up (shown in Figure 13.2) is a simple and primitive composite manufacturing technique. The steps involved in this technique are as follows:

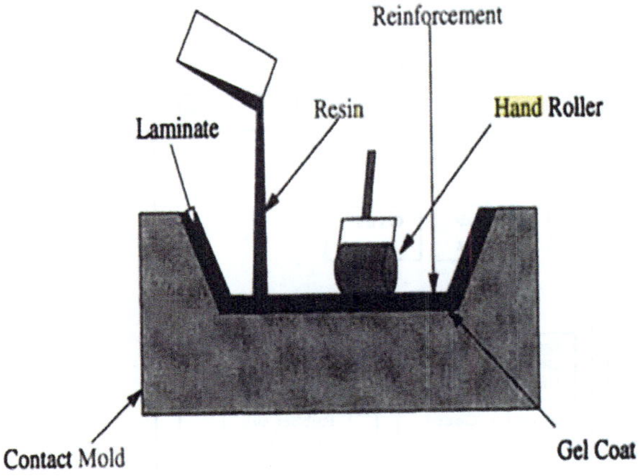

Figure 13.2: Hand lay-up process [23].

I. Applying gel coat and release agent on the surface of the mold
II. Placing manually the fiber reinforcement in the gel coated mold
III. Mixing the based resin with catalyst and applying accelerator by brushing, spraying, or pouring
IV. Using rollers or squeezers to wet the reinforcement with the resin to remove trapped air. Adding layers or reinforcement and resin to improve the thickness of the part [22].

This technique employs open mold fabrication method, which demands highly skilled labor and a large amount of styrene emission [24]. It is more commonly suitable for thermosetting polymer founded composites. Some of the commonly used matrix materials in hand-layup method include thermosetting polymers like epoxy, phenolic resin, unsaturated polyester and polyurethane, polyvinyl ester resin, and fiber reinforcements like all natural fibers (in fiber or mat form), aramid, glass and carbon fibers in uni-or bi-directional, woven, fabric, or random orientation. This type of manufacturing technique is used mainly for manufacture of components for aerospace, automobile, and marine industry. Its large use results from the extreme flexibility, which allows the manufacturing of a large variety of shapes. The variability of product composition, and the latitude in size, shape, and design possibilities, permits a lay-up part to be tailored to economically fit its desired function. Only the dexterity of hand lay-up techniques can provide the manipulations required to form the reinforcement without defects such as wrinkles or tears and despite the development of tape laying and other robotic approaches, hand lay-up is still widely used and remains in many cases the only realistic way to form the reinforcement on to the tool [25]. It is also a notable point that during the process of hand-layup, the emission of fumes is inevitable, which arises from the open mold, which causes some serious health concerns to the operator while the operator awaits the saturation of resin after laying the laminates by hand [26]. Some of the disadvantages associated with hand-layup method includes highly skilled labor, elevated labor cost, trained operator for laying, high cycle time but low output volume, and getting of only one smooth surface in the finished component, while the other one would be rough [27].

13.2.2 Pultrusion technique

Pultrusion can be an economic method for the production of composite materials in applications like beams, tubes, stiffeners, and other profiles having fixed cross-sectional area [27]. In most cases, this technique is much like the filament winding technique in which the resin is impregnated onto the reinforcement. In pultrusion technique, the fibers are subjected to a pulling force through a resin bath, mixed with the catalyst and then through a fixture to shape form the final product along with the removal of excess quantity of resin in the final product. Reinforcement fibers are then continuously pulled through the heat die by a frictional extraction system and the composite part gradually cures inside heat die through exothermic chemical reaction initiated by the heaters located on the die and solidifies. The cured and solidified profiles are cut into the desired length by a cut-off saw at the end of the process [28] as shown in Figure 13.3, schematic of the pultrusion process. Pultrusion renders a stock material and not exactly the finished components and the pultrusion product is much similar to the products obtained by conventional extrusion techniques [29].

Pultrusion technique is majorly used for the fabrication of high strength and continuous structured components. For instance, the reinforced composite beams that are used along with the cement concrete in the construction industry are usually made of pultrusion technique. It consists of manufacturing polymer composites, which are made from a thermoplastic or thermosetting resin matrix and glass fiber reinforcements, are generally referred to as glass fiber–reinforced plastic (GFRP) composites, and are used in structural applications, where a high specific mechanical strength and good chemical resistance are required. The reinforcement material used here can be a continuous filament mat (CFM) or unidirectional roving mat while the commonly used matrix materials include vinyl esters or polyester resins [30]. The merits of this process include low cost, simple tooling, and high productivity, while the disadvantages include the wastage of the material at a minimum quantity due to the process continuity [31].

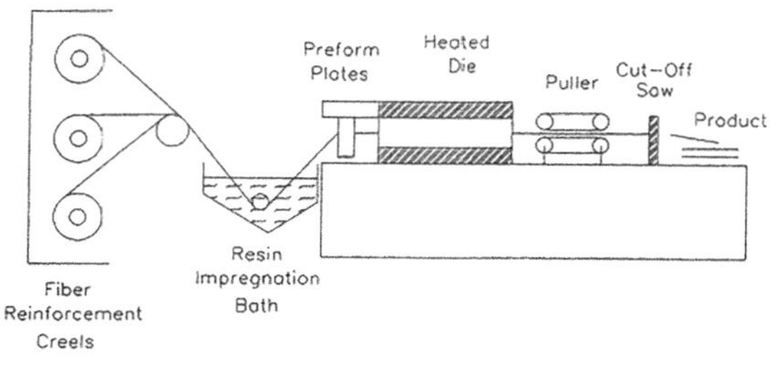

Figure 13.3: Schematic of the pultrusion process [29].

13.2.3 Filament winding technique

Filament winding technique renders products in lightweight and high strength. During this process, a continuous resin impregnated raving layer is coated over a monofilament round over a rotating mandrel, which was then postcured in an oven to room temperature. Usually carbon, glass, or aramid fibers are impregnated with thermosetting resins such as epoxies and vinyl esters [32]. There are two different filament winding methods: (I) wet winding, in which the fibers are passed through a resin bath and wound onto a rotating mandrel as shown in Figure 13.4 winding of prepreg in which the rotating mandrel is placed with the reimpregnated fiber tows. In this method, the wrapping of layers of fiber such as Kevlar, glass, carbon, or aramid can be done in two different angles such as at a perpendicular to axis of the cylinder and at helical layers at any angle to the axis of the cylinder. Among two filament winding methods, wet winding is more common and widely used for manufacturing fiber-reinforced thermosetting matrix composite cylinders, it is inexpensive and lets one control the properties of the composite. Compared with prepreg winding, wet winding has several advantages: low material cost; short winding time; and the resin formulation, which can be easily varied to meet specific requirements.

The process of filament winding as composite structures manufacturing technique can be characterized by considerable geometrical accuracy, constant product quality, and generally high fiber volume fractions. It is often used in automation, although the majority of filament wound products is rotationally symmetric and comprises objects like pressure vessels, shafts, containers, and pipelines. Nowadays, filament wound structures show an increasing tendency to become mass production items [33].

13.2.4 Compression molding technique

Compression molding technique is the most common and most widely used manufacturing technique for the fabrication of composite parts. This method is the extension of the hand lay-up process. Even though compression molding technique is most applied to thermoplastics, it can also be used for the manufacturing of composite parts made of thermosetting polymer composites. Generally, the raw material forms for the compression molding technique include putty-like masses, preforms, and granules. First, they are placed in an

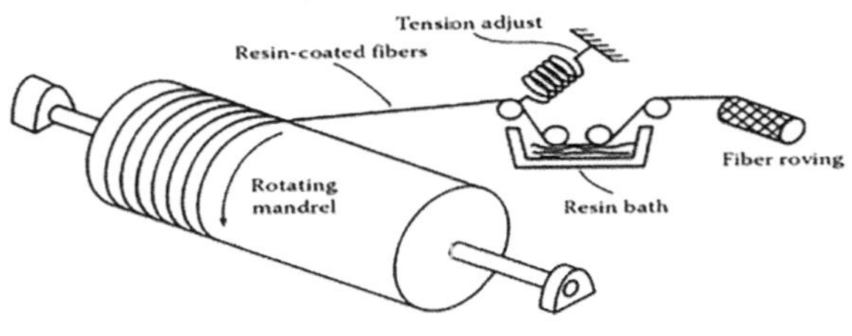

Figure 13.4: Filament winding process [32].

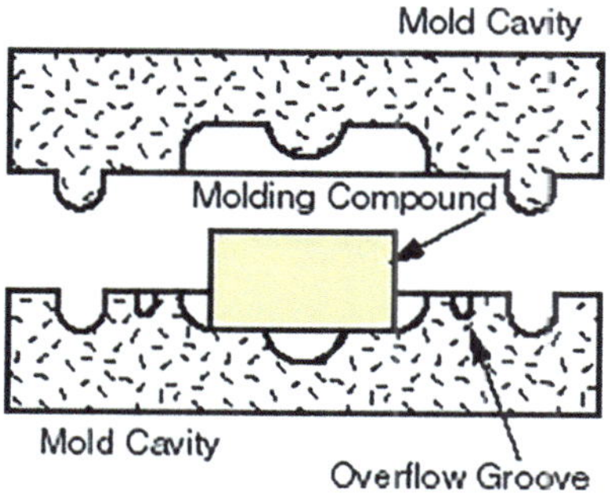

Figure 13.5: Schematic of composite material placed in the mold cavity [35].

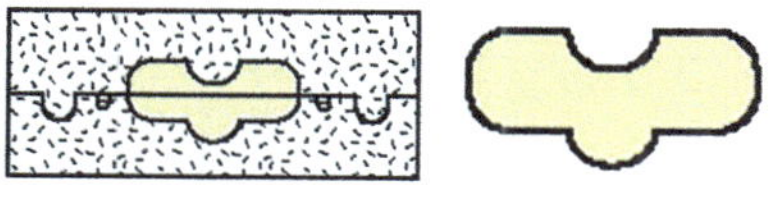

Figure 13.6: Mold is closed, and pressure is applied [35].

open, heated mold cavity, the mold is then closed, and pressure is applied to force the material to fill up the cavity. A hydraulic ram is regularly used deliver adequate force during the molding process. The heat and pressure are kept up until the plastic material is cured [34]). Figures 13.5 and 13.6 show the schematic of composite material manufacturing through compression moulding.

Bulk molding and sheet molding compounds (BMC and SMC) are the most used form of parts in any compression molding techniques. SMC can be used in any size that fits the mold shape by precutting but is not economic. SMC is also characterized by a higher consistency of the product owing to the homogenously distributed material onto the surface of the mold resulting in lower disorientation of the material [39]. Utilization of compression molding technique is most common for applications like gears, containers with large size, electrical components, housing appliance components, flatware, electronic device components, door handles, knobs, and buttons. Commonly used matrix materials in the compression molding technique include polyetherether ketone (PEEK), polyester, polyamide imide (PAI), and polyimide (PI) along with the commonly used natural and synthetic fibers [40]. Four significant parameters that strongly govern the compression molding technique include heating time and method, cooling time and method, quantity of material placed, and the force applied during the curing process.

13.2.5 Plastic injection molding technique

Injection molding is the most important commonly used manufacturing process for the fabrication of plastic parts [41]. The size, application, and complexity of the products vary

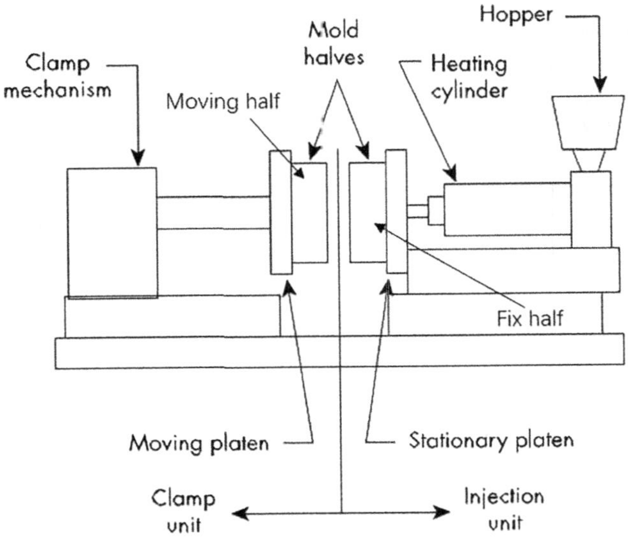

Figure 13.7: Plastic injection molding overview [36].

widely when manufactured using injection molding technique and the coverage of product range is enormous. Usually, the process of injection molding demands the plastic material to be manufactured, a mold and an injection molding machine [42]. The process of injection molding involves the heating of the precursor plastic in an injection molding machine, injection of the melt into a mold and cooling it and finally the solidification of the part. Figure 13.7 shows the overview of plastic injection moulding process.

The procedure cycle for plastic injection molding is short, commonly between 2 s and 2 min, and comprises the accompanying four stages [43]:

- **Clamping:** Clamping has to be carried out before the injection of the molten plastic into the mold for an air-free operation. Both the halves of the mold are attached with the injection molding machine while any one half can slide over the other. Both the mold halves are closed tightly by a hydraulic pressure clamping unit, and a sufficient force is applied onto them to be closed safely and securely before the process of injection of molten metal into the mold.
- **Injection:** Injection unit functions to progress the raw plastic material, which will usually be in pellet form, to the injection molding machine where the material is subjected a high pressure and temperature for melting and injecting. The molten metal is then rapidly transferred to the mold under a very high pressure applied for the material holding. The transferred material to the mold cavity is usually termed as "shot." Injection pressure and power and the shot volume can be used to determine the time of injection.
- **Cooling:** As the molten plastic material reaches the mold surface, it begins to enter the cooling cycle. During the cooling cycle, the plastic attains the required shape and

size adopting to the shape and size of the mold. Until the required cooling time is reached, it is impossible to open the mold. Various thermodynamic properties of the plastic material and the maximum wall thickness can be used to determine the cooling rate of the material.

- **Ejection:** Ejection of the cooled part can be done by using an ejector system, which is attached to the latter part of the injection molding machine after the required cooling time is attained. During the opening of the mold, this ejector mechanism is used to push away from the mold. Since the solidified part shrinks and sticks to the wall of the mold during the cooling process, a force is needed to eject the part out. In order to avoid the problem of the part sticking to the mold surface, a releasing agent can be coated or applied into the mold surface before the molten plastic is being injected into the mold cavity. The time of mold opening and ejection can be assessed based on the dry cycle time of the operation to the free fall of the mold after the process of ejection. After the ejection of the current part, the mold is shut closed for the next shot to be fed in.

13.2.6 Vacuum bagging technique

Vacuum bagging technique is a typically used method for manufacturing of laminated composites. In this technique, the atmospheric pressure is used to withheld the matrix and the laminates until the curing of the matrix to bond the laminates (as shown in Figure 13.8) [44].

Pressurizing a composite lamination serves several functions. First, it removes trapped air between layers. Second, it compacts the fiber layers for efficient force transmission among fiber bundles and prevents shifting of fiber orientation during cure. Third, it reduces humidity. Finally, and most important, the vacuum bagging technique optimizes the fiber-to-resin ratio in the composite part [45].

13.2.7 Spray up technique

When it is compared to wet lay-up process, spray up is more cost effective meanwhile due to the randomly oriented fiber spray up process produces lower mechanical properties. During the process of spray up, a special gun is used to chop the fed glass mat into small pieces and mix with the matrix prior to spraying on the substrate. In order to compact the layers after spraying, a metallic roller can be used. For quality improvement, vacuum bag

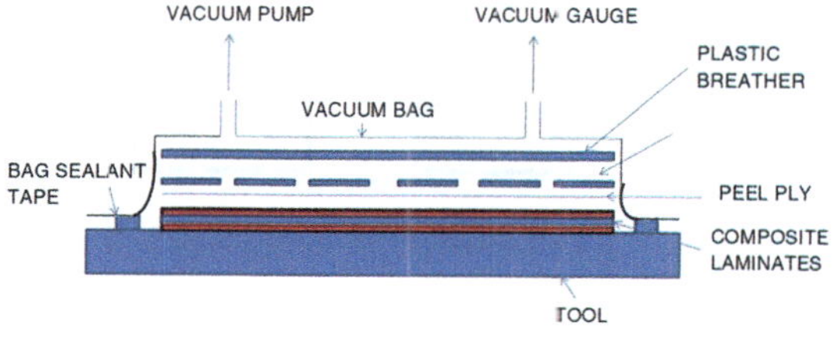

Figure 13.8: Vacuum bag molding [37].

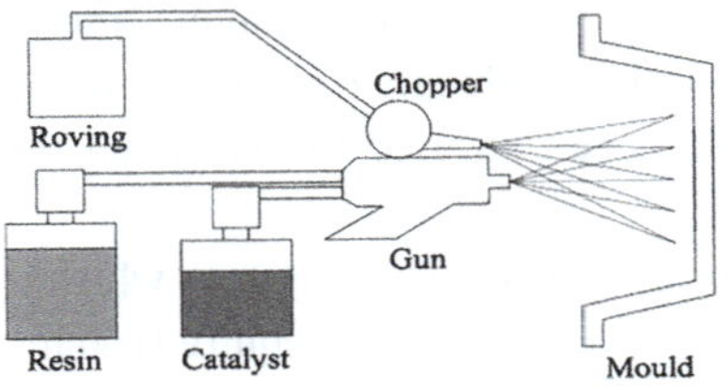

Figure 13.9: Spray up process [38].

cures might be used but not mostly used. Few parts cannot be made using this process such as load-bearing parts owing to the necessity of random orientation and short fibers [46]. The process of spray up technique is illustrated in Figure 13.9.

13.2.8 Resin transfer molding (RTM) technique

RTM is a technique in which the preform is directly placed within the resin mold to form the composites. The mold is closed and the matrix material is injected into the mold containing the preform till the fiber is completely impregnated within the matrix. After the curing of the composites within the mold, they were removed from the mold, and then the composites were subjected to postprocessing or postcuring process. During the RTM process, two ports are highly important, one for the matrix to be injected into the mold and the other to release the air at the time of matrix injection into the mold [47]. The composites manufactured using RTM finds their application in civil, aerospace, automotive, and sporting fields. Various process parameters such as volume fraction, temperature, pressure, viscosity of the matrix, permeability, and time of filling the mold are significant in determining the product quality of the finished composite material [48]. Figure 13.10 shows the sequence and process of RTM.

13.2.9 Vacuum infusion molding technique

Vacuum infusion molding is a suitable technique that can be used only for medium and low volume of production parts since its processing cycle is too long and it is difficult to

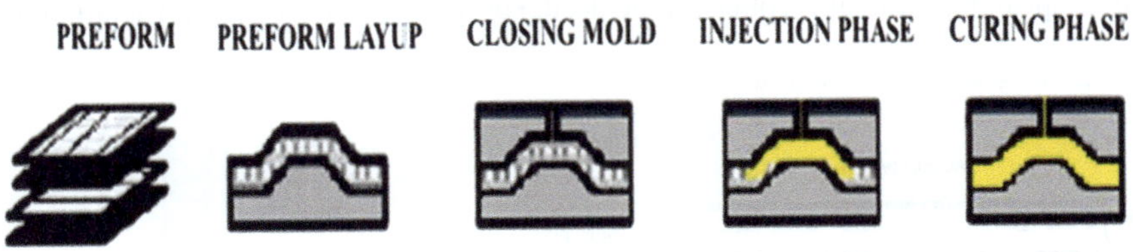

Figure 13.10: Sequence of the main steps in RTM [49].

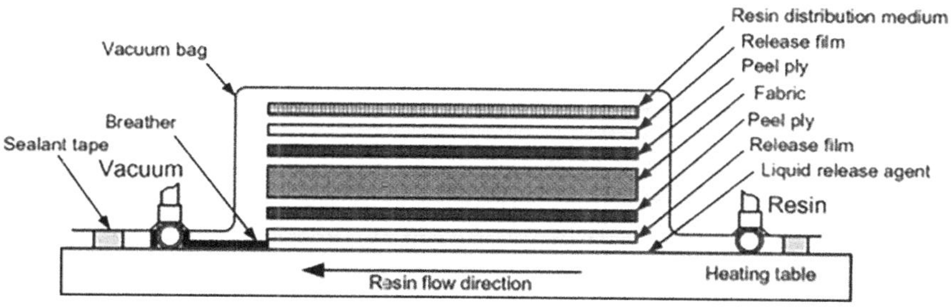

Figure 13.11: Vacuum infusion molding process [60].

reduce the cycle time of curing and by achieving these two can render better quality automotive products with a better process efficiency. The following are the major steps involved in vacuum infusion molding (as illustrated in Figure 13.11):

- Initially, the dry fabric or preform is laid onto the mold along with the releasing agent, peel plies over the surface of the tools.
- Then the vacuum bag containing the preform is sealed completely and a vacuum is created by means of a vacuum pump.
- Matrix material and hardener mixture is then sucked into the vacuum bag from an external reservoir by means of a pump.
- The impregnation is completed by infusing the matrix with hardener all through the preform.
- After the complete impregnation, demolding and curing processes are carried out.

13.3 Energy saving measures in composite industry through good manufacturing practice

After fibers are reinforced and polymer matrices are made, as we mentioned before in composite manufacturing, variable process methods are required preceding integration of polymer resins and fibers with necessary additional energy less than prior manufacturing methods. Also, during curing, necessary heat and pressure are provided by a considerable amount of energy. Owing to the energy intensity of each composite manufacturing process, the methods can be distinguished based on the energy intensity of each process. Various energy intensities for different process are enlisted in Table 13.2.

It is necessary to reduce the energy intensity of every composite manufacturing process to reduce the energy consumption during the process. It was stated in a study that the processing of natural fibers utilize 45 % less energy when compared with the energy utilized for processing synthetic fibers [51]. Specifically, natural fibers like China reed and flax fibers can be processed with lower energy intensities. Similarly, among the polymer

Table 13.2: Energy intensity of variable composite materials [50].

Material type	Material	Energy intensity, MJ/kg
Polymers	Polyester	63–78
	Epoxy	76–80
	LDPE	65–92
	PP	72–112
	PVC	53–80
	PS	71–118
	PC	80–115
Fibers	Glass fiber	13–32
	Carbon fiber	183–286
	China reed fiber	3.6
	Flax fiber	6.5

matrices, polyester and epoxy resin matrices can be processed with lesser energy intensities comparatively while manufacturing fiber-reinforced composites with these resins.

It was stated in some of the composite-related news published during late 70s that the phenolic resin reinforced with cotton fabrics were used as substitutes for phosphor bronze bearings in the steel and nonferrous neck rolling mills, and the energy savings for this process were saved about 25 % [52]. In order to save the energy consumption, also recycling technologies are being developed. In the recycling perform, the ideal maximization of energy recovery is achieved by closed-loop recycling, where the fibers used for the manufacturing of other fiber-reinforced composites such as short fiber-based composite materials and sheet molding compounds keeping their characteristics of performance. Recovery of energy from the composite can be made through the pyrolysis method of recycling in which matrix is decomposed slightly at low temperatures into short chain hydrocarbons to be used again in the chemical industry or as fuel, therefore, enable the reuse of the rest of fiber materials [50].

Moreover, the increase of energy absorption capability of the composite material is positive resultant in energy saving. S. Ramakrishna and D. Hull [53] in their experiments used knitted fabric composite tubes, and the results showed that and the specific energy absorption of knitted fabric composite tubes increased with increasing additional fiber content. Also, the larger the tow size of inlay fibers, the greater was the loads required to bend stiffer fronds, and this resulted in increase of energy-absorption capability.

13.4 Measures to save energy in a composite industry

In recent days, utilization of a certain class of materials might lead to the improvement in their environmental and economic aspects during their entire life cycle including the process of manufacturing and during recycling at the end of their life when compared with their predecessor materials. Many advanced materials are being invented these days in the composite manufacturing industry that aids in energy saving during their manufacturing [54].

13.4.1 Energy efficient smart windows using soft matter composites

Smart windows are made of soft matter containing polymer composites stabilized with cholesterol liquid crystals and these windows can be used as electrically switchable broadband infrared (IR) reflectors, which has a transparent vision in visible light region. Additionally, based on the environmental conditions, the infrared reflectors can have variation in transmission or reflection behavior through the variation in voltage supply [55].

In most cases, the interior temperature conditions are to be compatible with the use of heating, air-conditioning, and ventilation of the equipment, which is necessary for performing the intended function of the equipment. This utilization of such energy consuming form of interior air-conditioning results in high electricity bills and also in high carbon dioxide emissions along with some hazardous environmental effects.

Few experimenters developed an inorganic materials-based indium oxide nanocrystals doped with tin in niobium oxide glass for thermal responsive IR reflectors to reflect back the IR radiation even at elevated temperatures. These materials were also capable of transmitting at lower temperatures by using electrically switchable IR reflectors by using manual controls [56]. However, these reflectors might reflect the minor portion of solar light, which reduces the impact of radiation.

Based on the energy savings values, the performance efficiency of the switchable IR reflectors were evaluated in various studies. During the actual functioning of the IR reflector in the buildings, it serves for a wider scope in such a way that it reduces the building energy consumption, thereby regulating the emission of carbon dioxide and also make the air-conditioning system of the building to operate a very least powder consumption [57]. Additionally, when the solar gains are controlled in an efficient manner, it results in various advantages like improvement of environmental quality at the indoor due to an effective use of sunlight at the day, better outside landscape views, and decreasing the radiant window temperature, which in turn increases the summer thermal comfort at the buildings.

13.4.2 Electrical water heater using phase change materials for energy efficiency

Paraffin wax is the most prominently used phase change material (PCM), which can store the thermal energy by transforming itself into two different states of matter. The heat for the paraffin wax was gained from the concentrated solar dish. In many experiments, the

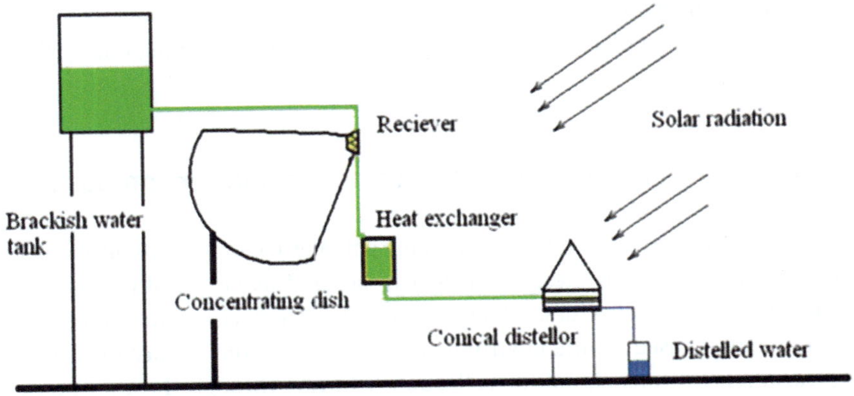

Figure 13.12: A schematic diagram for the recent studying [59].

paraffin wax was used as a PCM, and the system was attached with a solar tracker. It was found that the working time of the system without PCM was about 3 h as per the sun tracker, while it was increased to 5 h overall as per the sun tracker reading placed in the concentrated solar dish [58]. With the available sun tracker, it was found that various efficiency values such as heating, system, and concentrating efficiency have increased by 112.8 %, 307.6 %, and 65 %, respectively, when compared with the system without sun tracker. In the absence of sun tracker and PCM the heating, system and concentrating efficiency was about 41.65 %, 185 %, and 50.47 %, respectively [59]. This is because the phase change material storages are used to store energy in several practical application areas, and to balance temporary temperature changes. In applications with a small temperature swing, PCM storage is better to practical heat storage, because of its nearly isothermal storing mechanism and high storage density. Figure 13.12 shows a schematic of hot water generation from a solar heater.

13.4.3 Composite material for building energy efficiency

The initial assessment of building thermal efficiency was based on the parameters such as thermal insulation of building doors, walls, windows and external walls, and roofs. Further analysis was made through the parameters like use of optimum building materials, method, and quality of building construction and manufacturing of components. All these assessments were made with reference to the use of industrial composite panels used for construction these days. It was stated in various studies that the use of industrial composite panels in the field of building construction and materials usage had contributed to the enhancement of accuracy of the construction industry, which in turn enhances the thermal insulation performance of the buildings when compared with the traditional buildings.

Figure 13.13: Raw straw board [61].

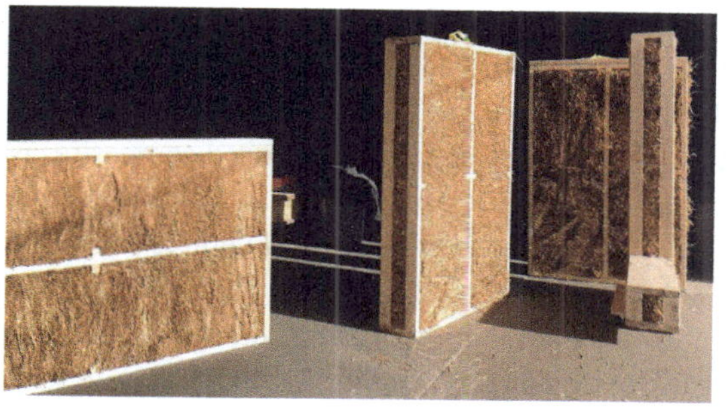

Figure 13.14: New straw board specimens [61].

Straw board is one such type of the composite panel used lately, which was usually made of natural wheat or natural straw as a main precursor, which gives it a clean texture. These materials are light in weight, possess better strength and better thermal insulation, and are environmental friendly materials that can contribute to green buildings. Such straw-based materials are mainly used for the construction of external and internal walls of roof of the building and other parts of the building. However, the straw-based construction materials are used as only support structures and not as main construction materials, which bear the building load [63]. Moreover, building applications demand some more properties such as fire-retardance and high temperature-resistant and moisture-resistant materials. Straw board materials are highly suitable for old buildings also which needs a rebuilding that includes temporary and storage buildings apart from which these materials can be used for load bearing building structures in light industrial buildings and civil constructions. The following Figures 13.13 and 13.14 show the raw straw board and the panels made of straw boards.

13.4.4 Soft magnetic composites

Soft magnetic composites (SMC) containing ferromagnetic powder encompassed within an electrically insulating film is a category of materials that are characterized by better soft magnetic properties with low hysteresis losses, high magnetic permeability even at low magnetic fields, minimal eddy current losses, and maximum flux density for saturation. Such soft magnetic materials finds their application mostly in electromagnetic DC and AC applications including sensors, low frequency filters, transformers, actuators, electromagnetic circuits and electromotors, magnetic field shielding, and magnetic seal systems, which demands the use of core with 3D ferromagnetic isotropic behavior [64–66].

13.5 Light weight composites for fuel and energy saving

Light weighting is an important for end-use energy efficiency strategy especially in transportation area, for example, a 10 % reduction in vehicle weight can improve fuel efficiency by 6 %–8 % for conventional internal combustion engines or increase the range of a battery-electric vehicle by up to 10 % [67]. The use of glass fiber-based composites is mostly found in the semi-structural or closed structures such as doors, brackets, roofs, and rear hatches of the vehicles, which contributed to the reduction of vehicle weight by 8–10 % of the entire vehicle weight. Besides, the use of glass fiber composites can also be extended to the application, which requires a combination of better damping behavior and corrosion resistance. Similarly, the use of carbon fiber composites in the commercial vehicle components has been done for the past four decades especially in roof and hood semi-structural parts and some nonstructural parts such as seat fabrics where they can be easily produced by low volume productions [68]. Use of both carbon and glass fibers in the structural and nonstructural parts is majorly supported by the advantage of lightweight offered by these fiber-based composites. When these fibers are used in the lightweight body of the vehicles, almost 23–28 % of the vehicle weight was saved [69].

In addition, the benefits of light weighting can also be extended to military vehicles, which results in capability to support functional operation, enhanced performance, increased fuel economy, and enhanced survival ability at extreme conditions. These factors were listed by the application of lightweight technology magazine published by National Research Council report in the year 2012.

By considering life cycle assessment (LCA) where a valuable technique for estimating the environmental performance of products, materials, and services from extraction of raw materials to final disposal, which includes extraction, materials processing, manufacturing, transport, use, reuse, maintenance, and recycling, can determine the type of material can be applied on the vehicle.

13.5.1 Zero emission

The vehicle structural material made components were used to evaluate and compare the total emissions when put into use. For instance, the comparison material on structural aircraft like aluminum, GLARE, and carbon fiber-reinforced polymer (CFRP) plates was performed to study the potential of lightweight composites in reducing the aviation

Figure 13.15: Ultra-low-drag platform vehicle for hypercar concept [62].

emissions. In the same time, the effect of aviation emissions, which consist mainly of carbon dioxide (71 %), water vapor (28 %), and nitrogen oxides in the atmosphere [70]. It is widely recognized that emission at high altitudes is more environmentally damaging than those at ground level, due to increased interaction between gases in the atmosphere. For instance, the concept of hypercar combines the ultra-low-drag and ultralight vehicle with a hybrid electric-based drive system. This combination results in drastic improvement in fuel economy and much reduced emissions in both lightweight trucks and cars, which also aids in exceeding or maintaining the performance, amenity, safety and affordability.

Because of their fuel efficiency, low emissions, and lifecycle benefits, hypercars should certainly be "green," (Figure 13.15) but their prospect for extensive adoption may depend less on environmental regulation like climate-protecting and equivalent-zero-emission vehicles (EZEV) regulations compared to the two powerful market forces, which are what customers want at superior products, and also desire of manufacturers competitive advantage [62]. In conclusion, hypercars have to be both useful and fun to drive, and fast, less-risk, and inexpensive in manufacturing.

13.6 Energy management system implementation in composites industry

In composite industry, energy is mostly consumed more in heaters, compressed air plant steam boiler, cooling water pumps, thermal oxidizer, and chiller. Different energy management systems used in different processes are explained in the next section [71]. As the use of carbon fiber reinforcement increased, there is a need for recycling these materials waste to save energy. Meanwhile there are still problems faced when trying to recycle these materials. A past study was done to validate the use of recycling for fiber reinforcement plastic. In order to qualify the above recycling method as an acceptable waste management solution, it has to be raised to industrial standards to comply with the government legislations, which ultimately decreases the accumulation of waste during the production stage and end of life stage [72].

13.6.1 Energy management in injection molding

There are two phases for the energy consumption in injection molding, which are high power that is used in a short term when the polymer is injected or the part is ejected as well as the low power over a long time used when the injected polymer is cooling [73]. Figure 13.16 shows the example of energy consumption in injection molding plant. Figure 13.17 shows where the energy is used in a molding machine.

To reduce energy used in injection molding, manufacturers use all-electric machines instead of hydraulic machines and hybrid machines. This is because all-electric machines can reduce energy consumption from 30 to 60 % according to the mold and the machine used, and all power used for all-electric machines are from direct electric drive. Moreover, all electric machines require low power consumption to start up. [74].

One of the ways to reduce energy consumption used in injection molding is the use of barrel insulation jackets, which mainly used to reduce the energy consumption of the heating elements up to 50 %. The most important advantages of installing barrel

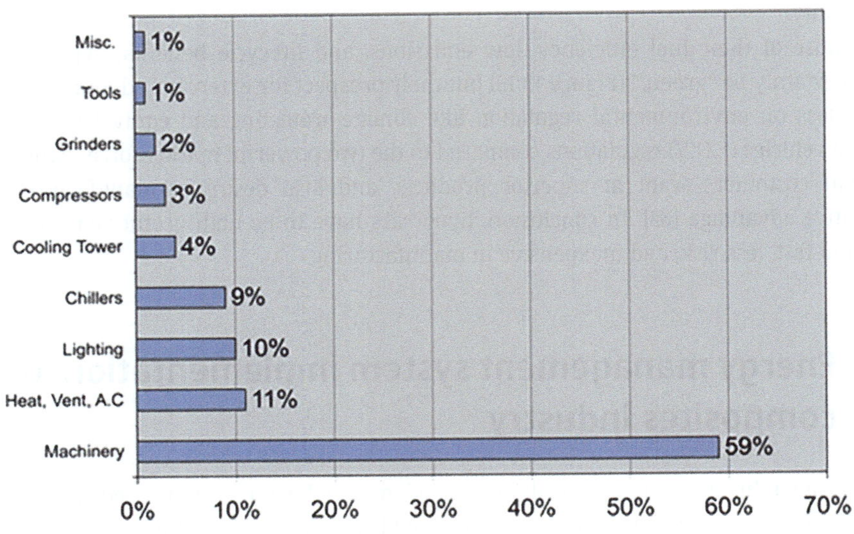

Figure 13.16: Example of energy consumption in injection molding plant [73].

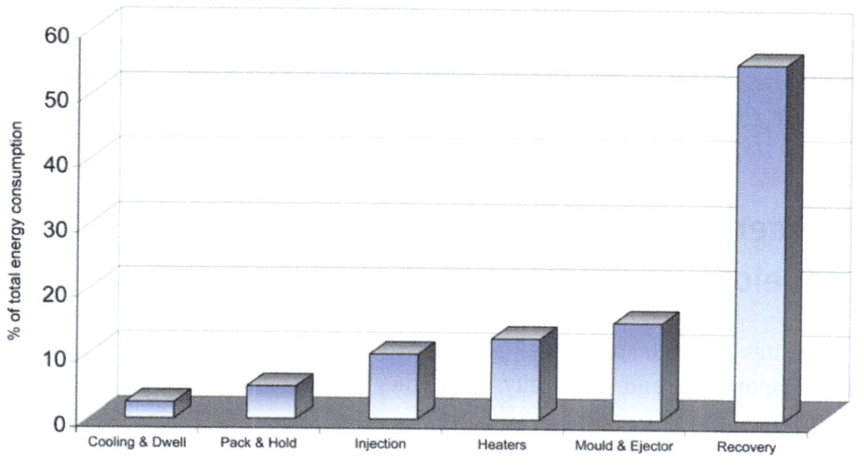

Figure 13.17: Energy used in a molding machine [74].

Table 13.3: Properties of cylinder heater bands.

Comparison of cylinder heater bands with and without insulation jackets			
Without insulation jackets		**With insulation jackets**	
Consumption kW	**Emission temp °C**	**Consumption kW**	**Emission temp °C**
33.67	110	26.28	55
Temperature emission difference: 50 %; energy saving: 22 %			
Source: Nickerson Europe Ltd.			

insulation jackets are short start up time, reducing electricity consumption, and reducing operation cost [75]. Table 13.3 shows the properties of cylinder heater bands without and with insulation jackets.

13.6.2 Energy management in extrusion process

During the process of extrusion components like cooling and heating systems, lighting systems, drives, and motors are the major components that consume most of the energy. By setting the temperature of the die to the lowest possible value without compromising the uniformity of melt flow can be one of the possible ways to reduce the consumption of energy. Moreover, to ensure an efficient energy usage, in cooling the profile the water should be checked so it is not colder than it needs to be. In extrusion, the application of superimposing longitudinal and transverse oscillations to the die was observed to increase the mechanical properties of the polymer; on the other hand, it decreases the pressure drop in the die section, which results in saving on the total power consumption per unit weight [76].

13.7 "Green composites" and sustainable development strategies

Green composites are the materials that are characterized with green attributes and are viable both economically and technically, where they act as minimizer of contamination era as well.

When compared with normal composite materials, the green composite material utilization has gained much attention in terms of environment and sustainability [77]. Transition toward the green materials and biopolymers based on natural fibers has become possible these days owing to the continuous depletion and over utilization of petroleum-based products and the gradually rising environmental hazards and their nonrenewability. The environmentally friendly and sustainable nature of natural fibers and biopolymers has pulled in significant consideration of researchers and industries [77]. Research trends in green composite materials have increased exponentially since these materials are considered to be a potential replacement for the currently used traditional composite materials. A consideration has to be given on the manufacturing and disposal phase of a composite material to account it as a green composite material [78].

Utilization of plant-based natural fibers is considered to be very cheap and has been widely used all around the globe for the past couple of decades. The natural fiber-based composite materials render some other advantages as well. Some of the advantages include their good acoustic behavior fueled by the cellulosic nature and hollow cellular structure, their CO_2 neutrality at the time of burning, nonabrasive behavior on the processing equipment, their disposal through incineration, and their lightweight because of their tubular hollow structure. Some of the prominently used nonbiodegradable thermoplastic polymeric resins for reinforcing various plant-based natural fibers include polyvinyl chloride, saturated and unsaturated polyester resin, low and medium density polyethylene, polypropylene, and nylon. The composites manufactured with the help of aforesaid reinforcements and matrices can be termed as green composites [79]. The life cycle of green composite material is shown in Figure 13.18.

Including the green composites for the manufacturing of various automotive components might aid toward the transition to go green initiatives. Mercedez-Benz was the very first car maker to manufacture the door panels by using natural fiber composites with jute fibers in 1990 followed by Toyota motors while they utilized natural biofiller-based composite materials for headrests, parcel shelves, roof upholsteries, and door

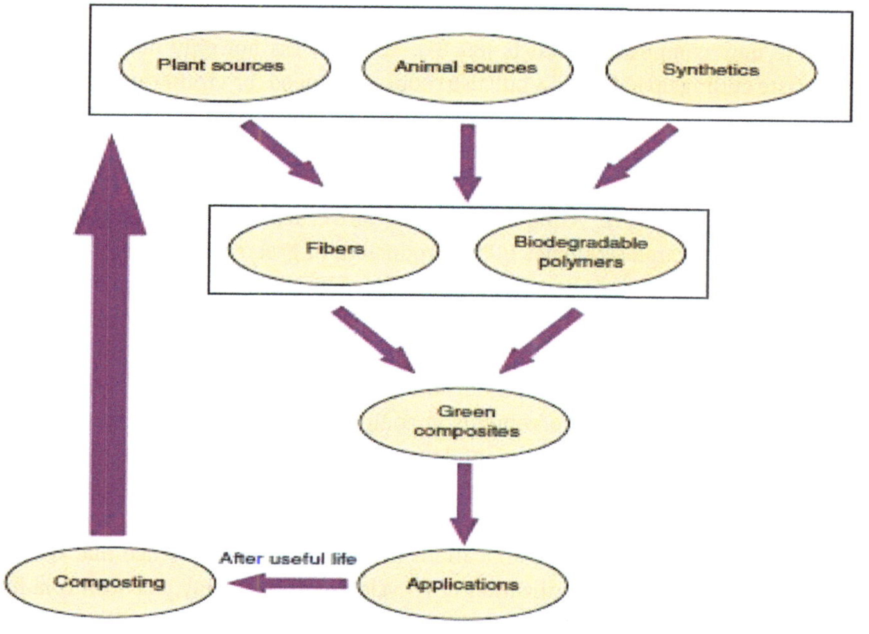

Figure 13.18: Typical life cycle of green composites [80].

panels in the year 2003 [81]. Novel applications in the field of medical devices, veterinary medicine, and healthcare applications have become possible by the use of various biomaterials-based green composites while their potentialities and importance in the field medical engineering and technologies [82].

13.7.1 Waste reduction

Ochi et al. reported that waste reduction is a fundamental aspect of green composites, as it focuses on minimizing environmental impact and conserving resources [83]. One of the key strategies employed in waste reduction is the utilization of recycled or recyclable materials. Green composites prioritize the use of materials that have undergone recycling processes or can be easily recycled at the end of their life cycle. By incorporating recycled materials into composite production, manufacturers significantly reduce landfill waste and alleviate the burden on natural resources as reported by Shang and Qu [84]. This approach not only mitigates environmental pollution but also contributes to a more circular economy, where composite waste is transformed into valuable resources for future composite manufacturing.

Previous researchers reported that recycling composite waste is an effective means of waste reduction and sustainability in the composites industry. Through innovative recycling techniques, composite waste is processed and reintroduced into the production cycle to create new composite products [85–87]. This process not only minimizes the amount of waste ending up in landfills but also reduces the need for virgin raw materials, which often have a high environmental impact. By embracing recycling practices, the composites industry demonstrates its commitment to resource conservation and environmental stewardship. Furthermore, the development of efficient recycling technologies and the establishment of recycling infrastructure are crucial in enabling widespread adoption of waste reduction strategies in the production of green composites.

Recent scientific publications have provided valuable insights and evidence regarding the effectiveness of waste reduction strategies in the composites industry. For instance, a study investigated the incorporation of recycled carbon fiber into epoxy-based composites, demonstrating its potential for waste reduction and improved sustainability. Another study explored the use of bio-based reinforcements in composite manufacturing, highlighting their recyclability and reduced environmental impact compared to traditional materials. Additionally, another research focused on the development of recycling techniques for thermoset composites, showcasing the feasibility of transforming composite waste into value-added products. These recent publications provide valuable evidence that waste reduction measures, such as the utilization of recycled materials, play a crucial role in achieving sustainability goals within the composites industry. By adopting such practices, manufacturers can contribute to a greener and more sustainable future.

13.7.2 Life cycle assessment (LCA)

Recent scientific publications have highlighted the importance of conducting life cycle assessments (LCAs) in the context of sustainable development in the composites industry. For example, a study conducted a comprehensive LCA of green composites made from bio-based polymers and natural fibers. The findings revealed that these materials have significantly lower environmental impacts compared to traditional composites, especially in terms of greenhouse gas emissions and energy consumption. Another study conducted by Smith et al. [88] focused on the LCA of recycled carbon fiber composites, showcasing the environmental benefits of incorporating recycled materials into composite production. These recent publications emphasize the role of LCA as a crucial tool for assessing and improving the sustainability of green composites, enabling manufacturers to make informed decisions throughout the entire life cycle of the product [88].

In line with the concept of sustainable development, recent scientific publications have explored the application of life cycle assessment (LCA) methodologies specifically tailored for green composites. For instance, a study by Chen et al. [89] proposed an LCA framework specifically designed for assessing the environmental impact of natural fiber-reinforced composites [89]. The study considered various life cycle stages, including fiber production, composite manufacturing, product use, and end-of-life scenarios. Another study conducted by Zhang et al. [90] focused on developing a streamlined LCA approach for bio-based polymer composites, taking into account the specific characteristics and environmental implications of these materials. These recent publications demonstrate the growing recognition of LCA as an essential tool in evaluating and optimizing the environmental performance of green composites [90]. By incorporating LCA into the decision-making process, stakeholders in the composites industry can effectively identify areas for improvement and implement sustainable practices, leading to a more environmentally friendly and resource-efficient composites sector.

13.7.3 Economic viability

The adoption of green composites in various industries offers not only environmental benefits but also economic viability. Recent studies have emphasized that incorporating green composites into manufacturing processes can lead to long-term cost savings and improved economic performance. For instance, research conducted by Dua et al. [91] analyzed the economic feasibility of using bio-based composites in automotive applications. The study demonstrated that despite the initial higher material costs, the overall lifecycle cost of these composites was lower than traditional materials due to reduced maintenance and fuel consumption [91]. Furthermore, a study by Eskander et al. [92] explored the economic impact of integrating recycled composites into the construction sector, revealing significant cost savings in waste management and material procurement. These findings highlight that green composites not only align with sustainability goals but also present economic advantages, fostering innovation, job creation, and enhancing market competitiveness [92].

The increasing demand for environmentally friendly products and the global shift toward sustainable practices have created a favorable market environment for green composites. Recent scientific publications have emphasized the economic potential of adopting green composites in various industries. For example, a study by Volk et al. [93] investigated the market outlook and growth potential of bio-based composites in the packaging sector. The research identified a growing consumer preference for sustainable packaging solutions, creating market opportunities for companies producing bio-based composites [93]. Similarly, a study by Sharma et al. [94] explored the economic implications of incorporating natural fiber composites in the furniture industry, demonstrating that these materials can enhance product value, attract environmentally conscious consumers, and improve brand reputation [94]. These recent publications illustrate that green composites not only contribute to sustainable development but also offer economic advantages by tapping into emerging markets and meeting consumer demands for eco-friendly products.

13.7.4 Positive environmental impacts

The utilization of green composites in industries has demonstrated positive environmental impacts, aligning with sustainable development goals and addressing climate change concerns. Recent scientific studies have highlighted the potential of green composites to reduce carbon emissions and decrease reliance on nonrenewable resources. For instance, a study by Rajeshkumar et al. [95] investigated the environmental performance of bio-based composites compared to traditional petroleum-based composites. The findings revealed that bio-based composites exhibited significantly lower carbon footprints due to the renewable nature of their raw materials [95]. Another study by Suriani et al. [96] assessed the environmental benefits of using recycled composites in construction applications, demonstrating a substantial reduction in energy consumption and greenhouse gas emissions compared to virgin composites [96]. These recent scientific references underscore the positive environmental impacts of green composites, paving the way for sustainable manufacturing practices and contributing to global efforts in mitigating climate change as reported by previous researchers [97–99].

In addition to reducing carbon emissions, the adoption of green composites also contributes to the reduction of waste generation, another crucial environmental concern. Recent scientific publications have emphasized the potential of green composites in waste management and circular economy practices. For example, a study by Arinze et al. [100] explored the utilization of agricultural waste-based composites, showcasing their ability to divert agricultural residues from landfills and transform them into valuable materials. This approach not only minimizes waste but also reduces the need for virgin resources [100]. Furthermore, a study by Colangelo et al. [101] investigated the recycling and upcycling potential of end-of-life green composites, highlighting the possibilities for closed-loop systems and resource recovery [101]. These recent scientific references demonstrate that green composites play a crucial role in waste reduction and circular economy strategies, offering sustainable alternatives that contribute to a more resource-efficient and environmentally friendly future.

13.8 Conclusions

The composite industry is one with a very high potential for growth and innovation. The concept of taking two constituent materials, making them one, and having the new material retain the characteristics that we desire is a very alluring one. This gives us the ability to make many types of composites with desirable traits. However, one serious aspect is the question of monetary and energy cost. Producing composites can be a very costly venture in some cases, as well requiring a great amount of energy. Therefore, we need to strive to discover new and better methods, as well as alternative energy sources for the composite industry.

"Green Composites" play a pivotal role in achieving sustainable development by addressing key aspects such as waste reduction, life cycle assessment, economic viability, and positive environmental impacts. Through the emphasis on using recycled or recyclable materials, green composites contribute to waste reduction, minimizing landfill waste and conserving resources. Conducting life cycle assessments enables informed decision-making throughout the entire life cycle of green composites, considering factors from raw material extraction to end-of-life disposal. Moreover, the economic viability of incorporating green composites into industries is evident, as the demand for environmentally friendly products creates opportunities for innovation, job creation, and market competitiveness. Lastly, the positive environmental impacts of green composites, such as reduced carbon emissions, decreased reliance on nonrenewable resources, and decreased waste generation, contribute to sustainable development goals and efforts to mitigate climate change. Overall, the adoption of green composites holds great promise in fostering a more sustainable and environmentally conscious future [102].

Acknowledgment: The authors would like to thank the editors S.M. Sapuan, Mohd Roshdi Hassan, Eris Elianddy Supeni, and Azizan As'arry for their guidance and review of this article before its publication.

References

1. Hamerton I, Mooring L. 7 – The use of thermosets in aerospace applications. In: Guo Q, editor. Thermosets. England: Woodhead Publishing; 2012:189–227 pp.
2. Linganiso LZ, Anandjiwala RD. 4 – Fibre-reinforced laminates in aerospace engineering. In: Rana S, Fangueiro R, editors. Advanced Composite Materials for Aerospace Engineering. England: Woodhead Publishing; 2016:101–27 pp.
3. Ramesh M, Maniraj J, Rajesh Kumar L. Biocomposites for energy storage. In: Abdullah MA, Khan A, Rangappa SM, Siengchin S, editors. Biobased composites: processing, characterization, properties, and applications. Germany: Wiley; 2021:123–42 pp.
4. Janarthanan P, Ashok Kumar P, Nurazzi NM, Norrrahim MNF, Naveen J. 17 – Structural composites nanofiller materials in automotive applications. In: Nurazzi NM, Ilyas RA, Sapuan SM, Khalina A, editors. Synthetic and Natural Nanofillers in Polymer Composites. England: Woodhead Publishing; 2023:387–98 pp.
5. Wazeer A, Das A, Abeykoon C, Sinha A, Karmakar A. Composites for electric vehicles and automotive sector: a review. Green Energ Intell Transport 2023;2:100043.
6. Zhang J, Lin G, Vaidya U, Wang H. Past, present and future prospective of global carbon fibre composite developments and applications. Compos B Eng 2023;250:110463.
7. Oloruntobi O, Mokhtar K, Gohari A, Asif S, Chuah LF. Sustainable transition towards greener and cleaner seaborne shipping industry: challenges and opportunities. Cleaner Eng Technol 2023;13:100628.
8. Perry J, Hulme CH, Rix L, McCarthy HS, Lan T, Roberts S, et al. Assessing allogeneic chondroprogenitor manufacture under good manufacturing practice (GMP) serum free and xeno-free conditions. Osteoarthr Cartil 2023;31:S231–2.

9. Abedi M, Hassanshahi O, Rashiddel A, Ashtari H, Seddik Meddah M, Dias D, et al. A sustainable cementitious composite reinforced with natural fibers: an experimental and numerical study. Constr Build Mater 2023;378:131093.
10. Jarrett W, Jeffs SP, Korkees F, Rawson M. The opportunities and challenges of hybrid composite driveshafts and their couplings in the aerospace industry: a review. Compos Struct 2023;320:117203.
11. Perera YS, Ratnaweera DAAC, Dasanayaka CH, Abeykoon C. The role of artificial intelligence-driven soft sensors in advanced sustainable process industries: a critical review. Eng Appl Artif Intell 2023;121:105988.
12. Rajeshkumar L, Sathish Kumar P, Ramesh M, Sanjay MR, Siengchin S. Assessment of biodegradation of lignocellulosic fiber-based composites – a systematic review. Int J Biol Macromol 2023;253:127237.
13. Rajeshkumar L, Sitharaj A, Arulmurugan B, Bhuvaneswari V. Life cycle assessment of mineral fibers and their composites. In: Sanjay MR, Vinod A, Gaurav M, Siengchin S, editors. Synthetic and mineral fibers, their composites and applications. England: Woodhead Publishing; 2024:631–55 pp.
14. Mushafiq M, Arisar MMK, Tariq H, Czapp SJE. Energy efficiency and economic policy: comprehensive theoretical, empirical, and policy review. Energies 2023;16:2381.
15. Ning Y, Cherian J, Sial MS, Álvarez-Otero S, Comite U, Zia-Ud-Din MJES, et al. Green bond as a new determinant of sustainable green financing, energy efficiency investment, and economic growth: a global perspective. Environ Sci Pollut Res 2022;1–16.
16. Dong S-C., Li C-G., Xian G-J., Zhao Z-J., Zhang X-F., Yun Q-W, J. J. O. Z. U.-S. A.. Environmental impact assessment of aircraft elevator made with new lightning protection material. J Zhejiang Univ Sci A 2022;23: 669–82.
17. Mehmood T, Hassan MA, Li X, Ashraf A, Rehman S, Bilal M, et al. Mechanism behind sources and sinks of major anthropogenic greenhouse gases. In: Climate Change Alleviation for Sustainable Progression. USA: CRC Press; 2022:114–50 pp.
18. Gomes D, Oliveira F. Combustível composto avançado UO2-UN com condutividade térmica melhorada e alta densidade. 65–66 Congr Brasileiro Cerâm 2022;1:1–10.
19. Sadergaski L, Graham S, Malmstead J, Miskowiec A, Klett J. Synthesis and characterization of silicon carbide ceramic composites with CeO2 powder. Oak Ridge: Oak Ridge National Laboratory; 2022. Technical Report. https://doi.org/10.2172/1874641.
20. Das S, Das P, Lal B. Development of microbial biopolymers: the eco-friendly sustainable products for environmental applications. In: Novel Polymeric Materials for Environmental Applications. Singapore: World Scientific; 2023:1–20 pp.
21. Tabassum Z, Mohan A, Mamidi N, Khosla A, Kumar A, Solanki PR, et al. Recent trends in nanocomposite packaging films utilising wasteindex: waste generated biopolymers: industrial symbiosis and its implication in sustainability. IET Nanobiotechnol 2023;17:127–53.
22. Razzaq MEA, Moma SE, Rabbi MSJMER. Mechanical properties of biofiber/glass reinforced hybrid compositesindex: composites produced by hand lay-up method: a review. Mater Eng Res 2021;3:144–55.
23. Sikder A, Mishra J, Krishna R, Ighalo JOJAJfS, Engineering. Evaluation of the mechanical and durability properties of geopolymer concrete prepared with C-Glass. Fibers 2022:1–16.
24. Hoffer A, Tóth Á, Jancsek-Turóczi B, Machon A, Meiramova A, Nagy A. Potential new tracers and their mass fraction in the emitted PM 10 from the burning of household waste in stoves. Atm Chem Phys 2021;21: 17855–64.
25. Fragassa CJFU. Lightening structures by metal Replace traditional gym equipment Adv fiber-reinforced Compos exoskeleton. Fact Univ: Ser Mech Eng 2021;19:155–74.
26. Nixon-Pearson O, Greaves P, Mamalis D, Stevenson L. WP4–D1. 1. WIND turbine blades design and manufacturing, current state-of-the art; 2022:1–94 pp. Report - Cornwall flow accelarator.
27. Panetta J, Isvoranu F, Chen T, Siéfert E, Roman B, Pauly MJATOG. Computational inverse design of surface-based inflatables. ACM Trans Graph 2021;40:1–14.
28. Defonseka C. Water-blown cellular polymers. In: Water-Blown cellular polymers. Germany: De Gruyter; 2019.

29. Teubler J, Weber S, Suski P, Peschke I, Liedtke CJJOCP. Critical evaluation of the material characteristics and environmental potential of laser beam melting processes for the additive manufacturingindex: manufacturing of metallic components. J Clean Prod 2019;237:117775.
30. Hurukadli P, Bharti G, Shukla BKJMTP. Behavior of fiber reinforced polymer laminates strengthening prestressed concrete beams. Mater Today: Proc 2023;78:731–7.
31. Hegab H, Khanna N, Monib N, Salem AJSM, Technologies. Design for sustainable additive manufacturingindex: manufacturing: a review. Sustain Mater Technol 2023;e00576. https://doi.org/10.1016/j.susmat.2023.e00576.
32. Somaiah A, Prasad BA, Nath NKJMTP. A comprehensive review: characterization of glass fiber reinforced polymer composites with fillers from a Thermo-mechanical perspective. Mater Today: Proc 2022;62:3226–32.
33. Quanjin M, Sahat IM, Mat Rejab MR Abu Hassan S, Zhang B, Merzuki MNJJORP. The energyindex: energy-absorbing characteristics of filament wound hybrid carbon fiber-reinforced plastic/polylactic acid tubes with different infill pattern structures. J Reinf Plast Compos 2019;38:1067–88.
34. Almeida Fd., Sousa VF, Silva FJ, Campilho RD, Ferreira LPJAS. Development of a novel design strategy for moving mechanisms used in multi-material plastic injection molds. Appl Sci 2021;11:11805.
35. Do TT, Uyen TMT, Minh PSJP. The feasibility of an internal gas-assisted heating method for improving the melt filling ability of polyamide 6 thermoplastic compositesindex: composites in a thin wall injection molding process. Polymers (MDPI) 2021;13:1004.
36. Bryce DM. Plastic injection molding: manufacturing process fundamentals. USA: Society of manufacturing engineers; 1996.
37. Salit MS. Tropical natural fibre composites: properties, manufacture and applications. Singapore: Springer; 2014.
38. Jacob AJRP. Spray-up offers process improvements. Reinf Plast 2002;46:32–6.
39. Thuan HD, Son TA, Minh PS. Estimate the melt flow length with internal induction heating for the injection molding process. Key Eng Mater 2020;863:97–102.
40. Kanhere SV, Bermudez V, Ogale AA. Carbon and glass fiber reinforced thermoplastic matrix composites. In: Fiber Reinforced Composites. Netherlands: Elsevier; 2021:273–306 pp.
41. Rosli M, Ishak MI, Jamalludin MR, Khor C, Nawi M, Syafiq AM. Simulation-based optimization of plastic injection molding parameter for aircraft part fabrication using response surface methodology (RSM). IOP Conf Ser Mater Sci Eng 2019;551:012108.
42. Singh N, Singh B. Polymer-based additive manufacturing. In: Additive and Subtractive Manufacturing Processes. USA: CRC Press; 2019:121–43 pp.
43. Meiabadi MS, Vafaei A, Sharifi FJJOOIE. Optimization of plastic injection molding process by combination of artificial neural network and genetic algorithm. J Optim Ind Eng 2013;6:49–54.
44. Mujahid Y, Sallih N, Mustapha M, Abdullah MZ, Mustapha FJP. Effects of processing parameters for vacuum-bagging-only method on shape conformation of laminated compositesindex: composites Processes 2020;8:1147.
45. Wiley NC. Process for vacuum bag molding. WIPO: Google Patents; 1975.
46. Zin MH, Razzi MF, Othman S, Liew K, Abdan K, Mazlan N. A review on the fabrication method of bio-sourced hybrid composites for aerospace and automotive applications. IOP Conf Ser: Mater Sci Eng 2016;152:012041.
47. Potter K. Resin transfer moulding. Netherlands: Springer Science & Business Media; 2012.
48. Torres MJCPAAS, Manufacturing. Parameters' monitoring and in-situ instrumentation for resin transfer moulding: a review. Compos Part A: Appl Sci Manuf 2019;124:105500.
49. Laurenzi S, Marchetti MJC, Properties T. Advanced composite materials by resin transfer molding for aerospace applications. Comp Prop 2012;197–226.
50. Song YS, Youn JR, Gutowski TGJCPAAS, Manufacturing. Life cycle energy analysis of fiber-reinforced composites. Comp Part A Appl Sci Manuf 2009;40:1257–65.

51. Wötzel K, Wirth R, Flake MJDAMC. Life cycle studies on hemp fibre reinforced components and ABS for automotive parts. Die Angew Makromol Chem 1999;272:121–7.
52. Aruniit A, Jaan K, Goljandin D, Saarna M, Kaspar T, Majak J, et al. Particulate filled composite plastic materials from recycled glass fibre reinforced plastics. Mater Sci 2011;17:276–81.
53. Khandelwal H, Loonen RC, Hensen JL, Debije MG, Schenning APJSR. Electrically switchable polymer stabilised broadband infrared reflectors and their potential as smart windows for energyindex: energy saving in buildings. Sci Rep 2015;5:11773.
54. Shan Z, Qin S, Liu Q, Liu FJIJOPE, Manufacturing. Key manufacturing technology & equipment for energy saving and emissions reduction in mechanical equipment industry. Int J Precision Eng Manuf 2012;13: 1095–100.
55. Gao Y, Yao W, Sun J, Zhang H, Wang Z, Wang L, et al. A novel soft matter composite material for energyindex: energy-saving smart windows: from preparation to device application. J Mater Chem A 2015; 3:10738–46.
56. Liang X, Guo C, Chen M, Guo S, Zhang L, Li F, et al. A roll-to-roll process for multi-responsive soft-matter composite films containing Cs x WO 3 nanorods for energyindex: energy-efficient smart window applications. Nanoscale Horiz 2017;2:319–25.
57. Zeyghami M, Goswami DY, Stefanakos EJSEM, Cells S. A review of clear sky radiative cooling developments and applications in renewable power systems and passive buildingindex: building cooling. Sol Energy Mater Sol Cells 2018;178:115–28.
58. Wang J, Xie H, Guo Z, Cai L, Zhang KJJOEHT. Using organic phase-change materials for enhanced energy storage in water heaters: an experimental study. J Enhanc Heat Transf 2019;26. https://doi.org/10.1615/jenhheattransf.2018029249.
59. Chaichan MT, Kazem HAJCSITE. Water solar distiller productivity enhancement using concentrating solar water heater and phase change material (PCM). Case Stud Therm Eng 2015;5:151–9.
60. Goren A, Atas CJAOMS, Engineering. Manufacturing of polymer matrix composites using vacuum assisted resin infusion molding. Arch Mater Sci Eng 2008;34:117–20.
61. Shijun S. Heating region retrofitting of eco-technology and strategy research [Master's thesis]. China: Harbin Institute of Technology; 2007;7.
62. Marani R, Palumbo D, Bono G, Cicirelli G, Galietti U, D'Orazio T. Analytical model approximation for defect classification in fiberglass composites inspected by long-pulse thermography. In: 2020 IEEE International Conference on Industrial Technology (ICIT). Argentina: IEEE; 2020.
63. Zhang J, Qin Q, Li G, Tseng C-HJJOEM. Sustainable municipal wasteindex: waste management strategies through life cycle assessment method: a review. J Environ Manag 2021;287:112238.
64. Pittini-Yamada Y, Périgo E, De Hazan Y, Nakahara SJAM. Permeability of hybrid soft magnetic composites. Acta Mater 2011;59:4291–302.
65. Meng B, Hou J, Ning F, Yang B, Zhou B, Yu R, et al. Low-loss and high-induction Fe-based soft magnetic compositesindex: composites coated with magnetic insulating layers. J Magn Magn Mater 2019;492: 165651.
66. Shokrollahi H, Janghorban KJJOM, Materials M. Different annealing treatments for improvement of magnetic and electrical properties of soft magnetic composites. J Magn Magn Mater 2007;317:61–7.
67. Li M, Pu Y, Thomas VM, Yoo CG, Ozcan S, Deng Y, et al. Recent advancements of plant-based natural fiber–reinforced compositesindex: composites and their applications. Compos Part B: Eng 2020;200:108254.
68. Agarwal J, Sahoo S, Mohanty S, Nayak SKJJOTCM. Progress of novel techniques for lightweight automobile applications through innovative eco-friendly composite materials. A Review 2020;33:978–1013.
69. Gillingham KT, Knittel CR, Li J, Ovaere M, Reguant MJJ. The short-run and long-run effects of Covid-19 on energy and the environment. Joule 2020;4:1337–41.
70. Mikinka E, Siwak MJJOMSMIE. Recent advances in electromagnetic interference shielding properties of carbon-fibre-reinforced polymer compositesindex: composites–a topical review. J Mater Sci Mater Electron 2021;32:24585–643.

71. Koumoulos EP, Trompeta A-F, Santos R-M, Martins M, Santos CMD, Iglesias V, et al. Research and development in carbon fibers and advanced high-performance compositesindex: composites supply chain in Europe: a roadmap for challenges and the industrial uptake. J Comp Sci 2019;3:86.
72. Jiang P, Li R, Liu N, Gao YJAE. A novel composite electricity demand forecasting framework by data processing and optimized support vector machine. Appl Energy 2020;260:114243.
73. Cardin O, Castagna P, Couedel D, Plot C, Launay J, Allanic N, et al. Energy-aware resources in digital twin: the case of injection moulding machines. Int Workshop Serv Orientat Holonic Multi-Agent Manuf 2020;9: 183–94.
74. Abeykoon C, McMillan A, Nguyen BKJR, Reviews SE. Energy efficiency in extrusion-related polymer processing: a review of state of the art and potential efficiency improvements. Renew Sustain Energy Rev 2021;147:111219.
75. Lohani SP, Dhungana B, Horn H, Khatiwada DJSET, Assessments. Small-scale biogas technology and clean cooking fuel: assessing the potential and links with SDGs in low-income countries – A case study of Nepal. Sustain Energy Technol Assess 2021;46:101301.
76. Gumaste A, Haridas RS, Gupta S, Gaddam S, Kandasamy K, McWilliams BA, et al. Solid Stir Extrusion: innovating friction stir technology for continuous extrusion process. J Mater Process Technol 2023;316: 117952.
77. Devarajan B, Lakshminarasimhan R, Murugan A, Rangappa SM, Siengchin S, Marinkovic D. Recent developments in natural fiber hybrid composites for ballistic applications: a comprehensive review of mechanisms and failure criteria. Facta Univ, Ser Mech Eng 2024;22:343–83.
78. Amara C, El Mahdi A, Akman PK, Medimagh R, Tornuk F, Khwaldia KJFS. Use of cellulose microfibers from olive pomace to reinforce green composites for sustainable packaging applications. Food Sci Nutr 2023;11: 5102–13.
79. Oladele IO, Omotosho TF, Ogunwande GS, Owa FAJAS. A review on the philosophies for the advancement of polymer-based composites: past, present and future perspective. Appl Sci Eng Prog 2021;14:553–79.
80. Tatar KV, Dhabarde S. Impact of nanomaterials on the properties of biocomposites material-review. Int J Creat Res Thought 2022;10:1–6.
81. Scaffaro R, Citarrella MC, Morreale MJP. Green composites based on mater-Bi® and Solanum lycopersicum plant waste for 3D printing applications. Polymers 2023;15:325.
82. Oliveira MS, Garcia FDC, Luz FSD, Demosthenes LCDC, Pereira AC, Colorado HA, et al. Evaluation of dynamic mechanical properties of fique fabric/epoxy compositesindex: composites. Mater Res 2019;22: e20190125.
83. Ochi D, Barbieri D, Reis AF, Severino P, Venturini AC, Pedroso Yoshida CM, et al. Chapter 1 - agro-industrial waste as fillers for green composites. In: Inamuddin, Altalhi T, Alrooqi A, editors. Green Sustainable Process for Chemical and Environmental Engineering and Science. Netherlands: Elsevier; 2023:1–26 pp.
84. Shang X, Qu N. Internal curing for high-performance concrete by a green composite capsule. Cement Concr Compos 2023;136:104867.
85. Bandyopadhyay S. Source composite curve for wasteindex: waste reduction. Chem Eng J 2006;125:99–110.
86. Tareq Noaman A, Abed MS, Al-Gebory L, Al-Zubaidi AB, Al-Tabbakh AA. Production of Agro-wasteindex: waste cement Composites: influence of nutshells on mechanical and hardened properties. Constr Build Mater 2023;394:132137.
87. Valenta RK, Lèbre É, Antonio C, Franks DM, Jokovic V, Micklethwaite S, et al. Decarbonisation to drive dramatic increase in mining waste–Options for reduction. Resour Conserv Recycl 2023;190:106859.
88. Smith L, Ali M, Agrissais M, Mulligan S, Koh L, Martin N. A comparative life cycle assessment of dental restorative materials. Dent Mater 2023;39:13–24.
89. Chen Z, Gu H, Bergman RD, Liang SJS. Comparative life-cycle assessment of a high-rise mass timber buildingindex: building with an equivalent reinforced concrete alternative using the Athena impact estimator for buildings. Sustainability 2020;12:4708.

90. Ramesh M, Deepa C, Rajeshkumar L, Sanjay MR, Siengchin S. Life-cycle and environmental impact assessments on processing of plant fibres and its bio-composites: a critical review. J Ind Text 2020;51: 5518S–42S.
91. Dua S, Khatri H, Naveen J, Jawaid M, Jayakrishna K, Norrrahim MNF, et al. Potential of natural fiber based polymeric compositesindex: composites for cleaner automotive component production -a comprehensive review. J Mater Res Technol 2023;25:1086–104.
92. Eskander SB, Saleh HM, Tawfik ME, Bayoumi TA. Towards potential applications of cement-polymer compositesindex: composites based on recycled polystyrene foam wastes on construction fields: impact of exposure to water ecologies. Case Stud Constr Mater 2021;15:e00664.
93. Volk M, Yuksel O, Baran I, Hattel JH, Spangenberg J, Sandberg M. Cost-efficient, automated, and sustainable composite profile manufacture: a review of the state of the art, innovations, and future of pultrusionindex: pultrusion technologies. Compos B Eng 2022;246:110135.
94. Sharma S, Asolekar SR, Thakur VK, Asokan P. Valorization of cellulosic fiber derived from wasteindex: waste biomass of constructed wetland as a potential reinforcement in polymeric compositesindex: composites: a technological approach to achieve circular economy. J Environ Manag 2023;340:117850.
95. Rajeshkumar G, Seshadri SA, Devnani G, Sanjay M, Siengchin S, Maran JP, et al. Environment friendly, renewable and sustainable poly lactic acid (PLA) based natural fiber reinforced composites: a comprehensive review. J Clean Prod 2021;310:127483.
96. Suriani M, Rapi HZ, Ilyas R, Petrů M, Sapuan S. Delamination and manufacturingindex: manufacturing defects in natural fiber-reinforced hybrid composite: a review. Polymers 2021;13:1323.
97. Adil S, Kumar B, Panicker PS, Pham DH, Kim J. High-performance green compositesindex: composites made by cellulose long filament-reinforced vanillin epoxy resin. Polym Test 2023;123:108042.
98. Sun Y, Bao L, Luo C, Xu A, Chen C. Development of high-strength green compositesindex: composites based on ramie sliver and cellulose acetate resin. Ind Crop Prod 2022;187:115525.
99. Syafri E, Jamaluddin J, Sari NH, Mahardika M, Suryanegara L, Sinaga R, et al. Effect of ultrafine grinding and ultrasonication duration on the performance of polyvinyl alcohol (PVA) agave gigantea cellulose micro fiber (CMF) bio-composite film. J Nat Fiber 2023;20:2192545.
100. Arinze RU, Oramah E, Chukwuma EC, Okoye NH, Chris-Okafor PU. Mechanical impact evaluation of natural fibres with LDPE plastic compositesindex: composites: waste management in perspective. Curr Res Green Sustain Chem 2022;5:100344.
101. Colangelo F, Farina I, Travaglioni M, Salzano C, Cioffi R, Petrillo A. Eco-efficient industrial wasteindex: waste recycling for the manufacturingindex: manufacturing of fibre reinforced innovative geopolymer mortars: integrated waste management and green product development through LCAindex: LCA. J Clean Prod 2021;312:127777.
102. Cheng L, Huang Y, Yin S, Chen M, Liu Y, Zhang Y, et al. Recent advances in cellulosic materials for aqueous zinc-ion batteries: an overview. Carbohydr Polym 2023;316:121075.

Murshiduzzaman, Azizan As'arry*, Lim Wei Keong,
Mohd Zuhri Mohamed Yusoff, Mohd Sapuan Salit, Eris Elianddy Supeni,
Wan Zuha Wan Hasan and Hanafiah Yussof

14 Analysis of an automated solar panel cleaning robot on photovoltaics (PV) module frames with composites materials

Abstract: Solar panels are essential for harvesting clean energy from the sun, but dust collection can dramatically limit their efficiency and power output. This challenge has led to the development of automated solar panel cleaning robots, a promising solution to ensure the continuous and optimal performance of solar arrays. One concern is that the cleaning method can cause long-term damage or degradation to the panels, such as cracking, if the cleaning robot is not positioned correctly on the solar panel. The risk of damage or degradation to solar panels during the cleaning process is a significant consideration for both manufacturers and operators of solar installations. Automated solar panel cleaning robots are valuable tools in maintaining the efficiency and longevity of solar arrays. However, their deployment requires a balanced approach that prioritizes effective cleaning while minimizing the risk of damage to the panels.

Keywords: solar panel; cleaning robot; photovoltaics (PV); modules

Abbreviations

PV photovoltaics

***Corresponding author: Azizan As'arry**, Department of Mechanical and Manufacturing Engineering, Universiti Putra Malaysia, 43400 UPM, Serdang, Selangor, Malaysia, E-mail: zizan@upm.edu.my. https://orcid.org/0000-0001-6804-3697

Murshiduzzaman, Lim Wei Keong, Mohd Sapuan Salit and Eris Elianddy Supeni, Department of Mechanical and Manufacturing Engineering, Universiti Putra Malaysia, 43400 UPM, Serdang, Selangor, Malaysia, E-mail: gs55223@student.upm.edu.my (Murshiduzzaman). https://orcid.org/0009-0002-5389-6552 (Murshiduzzaman). https://orcid.org/0000-0002-4598-6358 (M.S. Salit)

Mohd Zuhri Mohamed Yusoff, Department of Mechanical and Manufacturing Engineering, Universiti Putra Malaysia, 43400 UPM, Serdang, Selangor, Malaysia; and Laboratory of Biocomposite Technology, Institute of Tropical Forestry and Forest Products (INTROP), Universiti Putra Malaysia, 43400 UPM, Serdang, Selangor, Malaysia. https://orcid.org/0000-0002-1069-7345

Wan Zuha Wan Hasan, Department of Electrical Engineering, Universiti Putra Malaysia, 43400 UPM, Serdang, Selangor, Malaysia. https://orcid.org/0000-0003-4691-066X

Hanafiah Yussof, School of Mechanical Engineering, College of Engineering, Universiti Teknologi MARA, Shah Alam, Malaysia

As per De Gruyter's policy this article has previously been published in the journal Physical Sciences Reviews. Please cite as: Murshiduzzaman, A. As'arry, L. W. Keong, M. Z. M. Yusoff, M. S. Salit, E. E. Supeni, W. Z. Wan Hasan and H. Yussof "Analysis of an automated solar panel cleaning robot on photovoltaics (PV) module frames with composites materials" *Physical Sciences Reviews* [Online] 2024. DOI: 10.1515/psr-2024-0032 | https://doi.org/10.1515/9783111234762-014

14.1 Introduction

The sun is the main energy source of our galaxy and sunlight is by far the most important source of energy received by Earth. Solar energy is considered one of the best solutions for green and renewable energy. With contemporary advancements of solar energy technology, solar energy is not only converted into electricity but also other types of energy such as chemical and thermal energy [1]. Dust particles can vary in phase, kind, chemical, and physical characteristics based on numerous environmental factors. Sources of pollution and dust production are dust, sandstorms, soot, industrial gases, volcanic ash, fly ash, crop burning, and building development [2]. Subsequently, air, humidity, temperature, and wind speed all play a role in determining isolated dust and its accumulation on the PV cell [3]. According to study by [3]; the average daily energy loss caused by dust accumulated on the surface of PV modules is roughly 4.4 % in a year. This study aims to address this issue by focusing on the development of an automated solar panel cleaning system, which integrates advanced robotic technology to enhance cleaning efficiency and reduce manual intervention. By streamlining the cleaning process, the system ensures consistent performance for solar arrays, particularly in large-scale applications. An automated solar panel cleaning system is made up of an autonomous unit that uses sensors and controllers, as well as a cleaning mechanism unit that may be watered or waterless. Solar panels can be cleaned using a variety of ways like robotic cleaning, electrostatic cleaning, coating cleaning, vibrating cleaning, and forced-air cleaning [4]. Solar panels need to be cleaned every few weeks, which is especially difficult for huge solar-panel arrays to maintain optimal efficiency for power generation. The aging of mirrors' solar panels will increase if the solar panel is placed in humidity areas [5].

14.1.1 Solar panel robot cleaner (Single axis)

A robot's degrees of freedom often relate to the number of moveable joints. Dust soiling of PV panels is a physical process that is highly influenced by the operating area [6]. Cleaning using cleaning robot functions to ensure that the efficiency of solar power generation to electricity by making sure the surface of solar panel is kept clean without putting manual labor at risk. A single axis robot can move linearly in only one dimension. The robot would move in a single axis back and forth daily without requiring human intervenion as shown in Figure 14.1.

Based on the design below as illustrated on Figure 14.2, this design concept's main cleaning media is implemented rotary brush which will be fixed under the device. To provide normal force and let the device remain in contact with the end of solar panel, robot will be placed on both end side of the solar panel. Hence, the force will be only applicable located at the end of the solar panel. Meanwhile, this design is not only suitable for dry cleaning but is also applicable to wet cleaning. This device is installing a water tank on the above solar panel, the cleaning fluid will be stored in the tank and it will be

Figure 14.1: Single-axis robot [7].

flowed to the surface of the panel due to gravitational force. In terms of motion, the cleaning robot moves along one axis only. When the robot reaches until end of the solar panel, the levelling sensor will drop it from the surface of the solar panel to the frame of the panel. This robot will move backward from the panels to continue the cleaning process.

14.1.2 Solar panel frame

A solar panel frame also known as a solar panel mounting structure or solar panel racking system as shown in Figure 14.3. It is an essential component of a solar photovoltaic (PV) system. Its primary purpose is to support and secure the solar panels in the

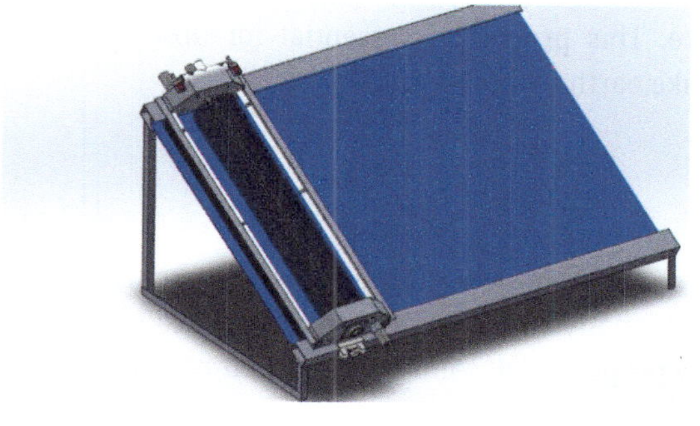

Figure 14.2: Concept-single axis robot.

Figure 14.3: Solar panel frame.

optimal position for capturing sunlight and converting it into electricity. Solar panel frames are designed to provide support and stability for the solar panels. They keep the panels elevated above the mounting surface, such as a roof or the ground, to prevent contact with moisture and debris. The frame also helps distribute the weight of the panels evenly, ensuring they can withstand environmental stresses like wind and snow loads. The selection of appropriate materials for these frames, such as structural steel, epoxy carbon UD, and epoxy carbon woven prepregs, is critical for ensuring durability, cost-effectiveness, and performance under load. The use of these materials in the simulation allows for a comparative analysis of their mechanical properties, including strength, deformation, and stress distribution, to optimize the frame design. To maximize the efficiency of solar panels, they need to be oriented and tilted at specific angles relative to the sun's position. Solar panel frames are adjustable structures that allow installers to set the desired tilt angle and orientation (usually facing south in the Northern Hemisphere) to ensure the panels receive the most sunlight throughout the day.

14.1.2.1 Structural steel

Structural steel is a specific type of steel that is used in the construction industry to create various structural components for buildings, bridges, towers, and other infrastructure projects. It is known for its high strength, durability, and versatility, making it a popular choice for supporting heavy loads and providing stability in construction projects. Besides, it is designed to withstand significant loads, both in tension (stretching) and compression (squeezing). This high strength-to-weight ratio allows for the construction of large and tall structures without excessive bulk. Structural steel is capable of withstanding deformation without failure. This property is essential for absorbing and redistributing energy during events like earthquakes or high winds.

14.1.2.2 Carbon fiber (290 GPa)

Carbon fiber is a composite material made from thin strands of carbon atoms bonded together. These strands exhibit exceptionally high tensile strength, typically ranging from 3,500 to 6,000 MPa, which is a key property that contributes to their performance in demanding applications. These strands are usually woven into sheets or fabrics and then combined with a polymer resin to create a composite material. This combination of carbon fibers and resin results in a material that exhibits unique properties. In the case of carbon fiber, its modulus of elasticity is exceptionally high at 290 GPa. This is significantly higher than most other materials commonly used in engineering and manufacturing, such as steel. Carbon fiber is renowned for its incredible strength-to-weight ratio. Its high modulus of elasticity allows it to withstand significant loads without deforming or breaking. This makes it an ideal material for applications where strength is crucial, such as in aerospace, automotive, and sports equipment. Additionally, carbon fiber demonstrates excellent fatigue resistance, low thermal expansion, and high chemical resistance, further enhancing its versatility.

14.1.2.3 Epoxy carbon UD (230 GPa) prepreg

Epoxy carbon UD (230 GPa) prepreg is a composite material used in various industries, including aerospace, automotive, and sports equipment manufacturing. Epoxy carbon UD (230 GPa) prepreg is a composite material made from unidirectional carbon fibers impregnated with an epoxy resin. The high stiffness (230 GPa) of the carbon fibers in this prepreg makes it suitable for applications where strength and stiffness in a specific direction are essential, such as in the aerospace industry for aircraft components or in high-performance sports equipment like tennis rackets or bicycle frames.

14.2 Methodology of simulation of load analysis

The CAD model's panels were imported into Motion SOLIDWORKS 2022, and ANSYS for load and motion simulation analysis. Figure 14.4 depicts a flowchart of the project process.

14.2.1 Pre-processing simulation of load analysis

Robotic systems with various mechanical configurations are now available for use in solar panel cleaning. Figure 14.5 shows the free body diagram analysis for robot-cleaning.

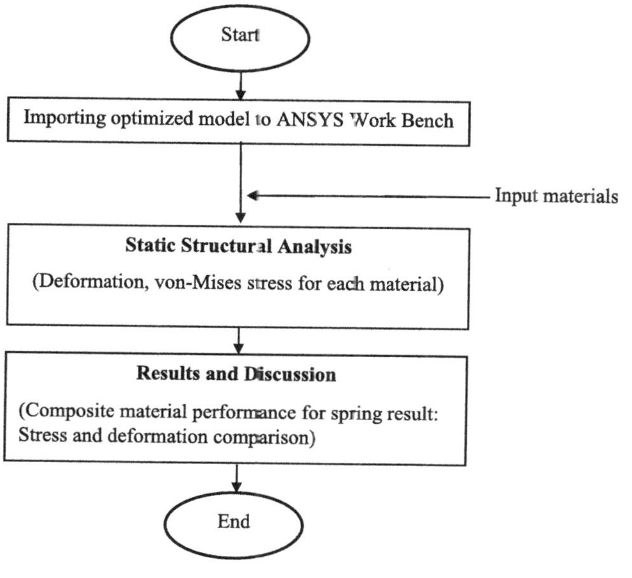

Figure 14.4: Flow chart of methodology.

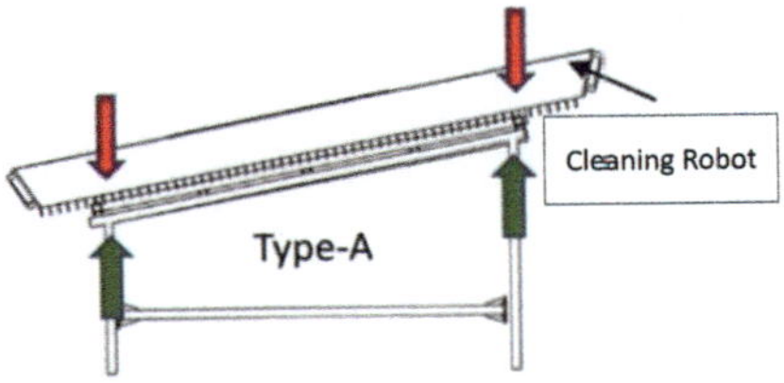

Figure 14.5: Free body diagram of PV frame.

To simulate the load, a mesh will be created on the PV surface using ANSYS 2022. The span angle center of the meshing will be set to coarse meshing, while the remaining parameters will be set to their default magnitudes, as shown in Figure 14.6.

Figure 14.7 depicts a complete meshing model. Within the model, the solver has successfully divided the components into small finite elements. The choice of coarse meshing is justified by the need to balance computational efficiency with result accuracy, given the size and complexity of the frame model. This approach captures critical stress and deformation regions without excessive computational overhead. Referring Figure 14.7, the meshing part of the frame will be more critical than the panel.

Additionally, the boundary condition was fixed on the PV frame as a fixed support, indicated by the blue color as illustrated in Figure 14.8. There are a total of three materials for frame that has been studied in this research. These boundary conditions were selected to realistically simulate the real-world constraints where the frame is anchored to the mounting structure, ensuring accurate representation of load distribution and stress behavior.

Meanwhile, Figure 14.9 displays the materials properties of solar panels. Silicon is the most common material used in solar cells to enhance the conversion of light into

Details of "Mesh"

Element Order	Program Controlled
Element Size	Default
Sizing	
Use Adaptive Sizing	Yes
Resolution	7
Mesh Defeaturing	Yes
Defeature Size	Default
Transition	Fast
Span Angle Center	Coarse
Initial Size Seed	Assembly
Bounding Box Diago...	3712.9 mm
Average Surface Area	82052 mm²
Minimum Edge Length	0.77878 mm

Figure 14.6: Setting meshing for the ANSYS load analysis.

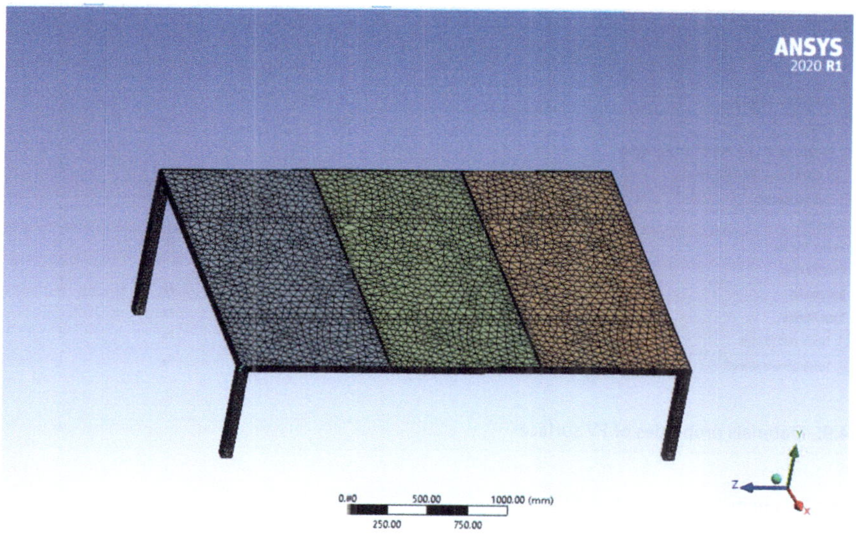

Figure 14.7: Meshing of the PV frame and PV surface.

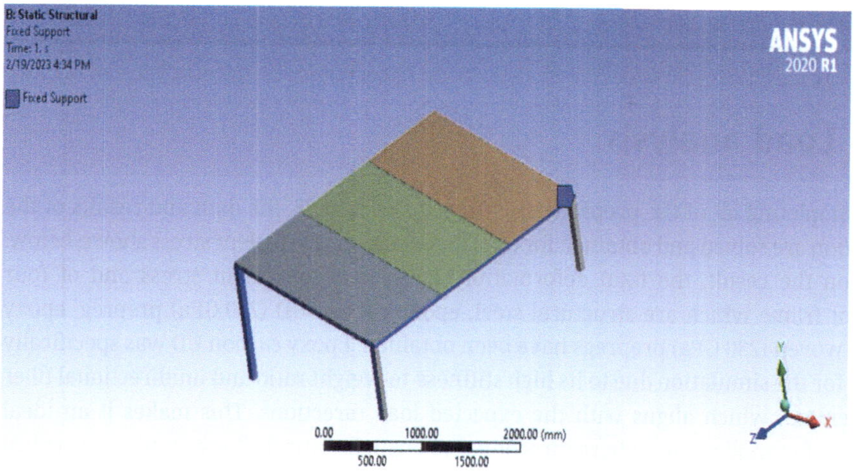

Figure 14.8: Boundary condition settings.

electricity, making it more efficient. The density of the solar panel is 2,600 kg m^{-3} and it has a tensile yield strength of 1.65×10^8 Pa.

In the following section, Table 14.1 shows the material properties of frame of panel that being going to be analyzed. The materials used for the solar frame are structural steel, epoxy carbon ud (230 GPa) prepreg, epoxy carbon woven (230 GPa) prepreg.

Properties of Outline Row 3: Solar Panel

	A	B	C
1	Property	Value	Unit
2	Material Field Variables	Table	
3	Density	2600	kg m^-3
4	Isotropic Secant Coefficient of Thermal Expansion		
5	Coefficient of Thermal Expansion	6E-06	C^-1
6	Isotropic Elasticity		
7	Derive from	Young's Modulus and Poisson's...	
8	Young's Modulus	1.1E+11	Pa
9	Poisson's Ratio	0.225	
10	Bulk Modulus	6.6667E+10	Pa
11	Shear Modulus	4.4898E+10	Pa
12	Tensile Yield Strength	1.65E+08	Pa
13	Tensile Ultimate Strength	1.7E+08	Pa

Figure 14.9: Materials properties of PV surface

Table 14.1: Data material tensile properties.

Material	Density	Tensile properties
Structural steel	7,850 kg m^{-3}	2.5×10^8 Pa
Epoxy carbon Ud (230 GPa) Prepreg	1,490 kg m^{-3}	9.69×10^6 Pa
Epoxy carbon woven (230 GPa) Prepreg	1,420 kg m^{-3}	3.69×10^6 Pa

14.3 Load analysis

After completing all of the preprocessing steps using ANSYS, the data and results of the simulation are solved and obtained for total deformation, equivalent stress shown below. Based on the result, the total deformation, maximum equivalent stress and of four different frame which are structural steel, epoxy carbon UD (230 GPa) prepreg, epoxy carbon woven (230 GPa) prepregs have been obtained. Epoxy carbon UD was specifically chosen for the simulation due to its high stiffness-to-weight ratio and unidirectional fiber arrangement, which aligns with the expected load directions. This makes it an ideal candidate for applications where directional strength is critical. These results are critical not only for evaluating the mechanical performance of the robot's structural frame under applied loads but also for understanding the long-term durability and reliability of the PV system. Lower deformation values, such as those exhibited by structural steel, indicate a more robust frame capable of maintaining alignment and support for extended periods, which is essential for minimizing maintenance and operational costs. These results are critical in evaluating the mechanical performance of the robot's structural frame under applied loads.

The load applied by the robot impacts only the structural frame, inducing stress and deformation primarily on the frame itself, rather than the solar panels. This ensures that the solar panels are shielded from excessive stress, which could lead to damage [8]. Protecting the solar panels from excessive stress ensures consistent energy generation efficiency, thereby extending the operational life of the PV modules and reducing replacement frequency. Deformation analysis provides insights into the behavior of the frames under load, considering displacement across all three coordinate axes (*X*, *Y*, *Z*). The simulation settings, including material properties, meshing parameters, and boundary conditions, were chosen to emulate realistic operational conditions for solar panel cleaning robots. The choice of simulation parameters ensures the results are highly applicable to practical scenarios, enabling informed decision-making regarding material selection and frame design. This ensures the derived results are applicable to real-world scenarios and aids in optimizing material selection (Figures 14.10, 14.11, 14.12, 14.13, 14.14, and 14.15).

A stress analysis is being conducted in this project in order to observe the maximum equivalent stress that being applied into the solar panel. From the above figures, it is noticed that for structural steel, the deformation is slightly higher near the top edge of the testbed panels. Whereas for epoxy carbon UD, the higher deformation happens both near the top edge and the lower-mid section of the testbed panels. On the other hand, for epoxy carbon woven, the deformation is only visible in the lower-mid section. The summary of the finite element method results are tabulated in Table 14.2. In structural analysis, deformation results can be added into the system in order to determine the total deformation of the panel frame and PV modules. Total deformation is the deformation results related to the model in three coordinates (*X*, *Y* and *Z*). For the results of the maximum deformation, tends to show a highest deformation which is 6.2827×10^{-6} mm compare to others robots. However, structural steel display a lowest deformation among

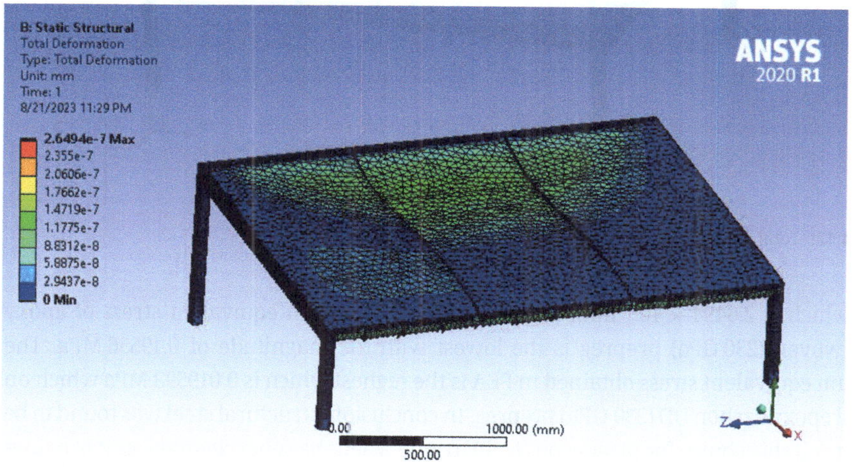

Figure 14.10: Total deformation for structural steel.

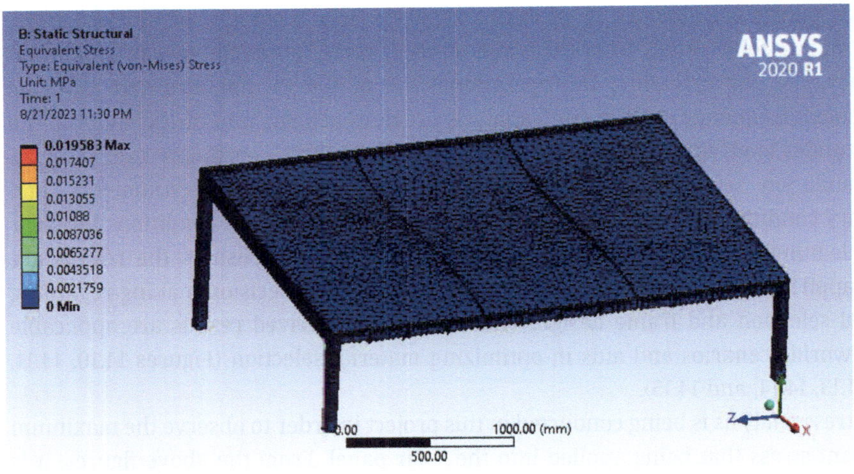

Figure 14.11: Equivalent stress for structural steel.

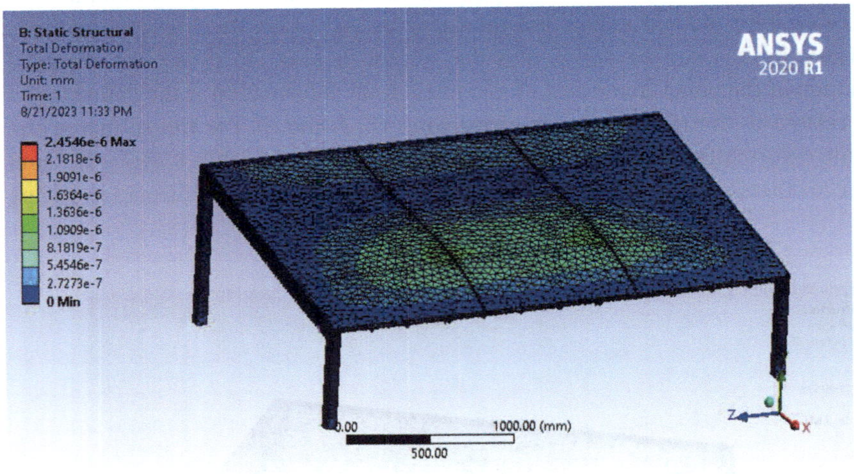

Figure 14.12: Total deformation for epoxy carbon UD (230 GPa) prepreg.

others which is 2.6494×10^{-7} mm. In addition, the maximum equivalent stress of epoxy carbon woven (230 GPa) prepreg is the lowest with the magnitude of 0.19556 MPa. The maximum equivalent stress obtained in FEA is the highest which is 0.019593 MPa which on material epoxy carbon UD (230 GPa) prepreg. In conclusion, structural steel was found to be the most suitable choice for placing on the PV frame when the robot cleaning, as it provides the highest support on the entire system. Its combination of low deformation and moderate stress resistance ensures the frame remains stable and aligned over time, protecting the

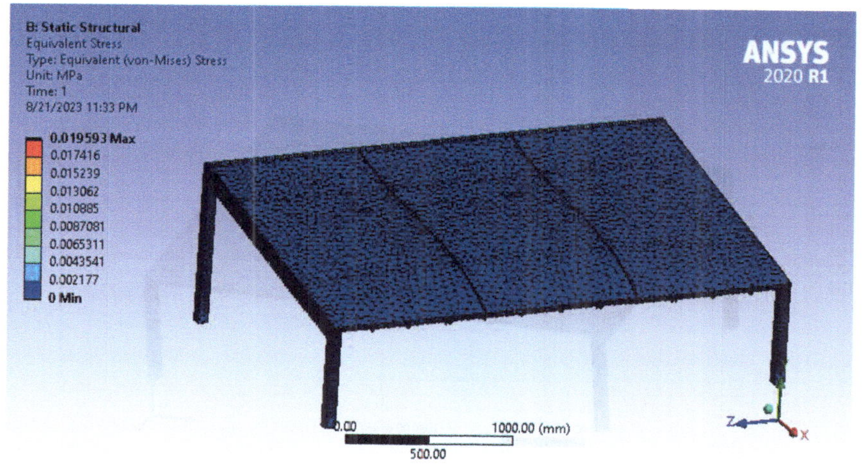

Figure 14.13: Equivalent stress for epoxy carbon UD (230 GPa) prepreg.

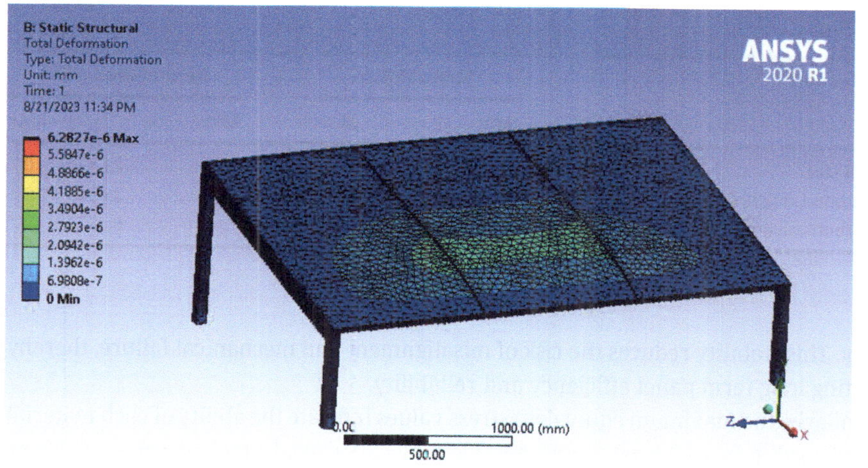

Figure 14.14: Total deformation for epoxy carbon woven (230 GPa) prepreg.

solar panels and maintaining energy generation efficiency. Due to the rising energy demand in the world, solar power is considered one of the most useful energies to be consumed not only in terms of sustainability but also in renewable and eco-friendly [9].

For the results of the maximum deformation, the highest deformation value, 6.2827×10^{-6} mm, occurs with epoxy carbon woven, which could impact the alignment of the panels under load over time. Structural steel shows the lowest deformation (2.6494×10^{-7} mm), demonstrating its superior rigidity and ability to maintain frame

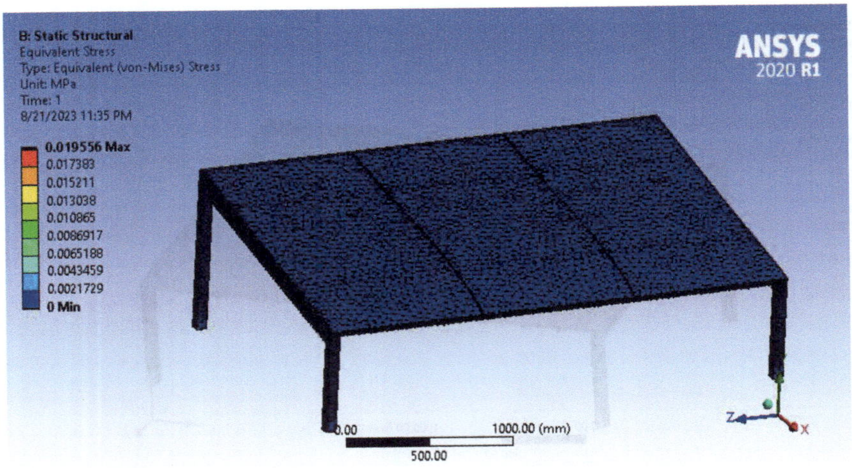

Figure 14.15: Equivalent stress for epoxy carbon woven (230 GPa) prepreg.

Table 14.2: Summarize of simulation results of load analysis.

Types of robots	Stress (MPa)		Deformation (mm)	
	Min	Max	Min	Max
Structural steel	0	0.019583	0	2.6494×10^{-7}
Epoxy carbon Ud (230 GPa) prepreg	0	0.019593	0	2.4546×10^{-7}
Epoxy carbon woven (230 GPa) prepreg	0	0.019556	0	6.2827×10^{-6}

stability. This stability reduces the risk of misalignment and mechanical failure, thereby supporting long-term panel efficiency and reliability.

Similarly, the maximum equivalent stress values indicate the ability of each material to resist load-induced stresses without yielding. Structural steel, despite its lower deformation, displays a comparable stress magnitude to epoxy carbon UD (0.019583 MPa vs. 0.019593 MPa), highlighting its balanced performance under load.

14.4 Conclusions

In Malaysia, the reduction in performance of solar panels can be obtained by up to 58.67 % which is equivalent of losing more than half of power output in a year [10]. To sum up, structural steel shows the best results in terms of the supporting with solar panels and cleaning robots. Moreover, structural steel is significantly more cost-effective compared to epoxy carbon UD and woven materials. Its widespread availability, ease of fabrication, and lower material costs make it an economically viable choice for large-scale implementation. Additionally, the ability of robot cleaning contributes to increased energy efficiency that can benefit the world. Structural steel also ensures durability and reduced maintenance requirements, which further enhances its cost-effectiveness over time. The research on solar panel cleaning robot enables to continue to produce more renewable energy by removing accumulated dust on the PV modules which aligns with SDG 7 where to ensure a clean and affordable energy to the environment. Effective cleaning, achieved through a better cleaning mechanism, can improve power generation, allowing for low cost and faster implementation. This supports SDG 7, which aims to increase the share of renewable energy in the global energy mix in a sustainable manner. This incentive holds the potential to inspire more industries to adopt solar energy in their respective areas. The introduction of this cleaning robot has effectively tackled one of the primary challenges linked to solar energy: the accumulation of dust on the surface, which hampers power generation.

Acknowledgement: The authors would like to thank the editors S.M. Sapuan, Mohd Roshdi Hassan, Eris Elianddy Supeni and Azizan As'arry for their guidance and review of this article before its publication. The authors would like to thank Universiti Putra Malaysia (UPM) and the Malaysian Ministry of Higher Education (MOHE) for their continuous support in this research work. This work was supported in part by the Geran Insentif Putra Siswazah (GP-IPS) fund (800-2/1/2021/GP-IPS/9697000).

References

1. Chen Z, Yiliang X, Hongxia Z, Yujie G, Xiongwen Z. Optimal design and performance assessment for a solar powered electricity, heating and hydrogen integrated energy system. Energy 2023;262:125453.
2. Katoch M, Kumar K, Dahiya V. Dust accumulation and reduction in electrical performance of solar PV panels. Mater Today Proc 2020;46:6608–12.
3. Hussain A, Batra A, Pachauri R. An experimental study on effect of dust on power loss in solar photovoltaic module. Renewables Wind Water Solar 2017;4:1–2.
4. Olorunfemi BO, Ogbolumani OA, Nwulu N. Solar panels dirt monitoring and cleaning for performance improvement: a systematic review on smart systems. Sustainability 2022;14:3–7.
5. Peinado Gonzalo A, Pliego Marugán A, García Márquez FP. A review of the application performances of concentrated solar power systems. Appl Energy 2019;255:113893.
6. Younis A, Onsa M. A brief summary of cleaning operations and their effect on the photovoltaic performance in Africa and the Middle East. Energy Rep 2022;8:2334–47.
7. Dawson, C. Solar panel cleaning systems and their pros and cons. Soilar technologies. 2022. https://soilar.tech/solar-panel-cleaning-systems-and-their-pros-and-cons/ [Accessed 13 January 2023].
8. Gabor AM, Lincoln J, Schneller EJ, Seigneur H, Janoch R, Anselmo A, et al. Compressive stress strategies for reduction of cracked cell related degradation rates in new solar panels and power recovery in damaged solar panels. In: 2018 IEEE 7th world conference on photovoltaic energy conversion, WCPEC 2018 - a joint conference of 45th IEEE PVSC, 28th PVSEC and 34th EU PVSEC, July. Piscataway: IEEE; 2018:2820–5 pp.

9. Akyazı Ö, Şahin E, Özsoy T, Algül M. A solar panel cleaning robot design and application. Eur J Sci Technol; 2019:343–8 pp. https://doi.org/10.31590/ejosat.638291.
10. Islam SZ, Izzati NS, Abdullah MN, Kamarudin MS, Omar R, Uddin J. Automated water recycle (AWR) method for dust removal from rooftop photovoltaic (PV) at Johor, Malaysia. SN Appl Sci 2022;4:1–2.

Belal Alemour*, Mohd Roshdi Hassan, Omar Badran and Hong Ngee Lim

15 Electrical conductivity and self-heating improvement of carbon fiber composite using functionalized graphene nanoplatelets

Abstract: The electrical conductivity of Carbon Fiber Reinforced Epoxy Composite (CFRE) is poor and need more improvement to be used in more applications such as deicing processes. Functionalized Graphene Nanoplatelets (FGNP) are used in present study to improve and enhance the electrical conductivity of CFRE by coating on its top surface. The results show that the electrical conductivity of CFRE is increased significantly by 11 times with percentage of improvement of 1,110 %. The effect of Joule heating in CFRE/FGNP specimen is also studied. The results show that the self-heating of CFRE is improved after coating it with FGNP. The self-heating of CFRE/FGNP specimen becomes more efficient and homogeneous, reaching high levels of temperature, more than CFRE neat at the same applied voltage. Tensile test is also performed by applying axial tensile load on the CFRE/FGNP specimen to investigate the effect of tensile load on its electrical conductivity. Therefore, it is found that the electrical resistivity of CFRE/FGNP is increased by increasing the applied tensile load.

Keywords: electrical conductivity; carbon fiber; composite material; graphene; deicing

15.1 Introduction

Conductive composite materials are made by adding conductive fillers such as carbon-based materials like graphene to a nonconductive matrix using certain techniques for creating composite materials having high electrical conductivity. These conductive composite materials can be utilized in many applications such as deicing and lightning strike for aerospace technologies. Conductive composites exhibit inherent features such

***Corresponding author: Belal Alemour**, Department of Mechanical Engineering, Bahrain Polytechnic University, Isa Town, Bahrain, E-mail: belalemour@gmail.com. https://orcid.org/0000-0002-7346-934X
Mohd Roshdi Hassan, Department of Mechanical and Manufacturing Engineering, Universiti Putra Malaysia, 43400 Serdang, Selangor, Malaysia, E-mail: roshdi_hassan@upm.edu.my
Omar Badran, Department of Mechanical Engineering, Al-Balqa Applied University, FET, Amman, Jordan, E-mail: omar_badran@bau.edu.jo
Hong Ngee Lim, Department of Chemistry, Universiti Putra Malaysia, 43400 Serdang, Selangor, Malaysia, E-mail: ganet@upm.edu.my

As per De Gruyter's policy this article has previously been published in the journal Physical Sciences Reviews. Please cite as: B. Alemour, M. R. Hassan, O. Badran and H. N. Lim "Electrical conductivity and self-heating improvement of carbon fiber composite using functionalized graphene nanoplatelets" *Physical Sciences Reviews* [Online] 2025. DOI: 10.1515/psr-2024-0033 | https://doi.org/10.1515/9783111234762-015

as light weight, electrochromic and photochromic nature, low cost, variable conductivity, corrosion resistant behavior, etc. Thus, these polymers are useful for a broad range of potential applications in fields such as energy storage, energy conversion, electrocatalysis, electroanalysis, microsystem technologies, microwave screening, electronic devices, corrosion protection, and sensors. The conducting polymer composites have even higher potential of applications in energy storage devices (batteries, supercapacitors), solar cells, fuel cells, catalysis, anticorrosive coatings, electrochromic devices, artificial muscles, biomedicine, and sensors [1, 2].

Carbon fiber reinforced composites (CFRE) are used nowadays in manufacturing of aircraft structures, due to their light weight and high strength. However, the electrical conductivity of CFRE is low, which means it cannot withstand the high currents from the lightning strike, also its heat conductivity will not be sufficient for deicing, eventually, CFRE can be damaged and need high-cost maintenance. Therefore, the electrical conductivity of CFRE must be improved, to withstand the lightning strikes and be useful for deicing process. So that, this conductive composite could be used as a heater for deicing in the aircraft structure to increase the performance of aircrafts. Nowadays the scientists found out that the thunderstorms are increasing significantly due to the global worming effect and icing effect is becoming severe in cold regions. Trapp et al. 2007 found that a doubling of greenhouse gases in the atmosphere would significantly increase the number of days that severe thunderstorm could occur in the southern and eastern United States [3]. The technology used by aircraft companies to protect the aircraft from lightning strikes is metal mesh, which is made of aluminum or copper; this metal meshes are attached to the top surface of CFRE composite, to increase the electrical conductivity of the aircraft's structure. On the other hand, metal mesh increases the weight of aircraft and need high-cost maintenance [4–9].

The researchers investigated high electrically conductive composite materials that can be attached to the aircraft's structure to replace the metal mesh due to their light weight and high electrical conductivity. These conductive composites contain conductive filler such as carbon-based materials like graphene and carbon nanotube [4, 10]. Imran, Bangwei et al. and Kumar et al. prepared a coating made of graphene that is applied on the surface of carbon fiber epoxy laminate to protect it from the lightning strike. Li et al. prepared a spray coating of carbon nanotubes and graphene nanoplatelets used to increase the electrical conductivity of CFRPs, and the results showed more improvement in the electrical conductivity that can be used later for heating purposes [10–14]. Thermal conductivity of CFRE also can be improved using graphene nanoplatelets, because graphene has very high thermal conductivity, and this will enhance the self-heating of CFRE to reach higher temperatures at same applied voltage. Therefore, these conductive graphene composites can be used as a deicing element heater in the aircrafts [10, 14].

The aim of this research is to develop a high electrical and thermal conductive carbon fiber reinforced epoxy composite material through the addition of FGNP. The results show that the electrical conductivity of CFRE is enhanced and increased significantly by about 11 times and with percentage of improvement of 1,110 %. The self-heating of all specimens by the Joule effect was analyzed. It is found that the self-heating of CFRE has been enhanced and improved after coating or printing it with FGNP.

15.2 Materials

Carbon fabric reinforced epoxy (HexPly 8552 prepregs) supplied by Hexcel Composites company has been used. The epoxy matrix is amine and thermoplastic toughened together, and the carbon fabric is manufactured with T300 fiber. In this research, FGNP shown in the Figure 15.1 is used to coat the top surface of CFRE.

Figure 15.1: Functionalized graphene nanoplatelets (FGNP).

15.3 Methodology

15.3.1 Preparation of FGNP

FGNP is prepared in the laboratory of chemistry department at faculty of science in University Putra Malaysia, using different processes. An experiment of 1 g of graphene nanoplatelets (GNPs) were mixed with 10 g of methacrylate and heated at 150 °C for 24 h. The GNPs functionalized with polymethyl methacrylate were decanted by filtration. Disperse 10 mg/ml GNPs functionalized with polymethyl methacrylate in a mixture of water, methanol, and polyvinyl alcohol at a mass ratio of 10:10:1.

15.3.2 Preparation CFRE and CFRE/FGNP composites

In this research, two specimens are prepared for testing. The first specimen is CFRE neat, which is carbon fiber reinforced epoxy composite without FGNP, which will be considered a control sample. The second specimen is CFRE/FGNP-P, where FGNP is printed on the top surface of CFRE using inkjet printer. CFRE neat is fabricated as shown in Figure 15.2(a). The 12 layers of HexPly 8552 prepregs were stacked one after another to form CFRE laminate, and then it is cured using hot press machine at 180 °C for 2 h. Specimen CFRE/FGNP-P is fabricated as shown in Figure 15.2(b), where FGNP paste is printed on the top surface of CFRE laminate after curing.

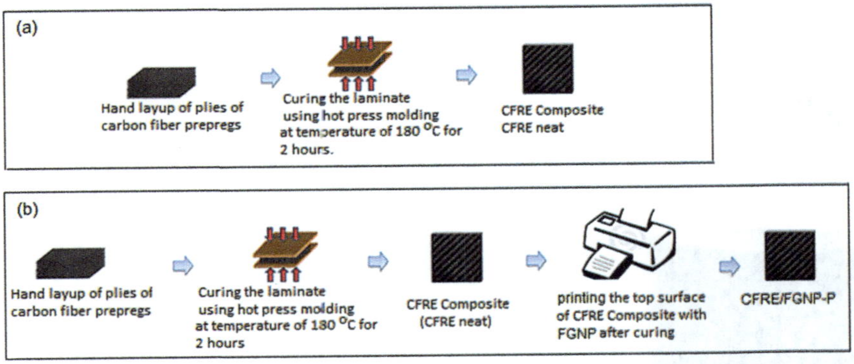

Figure 15.2: Procedures of preparing the testing specimens: (a) CFRE neat and (b) CFRE/FGNP-P.

15.3.3 Electrical conductivity test

Four-point probe device is used to find the electrical conductivity value for the specimens. The specimens were cut into a proper dimension (1 cm × 1 cm × 0.2 cm) to be used in the testing device. Four-point probe device was used to measure the voltage (V) and the current (I). Then, the electrical resistance (R) of the specimens is determined from equation (15.1). Finally, the electrical conductivity (σ) can be found from the equation (15.2) [2, 10]:

$$R = \frac{V}{I} \tag{15.1}$$

$$\sigma = \frac{RL}{A} \tag{15.2}$$

Where L is the length of specimen, and A is the area of specimen.

15.3.4 Self-heating test by the Joule effect

This test is performed to characterize the effect of self-heating caused by the transportation of electric current in the specimens (i.e., CFRE neat and CFRE/FGNP-P). The test consists of the determination of the temperature reached as a function of the applied voltage. All specimens are prepared and cut to a certain dimension (20 mm × 20 mm × 2 mm), and then four copper wires are attached to the top surface of each specimen, where two wires used for measuring the electric current through the specimen, and two wires for measuring the voltage across the specimen. Digital Infrared Thermometer is used to measure the temperature of each specimen due to self-heating [10].

15.3.5 Tensile test

The tensile test is performed on CFRE/FGNP-P specimen, to study the effect of tensile applied load on the electrical conductivity of CFRE/FGNP-P specimen. In this test, there is no failure to the specimen, only the specimen will be subjected to an axial tensile load in the elastic region, and there is two silver paste points at the specimen surface to measure the electrical resistance as shown in the Figure 15.3. The experiment was set up in accordance with American Society of Testing and Materials (ASTM) standards that best corresponded to the test specifications. The specimen is prepared and tested according to the ASTM D638 as shown in Figure 15.3 [2, 10, 15, 16].

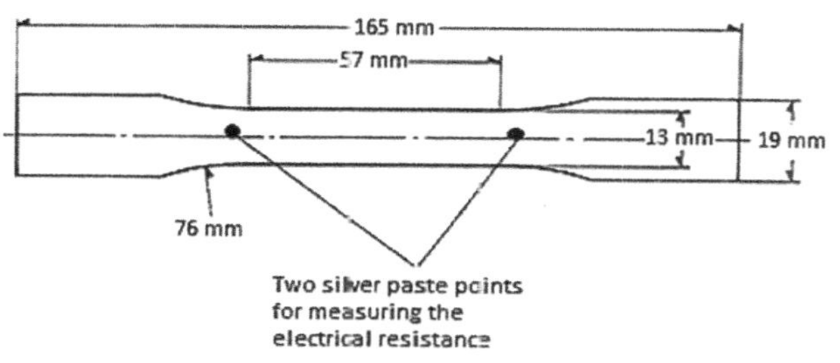

Figure 15.3: CFRE/FGNP-P specimen prepared for tensile test according to the ASTM D638.

15.3.6 Morphological characterization

The scanning electron microscopy (SEM) was used to characterize the top and side surfaces of three specimens (CFRE neat and CFRE/FGNP-P) to compare between them. Specimens were sputter-coated with gold to prevent electrical charging during the examination and obtain more clear images. An accelerating voltage of 20 kV was used to obtain the SEM images. In this research, the scanning electron microscope (brand: FEI, model: NOVA NANOSEM 230, Location: advanced technology (ITMA) at UPM, Malaysia) was used to characterize the top and side surfaces of the specimens. The Nova NanoSEM 230 is capable of magnifications up to 5,00,000× and can achieve a resolution of 0.8 nm at 15 kV.

15.4 Results and discussion

15.4.1 Electrical conductivity

The electrical conductivity value is measured for specimens CFRE neat and CFRE/FGNP-P, respectively. From the results, as shown in Figure 15.4, it is found that the electrical conductivity of CFRE is increased and enhanced significantly after coating it with FGNP by about 11 times with percentage of 1,110 %. Specimen CFRE/FGNP-P has the electrical conductivity of 3.1×10^4 (S/m), while the electrical conductivity of CFRE neat is 1.38×10^3 (S/m). It is found that the electrical resistance of CFRE is decreased significantly after coating it with FGNP, which indicates that the electrical conductivity of CFRE is enhanced. For example, specimen CFRE/FGNP-P has the lowest electrical resistance of 0.0169 (ohm), while CFRE neat has the highest electrical resistance of 0.36 (ohm).

Figure 15.5 shows the electrical conductivity versus temperature for all specimens (CFRE neat, and CFRE/FGNP-P). It is observed that the electrical conductivity of all specimens' decreases so slightly with increasing temperature from −10 °C to 27 °C; however, this change is almost constant.

The electrical conductivity of all specimens in this study is compared with copper, aluminum, and some previous studies as shown in the Figure 15.6. In this research, it is

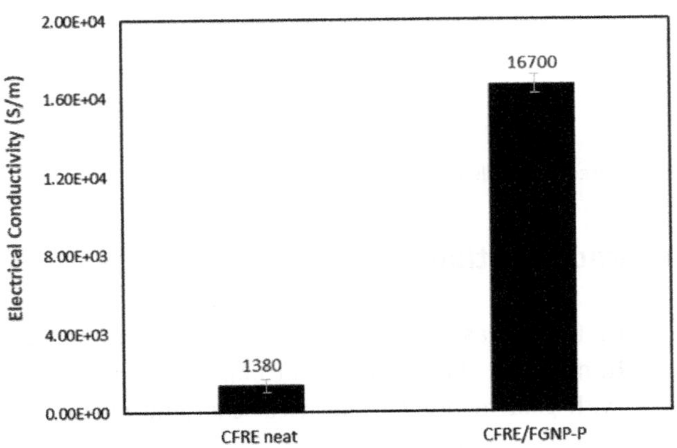

Figure 15.4: Electrical conductivity average values comparisons for all specimens.

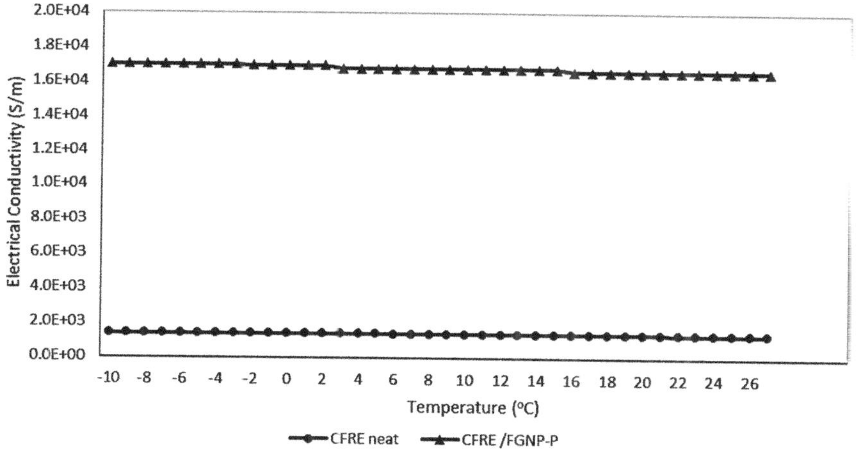

Figure 15.5: Electrical conductivity versus temperature comparisons for all specimens.

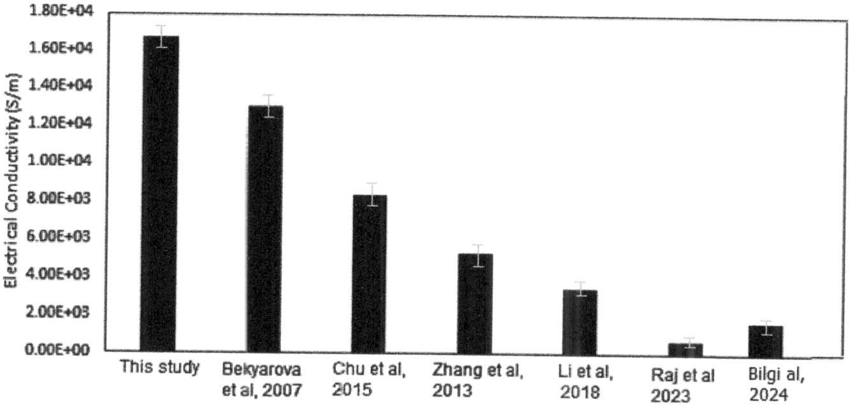

Figure 15.6: Comparison of electrical conductivity results with previous studies.

found that the electrical conductivity of CFRE has improved significantly after coating it with FGNP and reached to 1.7×10^4 (S/m), but it is still lower than the electrical conductivity of copper and aluminum. By comparing this study with other previous studies that used CNT [17, 18] as conductive filler, it is clear from Figure 15.6 that FGNP improved the electrical conductivity of CFRE more than CNT. Bekyarova et al. 2007 used CNT to coat CFRE surface to enhance its electrical conductivity, their results showed that the electrical conductivity was improved to 1.3×10^4 (S/m). While Chu et al. [18] reached a value of 8.3×10^3 (S/m) by inserting CNT based Bucky paper to improve electrical conductivity of CFRE. Figure 15.6 also shows the comparison of this study with other studies that used GNP [10, 19]. Zhang et al. obtained an electrical conductivity of 5.26×10^3 (S/m) by coating the carbon fiber surface with GNP [19]. While another research by Li et al. [10], a GNP/CNT hybrid was used to coat the surface of CFRE. Finally, Raj et al. [20] prepared a composite material made of carbon fiber reinforced with peek materials, but the electrical conductivity obtained was very small if compared with the other studies. The electrical conductivity of CFRE was improved and reached to 3.4×10^3 (S/m) [19]. Bilgi et al. [21] dispersed graphene nanoplatelets (GNPs) in carbon fiber reinforced polymers (CFRP), the results showed that FGNP as an additive achieved significant homogeneity and 50 % increase in electrical conductivity in the sample. It is noticed that the electrical conductivity of CFRE has been improved greatly after coating it with FGNP. Also, it is found that by increasing the percentage of FGNP content, the electrical resistance decreased significantly, which led to an increase in the value of electrical conductivity of CFRE several times. Therefore, the use of FGNP on CFRE is proved to enhance the conductivity than other materials.

15.4.2 Self-heating test

In this test, the self-heating of all specimens by the Joule effect was analyzed. The aim of this test is to make sure that the self-heating of CFRE has been improved after coating it with FGNP. It is found that the self-heating of specimen CFRE/FGNP-P is more effective, more homogeneous, and reaching higher temperatures for their high thermal conductivity as shown in Figure 15.7, also the value of applied voltage on the specimens is low, which reduces the energy consumption. The self-heating of CFRE/FGNP-P has been improved after coating it with FGNP. It was also confirmed that the self-heating is repetitive in several cycles, reaching the same temperature when the same voltage is applied.

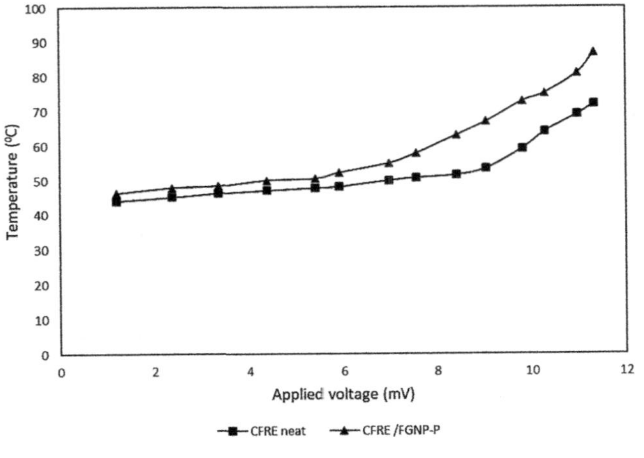

Figure 15.7: Temperature versus applied voltage curves of all specimens which obtained from the Joule effect.

Figure 15.7 shows the temperature versus applied voltage graph for the studied specimens (CFRE neat, and CFRE/FGNP-P). It is observed that the temperature of the specimens increases by increasing the applied voltage, because more electric current will pass through the specimen, for example, at applied voltage = 10 mV, the temperature of CFRE neat is reached to 64 °C due to self-heating, but the temperatures of CFRE/FGNP-P reached to higher value at same applied voltage, where the temperature of CFRE/FGNP-P reached to 75 °C, respectively. In addition to that, the addition of FGNP nanofiller induced an increase in the thermal conductivity of CFRE that is directly proportional to the nanofiller content. When the FGNP content is increased, the thermal and electrical conductivities of the samples are increased and then the self-heating due to joule effect is enhanced, eventually, the samples temperatures reached higher levels due to such augmentations.

From Figure 15.8, it is found that specimen CFRE neat took time of 17 s to reach the equilibrium temperature of 64 °C, at applied voltage of 10 mV. While the time taken to reach the equilibrium temperature for specimens CFRE/FGNP-P is 10 s at the same applied voltage. It is found also that the time taken to reach temperature of 64 °C for specimens CFRE/FGNP-P is 3.5 s at applied voltage 10 mV, which is 5–6 orders lower than time taken by specimen CFRE neat to reach the same temperature (17 s). This is due to the addition of FGNP to CFRE neat, which improved its thermal conductivity to higher values.

15.4.3 Tensile test

The tensile test was conducted on CFRE/FGNP-P, to find the influence of tensile load on the electrical conductivity of CFRE/FGNP-P. Figure 15.9 shows the electrical resistance versus

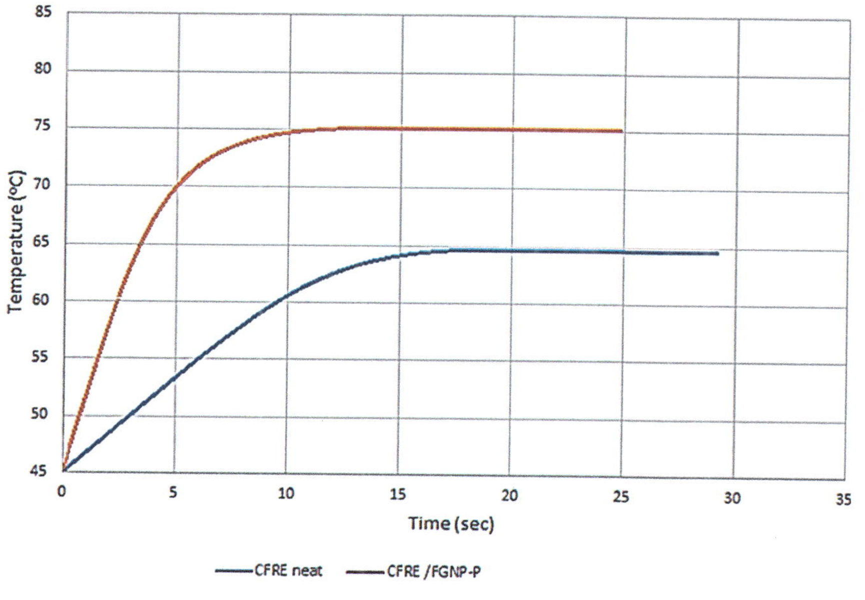

Figure 15.8: Temperature versus time of all specimens at applied voltage of 10 mV.

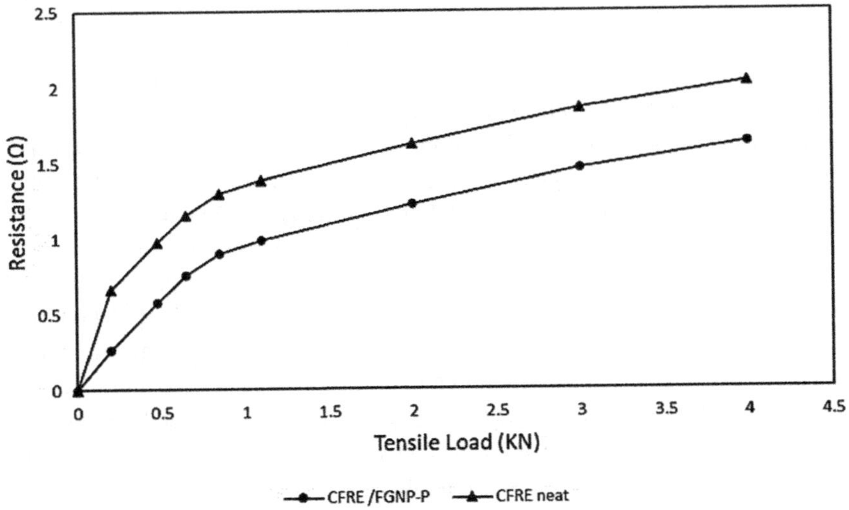

Figure 15.9: Electrical resistance versus tensile load curve of specimen CFRE/FGNP-P.

applied tensile load curve of specimen CFRE/FGNP-P. This test is performed by applying axial tensile load on the specimen CFRE/FGNP-P, to find out how the change in electrical conductivity of CFRE/FGNP-P will be. It is found from Figure 15.9 that the electrical resistivity of CFRE/FGNP-P is increased gradually by increasing the applied tensile load on the specimen; as the deformation increases in the specimen, the electrical resistivity of CFRE/FGNP-P also increases (decreasing the electrical conductivity); as the applied tensile load increase (the strain increases), the FGNP layer is stretched anisotropically, causing FGNP particles move away from each other (breakdown of the FGNP layer as it is stretched), leading to the rapid increase in resistivity of specimen CFRE/FGNP-P.

15.4.4 Material characterization

Figure 15.10 shows the characterization result of the top surface of two samples, CFRE neat and CFRE/FGNP-P, using scanning electron microscopy (SEM). Figure 15.10(a) shows the image of top surface of CFRE neat, which shows that there is no FGNP at top surface. While Figure 15.10(b) shows the FGNP coating over CFRE neat top surface. The image in Figure 15.10(c) shows the shape of FGNP sheets at magnification of 500 nm in the CFRE/FGNP-P sample, where the FGNP sheets are transparent and homogenous with less agglomeration. However, in Figure 15.11, the images show the side view surface of CFRE/FGNP-P.

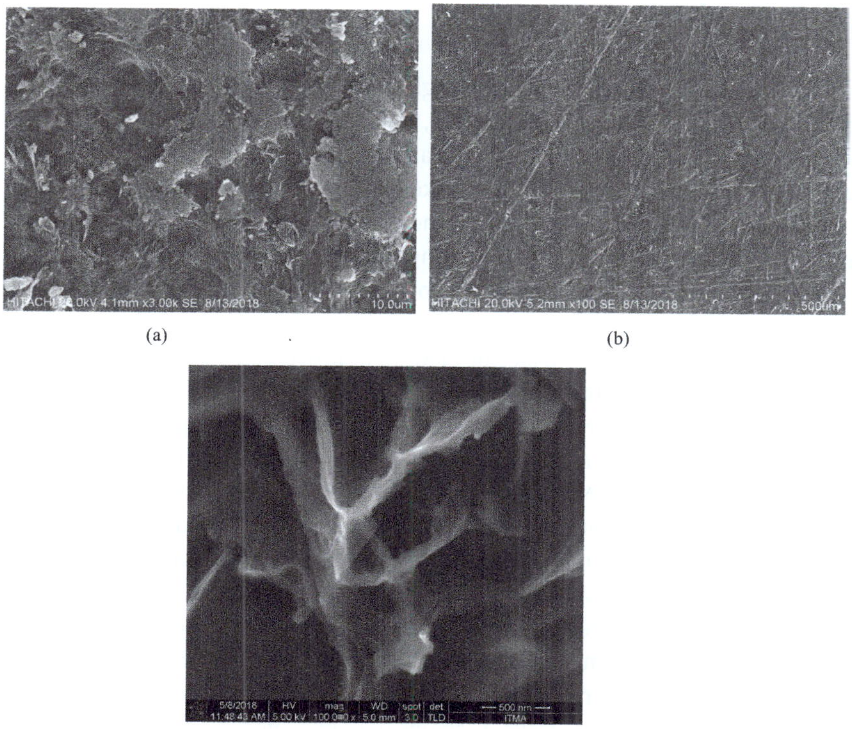

(a) (b)

(c)

Figure 15.10: Top view images of CFRE surfaces: (a) CFRE neat at magnification of 10 µm (b) CFRE/FGNP-P at magnification of 10 µm (c) top view CFRE/FGNP-P surface at magnification of 500 nm.

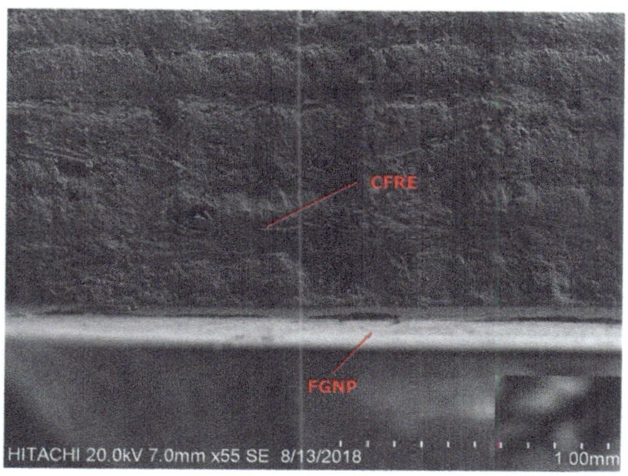

Figure 15.11: Side view image of CFRE/FGNP-P, at magnification of 1 mm.

15.5 Conclusions

The aim of this study is to develop a cheaper highly electrically conductive epoxy resins through the addition of functionalized graphene, which, when combined with conductive carbon fiber, resulted in a highly conductive carbon fiber reinforced epoxy composite material capable of withstanding lightning strike and can be utilized for deicing, which has lighter weights than metallic compounds. The development of such a material would result in safer aircraft, less weight, and cost savings from the elimination of the expensive metallic meshes as well as eliminating time associated with integrating the meshes into the structure.

FGNP is used to coat the top surface of CFRE to improve and enhance its electrical conductivity to be used for lightning strikes and deicing in the aviation applications. The electrical conductivity is measured for the specimens using four-point probe device. The results show that the electrical conductivity of CFRE is enhanced and increased significantly when it is coated with FGNP. The results show that the specimen CFRE/FGNP-P has the highest electrical conductivity of 1.67×10^4 (S/m), while CFRE neat has the lowest electrical conductivity of 1.38×10^3 (S/m), which means that the electrical conductivity of CFRE neat is increased by about 1,110 %.

The results showed that the self-heating of CFRE is enhanced and improved after coating it with FGNP. The self-heating of CFRE/FGNP-P is more effective and more homogenous, reaching higher temperatures, than CFRE neat. It was also confirmed that the self-heating is repetitive in several cycles, reaching the same temperature when the same voltage is applied.

The results also showed that the electrical resistivity of CFRE/FGNP-P is increased by increasing the applied tensile load on the specimen; as the deformation increases in the specimen, while the electrical conductivity decreased). When the applied tensile load increase (the strain increases), the FGNP layer is stretched anisotropically, causing FGNP particles move away from each other (breakdown of the FGNP layer as it is stretched), leading to the rapid increase in resistivity of specimen CFRE/FGNP-P and decrease in conductivity.

Acknowledgments: The authors would like to thank the editors S.M. Sapuan, Mohd Roshdi Hassan, Eris Elianddy Supeni, and Azizan As'arry for their guidance and review of this article before its publication.

References

1. Nayak NV. Composite materials in aerospace applications. Int J Scientific Res Publications 2014;4:1–10.
2. Alemour B, Yaacob MH, Lim HN, Hassan MR. Review of electrical properties of graphene conductive composites. Int J Nanoelectron Mater 2018;11:371–98.

3. Trapp R, Noah S, Harold E, Michael E, Eric D, Jeremy S. Changes in severe thunderstorm environment frequency during the 21st century caused by anthropogenically enhanced global radiative forcing. Proc Natl Acad Sci USA 2007;104:19719–23.
4. Alemour B, Badran O, Hassan MR. A review of using conductive composite materials in solving lightning strike and ice accumulation problems in aviation. J Aero Technol Manag 2019;11:1022.
5. Katunin A, Krukiewicz K, Turczyn R, Sul P, Bilewicz M. Electrically conductive carbon fiber-reinforced composite for aircraft lightning strike protection. In: IOP Conference Series: Materials Science and Engineering. Bristol, UK: IOP Publishing; 2017, vol 201:012008 p.
6. Gagné M, Therriault D. Lightning strike protection of composites. Prog Aero Sci 2014;64:1–16.
7. Gammon LM, Falcone A. Lightning strike damage in polymer composites. Adv Mater Process 2003;161: 61–2.
8. Morgan J. Thermal simulation and testing of expanded metal foils used for lightning protection of composite aircraft structures. SAE Int J Aerospace;6:371–7. https://doi.org/10.4271/2013-01-2132.
9. Alarifi IM, Khan WS, Rahman MM, Asmatulu R. Mitigation of lightning strikes on composite aircraft via micro and nanoscale materials. Adv Nanotechnology 2017;39:39–66.
10. Li Y, Zhang H, Liu Y, Wang H, Huang Z, Peijs T, et al. Synergistic effects of spray-coated hybrid carbon nanoparticles for enhanced electrical and thermal surface conductivity of CFRP laminates. Compos Appl Sci Manuf 2018;105:9–18.
11. Vishnyakov LR, Pereselentseva LN, Vishnyakova EL. Knitted soldered meshes and nanostructured carbon particles for lightning protection of composite wind turbine blades. Powder Metall Met Ceram 2014;53: 368–74.
12. Imran K. Enhancement of electrical conductivity of carbon/epoxy composites by graphene and assessment of thermal and mechanical properties [Doctoral dissertation]. Greensboro, USA: North Carolina Agricultural and Technical State University; 2016.
13. Bangwei Z, Soltani SA, Le LN, Asmatulu R. Fabrication and assessment of a thin flexible surface coating made of pristine graphene for lightning strike protection. Mater Sci Eng B 2017;216:31–40.
14. Kumar SS, Uddin MN, Rahman MM, Asmatulu R. Introducing graphene thin films into carbon fiber composite structures for lightning strike protection. Polym Compos 2019;40:517–25.
15. Alemour B, Yaacob MH, Lim HN, Badran O, Hassan MR. Improving the electrical conductivity of carbon fiber reinforced epoxy composite using reduced graphene oxide. Mater Res Express 2019;6:065607.
16. Ku H, Wang H, Pattarachaiyakoop N, Trada M. A review on the tensile properties of natural fiber reinforced polymer composites. Compos B Eng 2011;42:856–73.
17. Bekyarova E, Thostenson ET, Yu A, Kim H, Gao J, Tang J, et al. Multiscale carbon nanotube – carbon fiber reinforcement for advanced epoxy composites. Langmuir 2007;23:3970–4.
18. Chu H, Zhang Z, Liu Y, Leng J. Self-response multi-functional composite material base on carbon nanotube paper using deicing, flame retardancy, thermal insulation, and lightning-strike protection. In: SPIE Smart Structures and Materials+ Nondestructive Evaluation and Health Monitoring. Bellingham, Washington, USA: International Society for Optics and Photonics; 2015:94320S–8 pp.
19. Zhang B, Patlolla VR, Chiao D, Kalla DK, Misak H, Asmatulu R. Galvanic corrosion of Al/Cu meshes with carbon fibers and graphene and ITO-based nanocomposite coatings as alternative approaches for lightning strikes. Int J Adv Des Manuf Technol 2013;67:1317–23.
20. Raj R, Parasuram S, Kumar S, Bose S. Process dependent interface strengthening, de-icing and EMI shielding performance in PEEK/CF laminates. Mater Adv 2023;4:4180–9.
21. Bilgi C, Demir B, Aydın H, Üstün B, Kurtan Ü. Effect of functionalized graphene nanoplatelet dispersion on thermal and electrical properties of hybrid carbon fiber reinforced aviation epoxy laminated composite. Mater Chem Phys 2024;325:129702.

Index

https://doi.org/10.1515/9783111234762-016

Printed and bound by CPI Group (UK) Ltd, Croydon, CR0 4YY

24/08/2026

14947856-0001